Electronic Packaging for High Reliability; Low Cost Electronics

NATO ASI Series

Advanced Science Institute Series

A Series presenting the results of activities sponsored by the NATO Science Committee, which aims at the dissemination of advanced scientific and technological knowledge, with a view to strengthening links between scientific communities.

The Series is published by an international board of publishers in conjunction with the NATO Scientific Affairs Division

A. Life Sciences **B. Physics**	Plenum Publishing Corporation London and New York
C. Mathematical and Physical Sciences **D. Behavioural and Social Sciences** **E. Applied Sciences**	Kluwer Academic Publishers Dordrecht, Boston and London
F. Computer and Systems Sciences **G. Ecological Sciences** **H. Cell Biology** **I. Global Environment Change**	Springer-Verlag Berlin, Heidelberg, New York, London, Paris and Tokyo

PARTNERSHIP SUB-SERIES

1. Disarmament Technologies	Kluwer Academic Publishers
2. Environment	Springer-Verlag / Kluwer Academic Publishers
3. High Technology	Kluwer Academic Publishers
4. Science and Technology Policy	Kluwer Academic Publishers
5. Computer Networking	Kluwer Academic Publishers

The Partnerschip Sub-Series incorporates activities undertaken in collaboration with NATO's Cooperation Partners, the countries of the CIS and Central and Eastern Europe, in Priority Areas of concern to those countries.

NATO-PCO-DATA BASE

The electronic index to the NATO ASI Series provides full bibliographical references (with keywords and/or abstracts) to about 50,000 contributions from international scientists published in all sections of the NATO ASI Series. Access to the NATO-PCO-DATA-BASE is possible via a CD-ROM "NATO Science and Technology Disk" with user-friendly retrieval software in English, French, and German (©WTV GmbH and DATAWARE Technologies, Inc. 1989). The CD-ROM contains the AGARD Aerospace Database.

The CD-ROM can be ordered through any member of the Board of Publishers or through NATO-PCO, Overijse, Belgium.

3. High Technology – Vol. 57

Electronic Packaging for High Reliability; Low Cost Electronics

Edited by

Rao Tummala

Packaging Research Center (PRC),
Georgia Institute of Technology,
Atlanta, Georgia, U.S.A.

Marija Kosec

Department of Ceramics,
University of Ljubljana,
Ljubljana, Slovenia

W. Kinzy Jones

Department of Mechanical Engineering,
Florida International University,
Miami, Florida, U.S.A.

and

Darko Belavic

Department of Ceramics,
University of Ljubljana,
Ljubljana, Slovenia

Kluwer Academic Publishers

Dordrecht / Boston / London

Published in cooperation with NATO Scientific Affairs Division

Proceedings of the NATO Advanced Research Workshop on
Electronic Packaging for High Reliability; Low Cost Electronics
Bled, Slovenia
May 10–13, 1997

A C.I.P. Catalogue record for this book is available from the Library of Congress.

ISBN 978-90-481-5085-4

Published by Kluwer Academic Publishers,
P.O. Box 17, 3300 AA Dordrecht, The Netherlands.

Sold and distributed in North, Central and South America
by Kluwer Academic Publishers,
101 Philip Drive, Norwell, MA 02061, U.S.A.

In all other countries, sold and distributed
by Kluwer Academic Publishers,
P.O. Box 322, 3300 AH Dordrecht, The Netherlands.

Printed on acid-free paper

Table of Contents

Preface ix

Performance Predictions

"Possibilities of Thermal Resonance in Microcircuits," Gilbert De Mey 1

"Signal Integrity Analysis in Cost Sensitive Interconnections," Paul Svasta, Virgil Golumbeanu, Norocel-Dragos Codreanu, Willi Ciszkowski, Marian Vladescu 9

"The Use of Nonlinearity Measurements for the Reliability Assessment," P.J. Mach 17

"Effect of Plastic Package Geometry on its Propensity to Moisture Induced Failure," E. Suhir 25

"The Impact of Stress on the Noise, Quality and Reliability of Passive and Active Devices," Josef Sikula, Petr Vasina 53

Process and Materials Development

"Smart Unit for Technological Monitoring in Electrochemical Industry," Radu Ionescu, Sorin Ganciulescu, Dan Caravasile, Ovidiu Tol, Paul Svasta, Daniel Leonescu 61

"Laser Patterning of Printed Wiring Boards for MCM-Ls," ZS. Illyefalvi-Vitéz, J. Pinkola, M. Ruszinkó, P.J. Szabó 69

"Silver Filled Adhesives-Controlling Rheology and Electrical Conductivity," P. Sexton, K. McNeilly, S. Schlag 79

"Silicides Formation in Metal-Silicon Film Systems: The Role of Impurities," Yu Makogon, S.I. Sidorenko 91

"Mechanical Properties of Pb-Sn-In-Ag Solders from -100°C to 150°C," W. Kinzy Jones, Yanqing Liu, Milind Shah 99

MCM-D/L

"The Need for Low Cost, High Density Packaging and Interconnect Technology," Philip Garrou 107

Preface

This monograph will review recent advances in high performance, cost sensitive electronic packaging. The roadmap for integrated circuits (Ics) predicts an increase in input/output capabilities to exceed 2000 by the year 2001, with increasing speed and power. New packaging (chip scale, ball-grid array (BGA) and multi-chip modules) are required to maintain assembly of these devise within cost and size constraints.

Under the umbrella of the NATO Science Committee, an Advanced Research Workshop was held on Electronic Packaging for High Reliability, Low Cost Electronics, University of Ljubljana (host at Bled, Slovenia, May 10-13, 1997). This volume presents the majority of papers presented at this workshop.

This book addresses new hermetic packaging, new materials for thermal management and assembly, and new components which integrate multiple functions (embedded substrates and component arrays), while maintaining the reliability to previous levels found in high reliability applications.

There were many developments on fundamental materials and manufacturing processes of discrete components was occurring within the cooperating partner (CP) countries, while the strengths of NATO countries were the development of high speed, high integration packaging and more complex imbedded components technologies.

The organizers of the ARW gratefully acknowledge the financial support by the NATO Scientific Committee and the International Microelectronics And Packaging Society (IMAPS). Special thanks are due to Soren Norlyng, Dr. Zsolt Illyefalvi Vitez, Iztok Sorli, and Darko Belavic for their organization of an excellent program. Special thanks to Dr. Kinzy Jones and Dr. Marija Kosec for their support and guidance in organizing the conference and providing support for the manuscript.

Dr. Rao Tummala

next section, the classical diffusion equation has to be replaced by a more "wave-like" one. Phenomena such as resonance, wave propagation, wave reflection,... are then possible. In order to get resonance, dimensions comparable to the wavelength are necessary. As already pointed out, devices are becoming smaller and smaller, so one should investigate whether or not thermal resonance phenomena can be expected if integration densities are getting higher.

The literature on non Fourier thermal conduction in electronics is rather limited [3]. For the dimensions and time scales used in present day electronics, the differences between Fourier and non Fourier conduction are negligible [4]. However, research resulted in very small devices with dimensions in the range of $50nm$ [5]. For these kinds of components, which will operate at much higher frequencies, non Fourier thermal conduction will be necessary for appropriate device modelling.

2 FUNDAMENTAL EQUATIONS

The heat flux density $\overline{q}$ in a solid is related to the temperature gradient ∇T by the relation [6][7][8]:

$$\tau \frac{\partial \overline{q}}{\partial t} + \overline{q} = -k \nabla T \tag{1}$$

where:

- $\overline{q}$: heat flux density (W/m^2)
- τ : relaxation time constant (s)
- k : thermal conductivity (W/mK)
- $T(x,y,z)$: temperature distribution (K)

Fourier's lw is obtained by inserting $\tau = 0$ in (1). Physically, (1) means that if a uniform temperature gradient is sudenly applied in a solid, the heat flux density will increase exponentially with a time constant τ.

The divergence of $\overline{q}$ is, like in classical heat conduction theory, is equal to the heat released from cooling and the generated heat:

$$\nabla \cdot \overline{q} + C_v \frac{\partial T}{\partial t} = p \tag{2}$$

where:

- C_v : thermal capacity per unit volume $(J/m^3\ K)$
- p : dissipated power per unit volume (W/m^3)

Taking the divergence of both members of (1) and using (2), one obtains:

$$-\tau C_v \frac{\partial^2 T}{\partial t^2} + \tau \frac{\partial p}{\partial t} - C_v \frac{\partial T}{\partial t} + p = -k \nabla^2 T \tag{3}$$

or:

$$\nabla^2 T - \frac{C_v}{k}\frac{\partial T}{\partial t} - \frac{\tau C_v}{k}\frac{\partial^2 T}{\partial t^2} = -\frac{1}{k}(p + \tau\frac{\partial p}{\partial t}) \tag{4}$$

The first two terms remember us the classical heat diffusion equation. If one disregards the second term for a moment, one gets the wave equation. This identification gives us the wave propagation velocity c:

$$c = \sqrt{\frac{k}{\tau C_v}} \quad or \quad \tau = \frac{k}{C_v c^2} \tag{5}$$

Because temperature is by definition the mean kinetic energy of the particles, heat conduction can be seen as a propagation of mechanical energy. This is limited by the velocity of sound. Some textbooks derive (4) by using this point of view rather than starting from (1) [9]. However, only the third term of the left hand member is added in this way. The term $\partial p/\partial t$ is missing then. Remark that omitting $\partial p/\partial t$ will have the consequence that propagating temperature peaks as calculated by Guo and Xu will not be found [3]. Starting from (1) provides us also a better physical background.

For Si, Ge and $GaAs$ one has [10][11]:

	Si	Ge	$GaAs$
k (W/mK)	124	64	56
C_v (MJ/m^3K)	1.6	1.7	1.6
c (m/s)	5000	5000	5000
τ (ps)	3.03	1.49	1.4

The value $c = 5000\ m/s$ has been estimated using data for other solids. The values for τ are all calculated from this estimated c value. Nevertheless, the τ-values are all in the ps-range, in agreement with other authors [3][8].

If one deals with time harmonic problems, phasor notation can be used. The equation (4) is then rewritten as:

$$\nabla^2 T - \frac{C_v}{k} j\omega(1 + j\omega\tau)T = -\frac{1}{k}(1 + j\omega\tau)p \tag{6}$$

It is then assumed that the power density varies with the same frequency. This is a normal situation in microelectronics if a small AC input signal is superimposed on a DC level. It is evident that the temperature variations found by solving (6), should be added to the mean DC temperature caused by the DC power.

3 MATHEMATICAL MODEL

In order to have thermal resonance phenomena, one should have a device with at least one very short dimension, which gives rise to two thermal boundary conditions. In integrated circuits, this condition occurs when a 2-D periodic array of transistors is placed on the surface. If the power density is the same for every transistor, a regular pattern of heat sources is obtained. This phenomenon can be modelled by considering one unit cell around a transistor. Due to the periodicity of the problem, adiabatic boundary conditions can be applied at the boundaries of the unit cell. Similarly, the heat propagating from the semiconductor surface towards the gate electrode also propagates over a very small distance. Of course, these thermal phenomena have always to be put in parallel with het heat path to the bottom contact on the silicon, where no small dimensions exist.

In this contribution we will limit ourselves to a one dimensional aproach. The reason is that a full analytical treatment is possible, making physical interpretations of the results more obvious. A simplified one dimensional analysis also gives a good estimation of what can be expected from a full 3-D analysis.

3.1 One dimensional solution

We consider now a one dimensional array of heat sources (fig.1). In all points ... $x = -3a$,

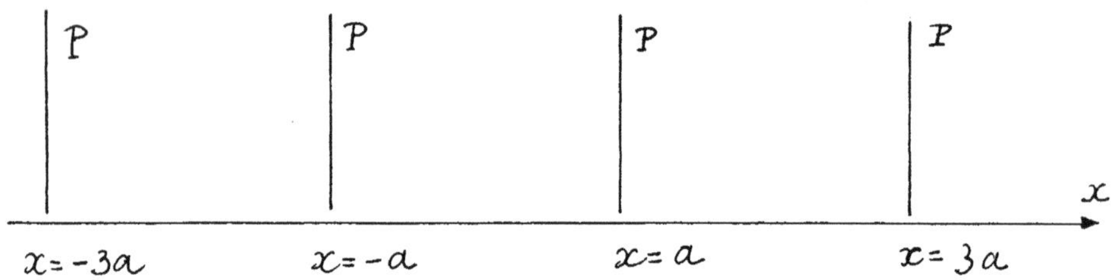

Figure 1: One dimensional array of heat sources

$x = -a$, $x = a$, $x = 3a$, ... a power density P (W/m^2) is dissipated in an infinite plane perpendicular to the x-axis. Due to the periodicity, we only have to consider one interval $[0, a]$. The equation we have to solve, is now:

$$\frac{d^2T}{dx^2} - \frac{C_v}{k} j\omega(1 + j\omega\tau)T = 0 \tag{7}$$

with the boundary conditions:

$$\frac{dT}{dx} = 0 \qquad at \qquad x = 0 \tag{8}$$

and:

$$k\left(\frac{dT}{dx}\right)_{x=a^-} = \frac{1}{2}P(1 + j\omega\tau) \tag{9}$$

The condition (9) can de derived from (6) by inserting $p = P\delta(x - a)$ and integrating both members from a^- to a^+.

The condition (8) is automatically fullfilled by the following solution:

$$T(x) = A\cosh\left[\sqrt{\frac{C_v}{k} j\omega(1 + j\omega\tau)}x\right] \tag{10}$$

The integration constant A is easily determined by applying (9) to (10):

$$T(x) = \frac{P\sqrt{1 + j\omega\tau}}{2\sqrt{kC_v j\omega}} \frac{\cosh\left[\sqrt{\frac{C_v}{k} j\omega(1 + j\omega\tau)}\ x\right]}{\sinh\left[\sqrt{\frac{C_v}{k} j\omega(1 + j\omega\tau)}\ a\right]} \tag{11}$$

We are not interested in the temperature distribution itself, bu rather in the thermal impedance as seen from the heat source at $x = a$. This is then:

$$Z_{th} = \frac{2T(x=a)}{P} = \frac{\sqrt{1+j\omega\tau}}{\sqrt{kC_v j\omega}} \frac{1}{\tanh\left[\sqrt{\frac{C_v}{k} j\omega(1+j\omega\tau)}\, a\right]} \tag{12}$$

The factor 2 reflects the fact that only half of the heat dissipated in $x = a$ is conducted to the region $[0, a]$.

By putting $\tau = 0$ in (12), one gets the thermal impedance according to the classical Fourier conduction law:

$$Z_{th,F} = \frac{1}{\sqrt{kC_v j\omega}} \frac{1}{\tanh\left[\sqrt{\frac{C_v}{k} j\omega}\, a\right]} \tag{13}$$

In order to simplify the equations, some dimensionless parameters are introduced now:

$$a' = \frac{a}{\sqrt{\frac{k\tau}{C_v}}} = \frac{a}{c\tau} \tag{14}$$

and:

$$\omega' = \omega\tau a'^2 = \omega\frac{a^2 C_v}{k} = \frac{\omega}{\omega_0} \tag{15}$$

For silicon one finds $c\tau = 15\ nm$ and $f_0 = \omega_0/2\pi = 4.9\ GHz$ for $a = 50\ nm$. The relations (12) and (13) can then be replaced by:

$$Z = \frac{Z_{th}}{a'\sqrt{\frac{\tau}{kC_v}}} = \frac{\sqrt{1+j\omega'/a'^2}}{\sqrt{j\omega'}\ \tanh\sqrt{j\omega'(1+j\omega'/a'^2)}} \tag{16}$$

and:

$$Z_F = \frac{Z_{th,F}}{a'\sqrt{\frac{\tau}{kC_v}}} = \frac{1}{\sqrt{j\omega'}\ \tanh\sqrt{j\omega'}} \tag{17}$$

The 1-D model can also be used for the heat transfer through the oxide and gate electrode in a MOS transistor. In this case, a two layer model should be set up. The boundary condition (8) can be maintained because the heat transfer by convection at the top of the gate electrode is negligible.

3.2 Poles and zeros

The poles and zeros of (16) and (17) can be determined by using the product expansions of the hyperbolic functions [12]

$$\sinh z = z\prod_{q=1}^{\infty}(1+\frac{z^2}{q^2\pi^2}) \tag{18}$$

$$\cosh z = \prod_{q=1}^{\infty}(1+\frac{4z^2}{(2q-1)^2\pi^2}) \tag{19}$$

One obtains then, after replacing $j\omega$ by the Laplace variable s:

$$Z = \frac{1}{s}\frac{\prod_{q=1}^{\infty}[1+\frac{4s(1+s/a'^2)}{(2q-1)^2\pi^2}]}{\prod_{q=1}^{\infty}[1+\frac{s(1+s/a'^2)}{q^2\pi^2}]} \tag{20}$$

and:

$$Z_F = \frac{1}{s}\frac{\prod_{q=1}^{\infty}[1+\frac{4s}{(2q-1)^2\pi^2}]}{\prod_{q=1}^{\infty}[1+\frac{s}{q^2\pi^2}]} \tag{21}$$

As expected Z_F has only poles and zeros on the negative real axis. The function (16) has the poles:

$$p_q = a'^2\frac{-1 \pm \sqrt{1-4q^2\pi^2/a'^2}}{2} \tag{22}$$

and the zeros:

$$n_q = a'^2\frac{-1 \pm \sqrt{1-(2q-1)^2\pi^2/a'^2}}{2} \tag{23}$$

If $a' < 2\pi$, poles and zeros are all located as shown on fig.2. resonance phenomena are

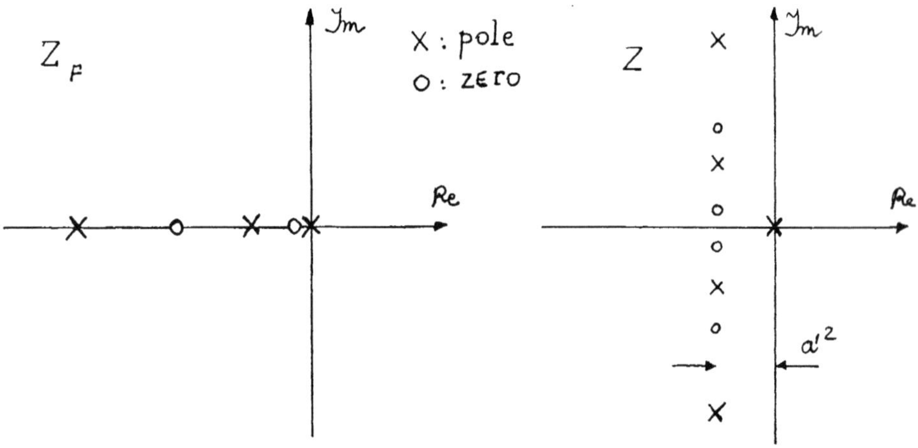

Figure 2: Poles and zeros of the thermal impedances Z and Z_F

then possible. If a' gets higher, the low order poles are on the negative real axis. The poles corresponding to high q-values are aligned parallel to the imaginary axis, but further away. The smaller a' the more pronounced resonance phenemena can be expected.

For Fourier conduction, a theoretical work on thermal impedances by means of poles and zeros has been carried out by Székely [13].

For silicon, the condition $a' < 2\pi$ means that the distance $a < 2\pi c\tau = 95.3\ nm$. Recently, an experimental FET transistor with a gate length of 50 nm has been reported [5]. Oxide thicknesses are usually in the range 50 to 100 nm. The condition $a < 95.3\ nm$ is not so unrealistic any more.

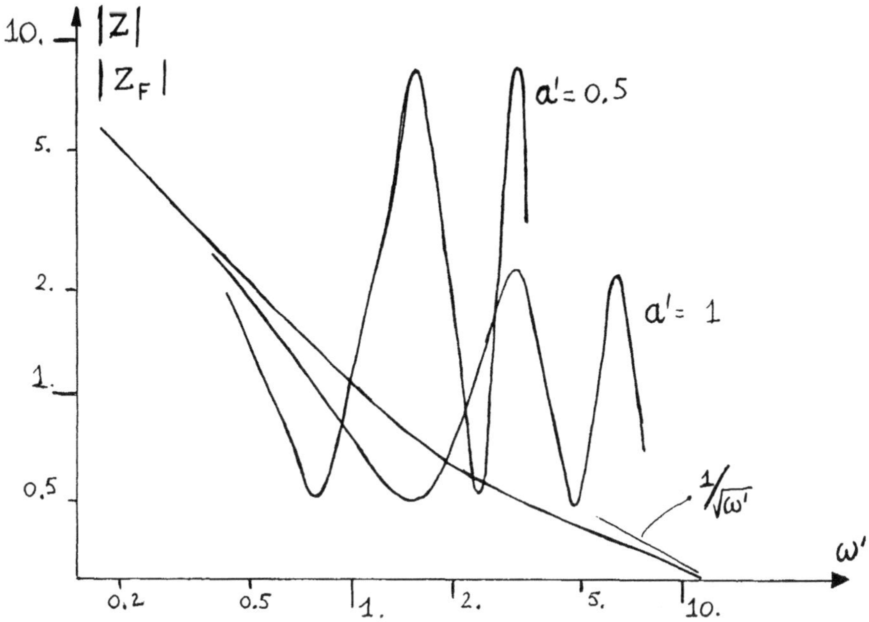

Figure 3: Bode plot of the thermal impedances Z and Z_F

3.3 Bode plot

On fig.3 one observes the amplitude Bode plot of the function (16) for several values of a'. The smaller a', the sharper the resonance peaks, because poles and zeros are then closer to the imaginary axis. For $a' = 1$ resonance is shifted to higher frequencies. For comparison, a Bode plot of (17) has also been drawn. At low frequencies (16) and (17) converge to each other.

4 CONCLUSION

In this contribution, thermal resonance phenomena using non Fourier thermal conduction have been analysed using a one dimensional model. Bode plots showing quite sharp resonance peaks, the results should be interpreted carefully. First of all, thermal impedances involving very small dimensions are in practice always part of a larger network of thermal impedances. The resonance peaks will then be less pronouced. Secondly, these phenomena only occur at very high frequencies. Experimental detection will only be possible through thermal feedback, i.e. due to the change of some electrical characteristics with temperature like in the work of Cavaglia and Iliadu [2].

5 REFERENCES

1. J. E. SOLOMON:
"The monolithic OPAMP: a tutorial study"
IEEE Journal on Solid State Circuits, 1874, vol.SC-9, p.314-322.

2. A. L. CAVAGLIA and A. A. ILIADU:
"Linear dynamic self heating in SOI MOSFET's"

- the signal caused by crosstalk (capacitive and inductive coupling) of the line with neighbouring lines; the forward (K_f) and backward (K_b) coefficients are calculated as follows[1]:

$$K_{bij} = \frac{1}{4}\left(\frac{|C_{ij}|}{\sqrt{C_{II} * C_{jj}}} + \frac{|L_{ij}|}{\sqrt{L_{ii*} * L_{jj}}}\right) \tag{1}$$

$$K_{fij} = \left(\frac{|C_{ij}|}{\sqrt{C_{ii} C_{jj}}} - \frac{|L_{ij}|}{\sqrt{L_{ii} L_{JJ}}}\right) \frac{\sqrt{\sqrt{L_{ii} L_{jj}} / \sqrt{C_{ii} C_{jj}}}}{2} \tag{2}$$

where, j is the line causing the disturbance and i is the victim line; C_{ij}, respectively L_{ij}, it is mutual capacitance, respectively mutual inductance; C_{ii} and L_{ii} are the self-capacitance and self-inductance of the i line.

- reflected signals, due to the difference that may appear between the line characteristic impedance and the input and output impedances of interconnected circuits; the line voltage on any point x °i and at any moment t is [2],

$$V(x,t) = V_A(t)\,[U(t - t_{pd}x) + K_s U(t - t_{pd}(2l - x)) + K_g K_s U(t - t_{pd}(2l + x)) + K_s^2 K_g U(t - t_{pd}(4l - x)) + K_g^2 K_s^2 U(t - t_{pd}(4l + x)) + \ldots\ldots..] + V_{CC} \tag{3}$$

- signal delay time on the interconnection line, that may cause undesirable phase displacement;
- difference between the propagation times (t_{PLH}, t_{PHL}) of interconnected circuits;
- parasitic signals through the common impedance of the supply line [3];
- electric, magnetic and electromagnetic reception of fields existent in the environment.

Starting with the design stage of the interconnection structure, an analysis of the parasitic signals should be realised, parasitic signals that may appear according to the manufacturing technology, of the used lines type and of circuits.

A few solutions that ensure an acceptable distortion of the digital signal transmitted on the line (used for electronic modules) are presented:

- the use of circuits with high noise immunity and with smaller currents consumed from the supply line;
- the decrease of the line length below the critical length;
- the decrease of the surface limited by the interconnection line routes;
- the increase of the distance between the parallel signals routes;
- matching the signal lines for line lengths above the critical length;
- the use of shielding;
- the reduction to minimum of the supply line impedance;
- decoupling the supply.

3. Optimisation methods - virtual costs

For a PCB electronic module achievement one can consider that the costs may be classified in two groups: real manufacturing costs and virtual manufacturing costs.

The real costs represent the union of all the expenses necessary for the manufacturing of a printed circuit board. These costs have to be considered by the PCB designer because they affect directly the final price of the product. Unlike the real costs, the virtual costs represent an optimisation method based on criteria connected with various items (vias, pads, segments, connections) and strategies (placing, routing, smoothing) available for the PCB designer [4]. The design system CADSTAR\WINDOWS offers the user an software tool of the virtual costs problem. This tool is based on the minimisation of vias number, of the total length of interconnection routes, of the segments number and of the length of segments routed in a different direction from the default one. These costs allow an optimisation of the items belonging to the interconnection structure and offer to the user the possibility to try to minimise a priori the manufacturing costs. The virtual costs may be classified in three domains: costs for the routes, costs for vias and costs connected with technological restrictions violation. The values are between 0 and 100. We have to keep in mind the fact that products increase leads to the increase of running time.

In our analysis we applied optimisation methods considering the vias criteria. The vias costs are divided in 3 types: Routing, Smoothing and Power Plane. The first type represents the virtual costs of added holes in the routing stage. It is clear that the increase of its value (default value is 1) leads to the decrease of the added vias. A secondary effect is the increase of the route distribution, at this stage the autorouter tries to use the existing routings and the through-hole pads in order to change the layers (in the case of tree routing). Figure 1 present the situation in which **Via routing cost** factor (F_r) is variable [5]. The second type is the cost of added vias in the smoothing stage of the routes, after routing. The observation regarding the cost applies here too, the implicit value being 1.7. The secondary effect is not present in this case. Power plane refers to the setting of the vias costs due to vias introduction towards the supply and/or ground planes in the case of surface mounted components.

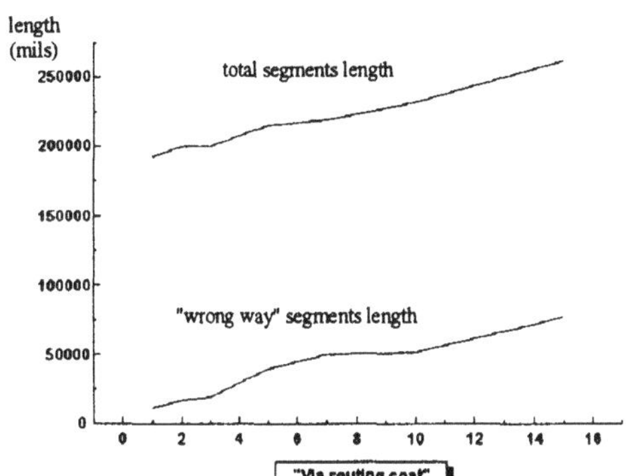

Figure 1. The dependence of the "via routing" factor

In the following, the costs and signal integrity analysis of some PCB modules with SMD components will be approached.

4. Practical example for the establish of a low cost interconnection structure

In order to illustrate the aspects presented in paragraphs 2-3, the circuit from figure 2 was analysed. The circuit has two multiplexing stages and it is analysed the signal at terminals 6 and 7 of U_1 circuit (figure 2). There were analysed nine variants with different package types (SMD and through-hole) and working areas. In this paper only two PCBs (layouts designed in CADSTAR - figures 3 and 4) are presented. The board has an 5150 x 3300 mils area. In the first case we want a routing with a small number of vias. In these conditions, it is accepted a large value for the item cost (100) in the routing stage. The item being “expensive”, the system will try to find other solutions for a complete generation of interconnection structure. One of these solutions is the monolayer routing with routes in a “bus” type configuration. The supply network was manually created in PCB environment. After a first processing stage, the route editor left 15 unfinished connections. These are due to the SMDs, the routes being forced to finalise only on one layer. After that, a smoothing stage was applied and other vias reduced costs (50, 20, 10). After the routing process the number of vias was 90. In the second case we want a routing with a large number of vias. As in the “through-hole” case it was introduced a null cost for vias at routing. The routing was done with effort 10 in the XOY co-ordination system, the orthogonal strategy was applied. The obtained number of vias was 278; after routing the interconnection structure was submitted to smoothing [4]. The parasitic elements of these layouts were extracted with POLARIS and presented in figures 5 and 6. These parasitic elements were used for the computations [1] of the new data set. The signals analysed with PSPICE are presented in figures 7 and 8.

5. Conclusions

The reducing of manufacturing costs must be realized in corelation with the keeping with the signal integrity. In the interconnection structure design one must take into account all the parasitic influences. Short routes (tracks) are preferable. These could lead, however, to a large number of vias (high cost). In our examples the PCB with a reduced number of vias (90) presents larger values of parasitic elements due to the increased length of the routes and of the parallel route segments ("bus" route structure). In the case of PCBs with large number of vias (278), the values of parasitic elements decrease due to the decreased number of routes length and of the large distances between routes. In some practical situations for the reduceing of costs one must be realized a compromise have in mind the vias number on one side and the EMC components used on the other side.

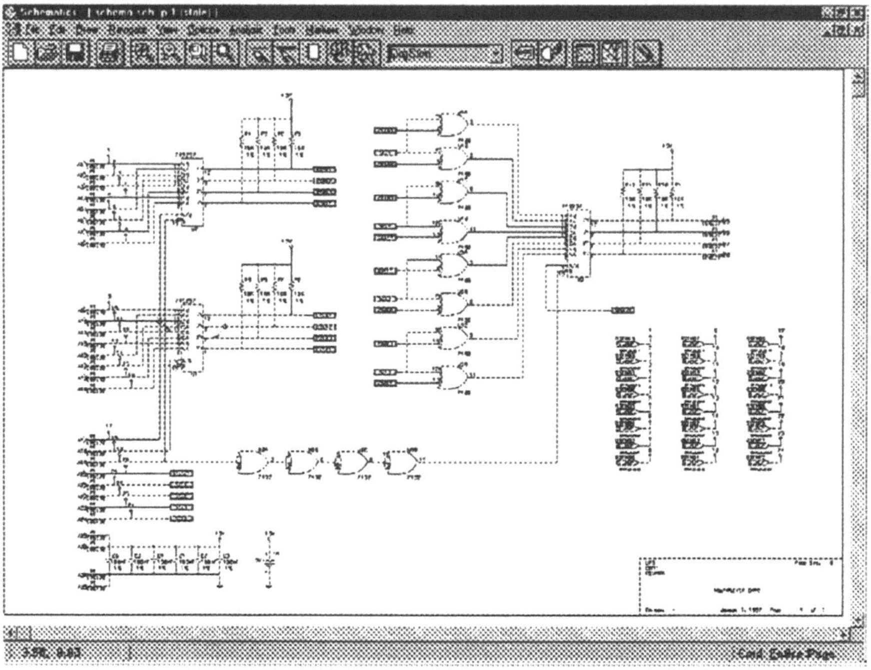

Figure 2. The schematic drawing of the analysed circuit

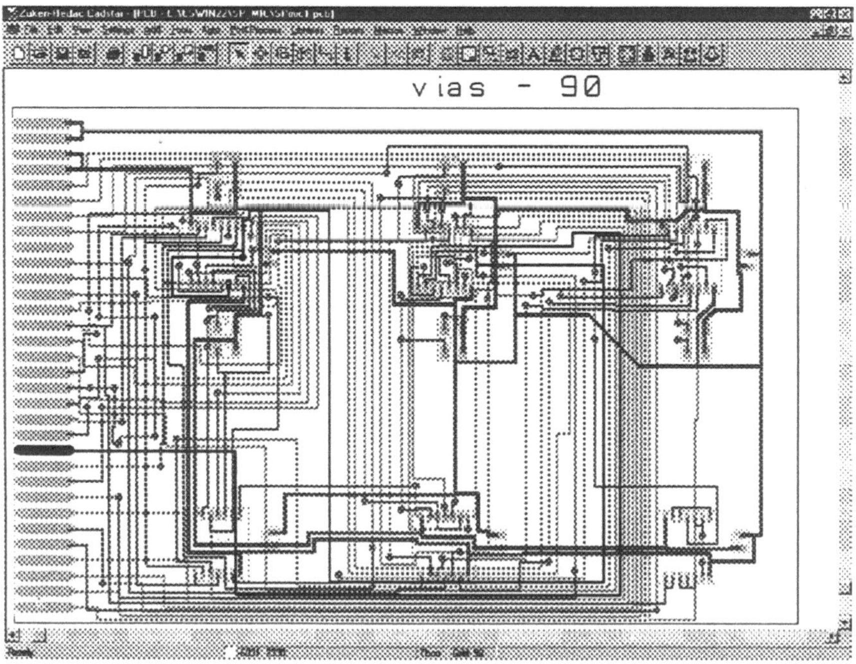

Figure 3. PCB with a small number of vias

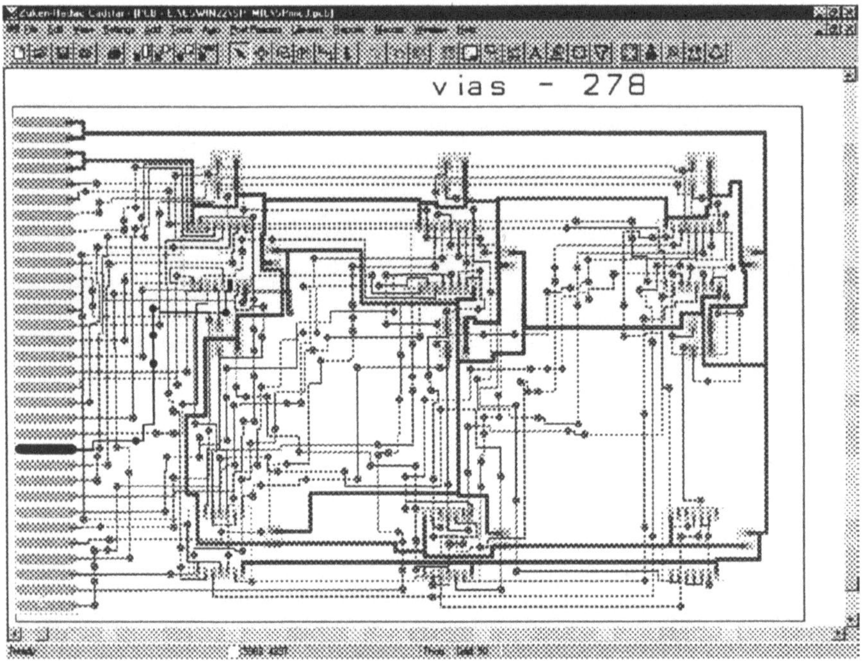

Figure 4. PCB with a large number of vias

Select signals

Name	Coupling	Length	L (nH)	R	C (pF)	C_via	Z0	Tpd (ns)
3	2.977465	0.083007	151.18	0.307358	5.65	3.00	163.52	0.924530
5	2.945046	0.089459	164.18	0.331247	3.96	4.00	203.72	0.805900
4	2.543746	0.088500	161.31	0.327698	5.72	3.00	167.93	0.960566
NN_0002	2.398068	0.076606	140.18	0.283657	4.10	4.00	184.97	0.757860
7	2.379533	0.084353	153.89	0.312343	5.10	3.00	173.79	0.885469
23	2.211136	0.101524	186.39	0.375921	4.19	7.00	210.90	0.883797
17	2.196730	0.091948	168.07	0.340464	5.09	3.00	181.67	0.925123
CCLK2	2.163257	0.079728	146.37	0.295216	3.37	5.00	208.28	0.702743
11	2.102672	0.172314	315.25	0.638041	9.00	3.00	187.18	1.684249

Figure 5. Results of the extraction with POLARIS of the PCB from figure 3

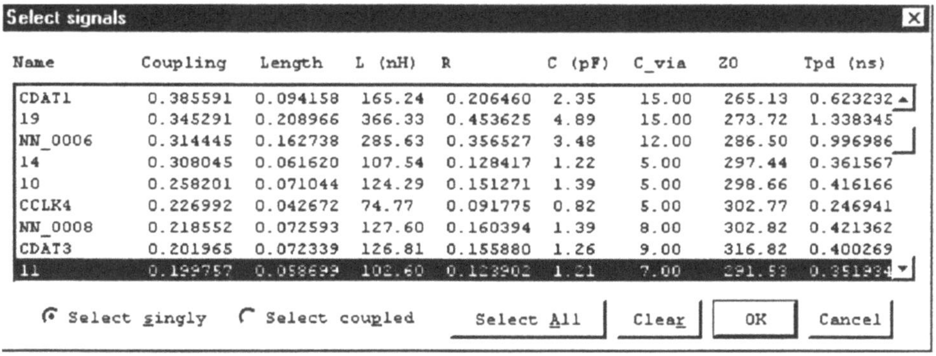

Select signals

Name	Coupling	Length	L (nH)	R	C (pF)	C_via	Z0	Tpd (ns)
CDAT1	0.385591	0.094158	165.24	0.206460	2.35	15.00	265.13	0.623232
19	0.345291	0.208966	366.33	0.453625	4.89	15.00	273.72	1.338345
NN_0006	0.314445	0.162738	285.63	0.356527	3.48	12.00	286.50	0.996986
14	0.308045	0.061620	107.54	0.128417	1.22	5.00	297.44	0.361567
10	0.258201	0.071044	124.29	0.151271	1.39	5.00	298.66	0.416166
CCLK4	0.226992	0.042672	74.77	0.091775	0.82	5.00	302.77	0.246941
NN_0008	0.218552	0.072593	127.60	0.160394	1.39	8.00	302.82	0.421362
CDAT3	0.201965	0.072339	126.81	0.155880	1.26	9.00	316.82	0.400269
11	0.199757	0.058699	102.60	0.123902	1.21	7.00	291.53	0.351934

Select singly   Select coupled   Select All   Clear   OK   Cancel

Figure 6. Results of the extraction with POLARIS of the PCB from figure 4

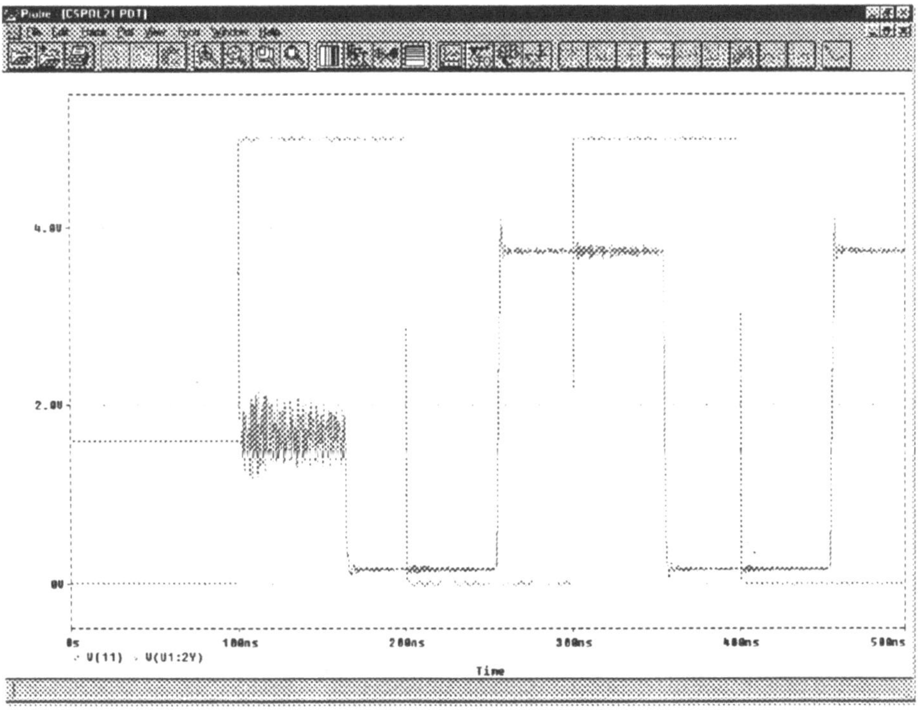

Figure 7. Results of the simulation with PSPICE for the PCB from figure 3

measurement of nonlinearity of nominally linear components, such as resistors and capacitors. The comparison of the results of the life-time tests carried out as the classical accelerated environmental tests with the results of the measurement of nonlinearity shows a comparable correlation like results of accelerated life-time test and noise measurement in many cases. Experience showed that the components which exhibit a relatively high nonlinearity are less stable a have the shorter life-time than the others usually. The relationship between reliability of the component and its nonlinearity correspondents very closely with the already accepted relationship between current noise of the resistor and its reliability.

2. Origin of Nonlinearity

The nonlinearity is closely connected with the noise of components. There is as yet no generally accepted complete explanation for the occurrence of current noise but many theories have centered around the modulation in conductivity which arises when current carriers pass into and out of the conduction band. However, a similar general explanation can not be utilized to explain nonlinearity as there is no reason why any such interchange of carriers should be dependent upon the level of the applied voltage. It is assumed that the structure of the thick film resistive and conductive layer consists of particles of a conductive material which are located in insulating matrix. The particles are in a direct mutual contact. Inside the conductive material the conductivity depends upon a phonon-electron interaction. Electron overcomes the boundary of the particle, leaves one particle and enter a neighboring one, by different mechanisms, e.g. by tunneling, Schotky emission, hoping etc. These mechanisms of conductivity can be near to linear, slightly nonlinear or strongly nonlinear. It depends on material parameters and technology of processing. The C-V characteristic of the resistor is influenced by all these mechanisms including the main mechanism of phonon-electron interaction inside the conductive particles. Therefore it can be less or more nonlinear. When the applied voltage increases, new channels with mostly nonlinear mechanisms of conductivity are opened. According to the type of the channels the nonlinearity of the component changes.

Another idea describes the mechanism of nonlinearity as follows: when different energy levels or different carrier concentration between two regions of material exist a junction between these regions exists. Such the „junctions“ are distributed throughout the material and may clearly involve ohmic nonlinearity. This type of the junction has an assymmetrical characteristic. However, the resistors exhibit a symmetrical form of nonlinearity. This situation is caused by the random distribution of the junctions which are connected in series and in parallel in material. So, the basic idea about the origin of the nonlinearity of resistors is that the basic source of the nonlinearity is the complex random set of quasi-junctions distributed throughout the material of the resistive layer.

3. Measurement and Evaluation of Nonlinearity

The nonlinearity of the current vs. voltage characteristics of a good resistor is very low and therefore it usually can not be observed by the direct measurement of the C-V characteristic. The first method of the measurement of nonlinearity was based upon the measurement of the dynamic resistance at two levels of voltage. Using this method a „voltage coefficient of resistance“ was found. The measurement had to be carried out at the very low level of measuring voltage because a higher level could cause heating of the resistor. To avoid to heating of a component during measurement of nonlinearity the measuring voltage was applied for a very short time, the time 0.2 sec was utilized in commercial apparatus for measurement of the voltage coefficient. Serious disadvantage of this method was a very considerable dependence of the value of the voltage coefficient on the levels of applied voltage.

An alternative method for evaluation of the nonlinearity of the resistors is the method based on the measurement of distortion of the periodical voltage which occurs across the resistor when it is powered by the sinusoidal current. The basic arrangement of this method is presented in Fig. 1.

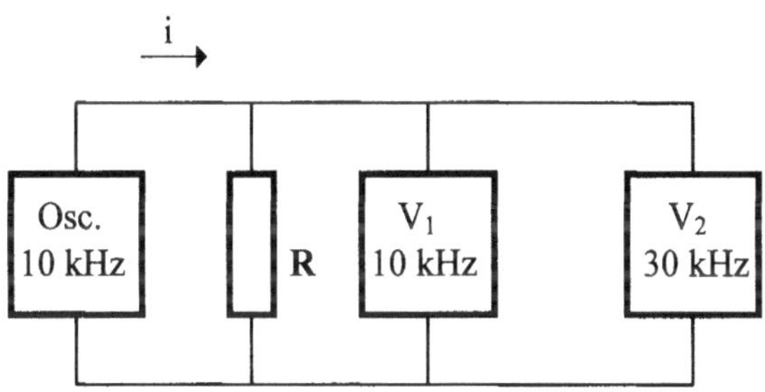

Figure 1. Schematic diagram of an equipment for indication of nonlinearity by the measurement of third harmonics which appears across the nonlinear resistance R powered by the sinusoidal current i

The quality of the oscillator has to be very high to avoid a systematic error which would be generated by a distortion of a the powering sinusoidal current. The voltmeters V_1 and V_2 are the selective voltmeters with a high selectivity. The voltmeter V_1 is tuned for frequency 10 kHz and is used for measurement of a voltage of the first harmonics which appears across the resistor under test. The voltmeter V_2 is tuned for the frequency 30 kHz and is used for measurement of the voltage of third harmonic component. Both the voltmeters have a linear and logarithmic scale. The equipment is completed by a matching transformer, the component can be matched to the 10 kHz generator and 30 kHz voltmeter within a very wide impedance range. The matching transformer is a very special design inasmuch as its residual distortion must be very low in order not to influence the quality of the measurements.

The nonlinearity of the component is expressed using THI (Third Harmonics Index) which is calculated according to the equation:

$$THI = 20 * \log \frac{U_3}{U_1^n} \qquad (dB, \mu V, V) \qquad (1)$$

Where U_3 ... voltage of third harmonic component, U_1 ... voltage of basic harmonic component, n ... coefficient which is near to 3.
It is necessary to accept for all measurements of nonlinearity the maximum power which can be dissipated by tested component to avoid to distortion of the measurement caused by heating of the resistor. It is assumed that maximum power must not overcome 10 % of the nominal power value of the component for the short-time measurement (time of the measurement $\leq$ 1 s). The influence of heating on measured characteristic of the component must be observed during the long-time measurement (time of measurement $\geq$ 1 s). When the measured value of third harmonics depends upon the time of measurement the level of the sinusoidal current is too high and must be decreased. This situation is often observed when components with low resistance (e.g. contacts) are measured.

4. Assessment of Reliability

The method of reliability assessment based on the evaluation of the nonlinearity of passive electronic components such as resistors and capacitors is based on the assumption that the component which exhibit the higher level of nonlinearity has more elementary „junctions" distributed throughout the active material and therefore it could be less stable and probably will have a shorter life-time in comparison with the rest of the batch. This relationship between the component nonlinearity and reliability corresponds with the already accepted relationship between noise and reliability, in spite of the ideas about the mechanisms of noise and nonlinearity differ.
As previously mentioned, the established measurement of nonlinearity is taken as a method of assessment of reliability of a component. The basic question is therefore to fix a reasonable limit for rejection of components.

5. Experimental Technique

Investigation of thick film resistors of different types was carried out. The changes of their nonlinearity with the changes of other typical parameters (noise, resistance, contact resistance) were investigated during thermal ageing. The resistors were made of three types of resistive pastes (100 Ω/□, 1 kΩ/□, 10 kΩ/□), contacts was made using two types of conductive pastes (Au, AgPd). Six different combinations resistive paste-conductive paste were used, sixty resistors were made for every combination of resistive-conductive paste. Nonlinearity was measured using the equipment CLT 1 (Radiometer Copenhagen), the amplitude of the sinusoidal current was very low to minimize heating of the resistor during the measurement. The same rule was respected for all types of measurements. Resistance was measured by the four-point method to avoid problems with the contact resistance between resistive and conductive layer. Contact resistance between conductive and resistive layer was measured by the three-point method. The schematic diagram of this method is shown in Fig. 2.

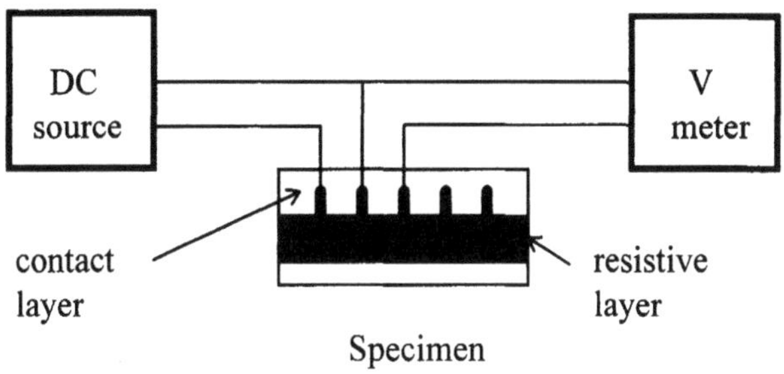

Figure 2. Three-point method for measurement of the contact resistance

Noise of the resistors was measured by the use of the equipment Quan-Tech, model 315. The measured voltage is very low and therefore the specimen has to be carefully screened during the measurement. The noise level is described by a noise index NI in dB:

$$NI = T - \left(f\{T - S\}\right) - D \qquad \text{(dB)} \qquad (2)$$

Where T ... total noise, S ... system noise, D ... noise equal to the powering dc voltage, f ... function for calculation of a correction factor

The noise index in dB may be converted to the noise voltage in μV/V.

The samples were aged in an oven with controlled temperature. Temperature of ageing was 120 ^{O}C, time of ageing was 4 weeks. The measurements were carried out to the end of every week of ageing. The samples were took out, cooled at the room temperature and measured. Then they were given back to the oven.

Some results are shown in Fig. 3 ... Fig 7 where the measurements of resistors made of paste 10 kΩ/□ are presented. Similar results were found for resistors made of paste 1 kΩ/□ and paste 100 Ω/□.

All points of presented characteristics were found by processing of the group of 21 measured values. The lowest and the highest value were neglected and the value of the point was calculated as an average of 19 values.

6. Measured Results and Discussion

The averaged values of THI are presented in Fig. 3. THI depends on amplitude of the input voltage. This dependence is not too strong usually and is influenced by the material of the resistor and by its technological processing. The averaged value of THI was calculated of all points of THI vs. U_1 characteristic. The range of the input voltage U_1 was chosen to avoid heating of the resistor. As for the resistors made of paste 10 kΩ/□ the voltage in the range of 0.1 - 10 V was used. Typical course of THI for U_1 = 10 V is presented in Fig. 4.

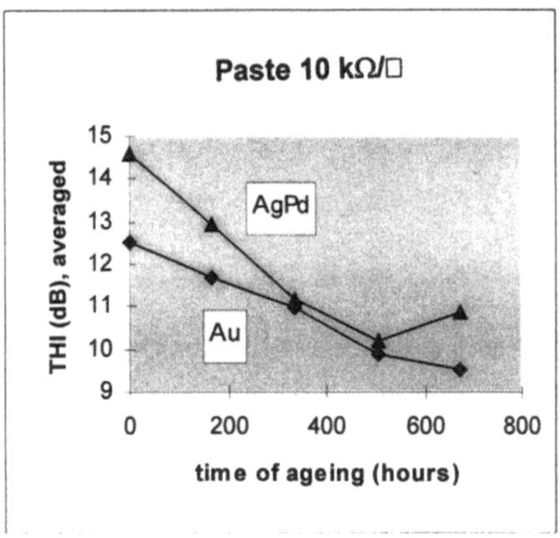

Figure 3. Averaged THI of the resistors

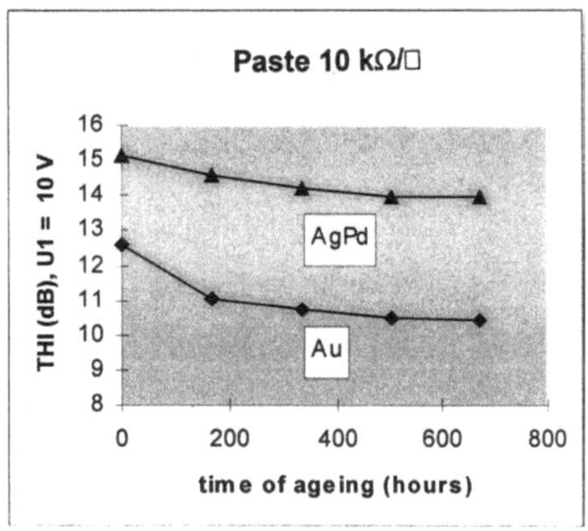

Figure 4. THI for U_1 = 10 V

The difference between both the THI characteristics is caused by the anomalous value of the last point of the characteristic of resistors contacted by layer made of AgPd paste. This anomaly was found for a part of the group of tested resistors at low amplitude of the voltage U_1 after the time of ageing 672 hours. It could be explained by a substantial change of the structure of resistive or conductive layer caused by ageing. It is interesting that this change did not appear in the other characteristics.

The change of the noise index during ageing is shown in Fig. 5. The contact resistance and its change is very low, as shown in Fig. 6. The change of resistance is presented in Fig. 7.

It was found that the correlation between the noise index and THI is high. Therefore it is possible to assume that the significance of THI for assessment of reliability of resistors is comparable with the significance of the noise index.

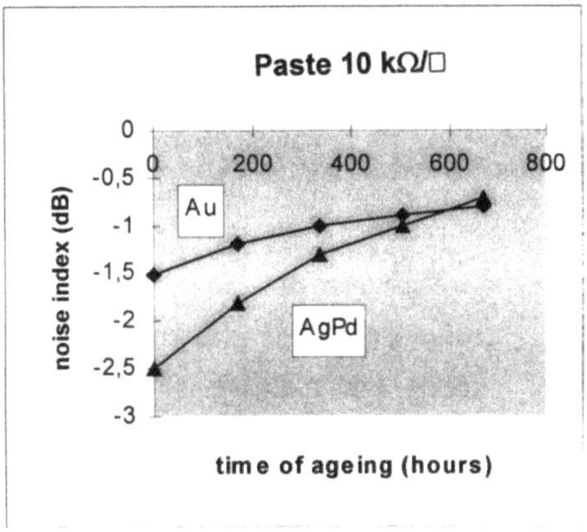

Figure 5. Noise index of the resistors

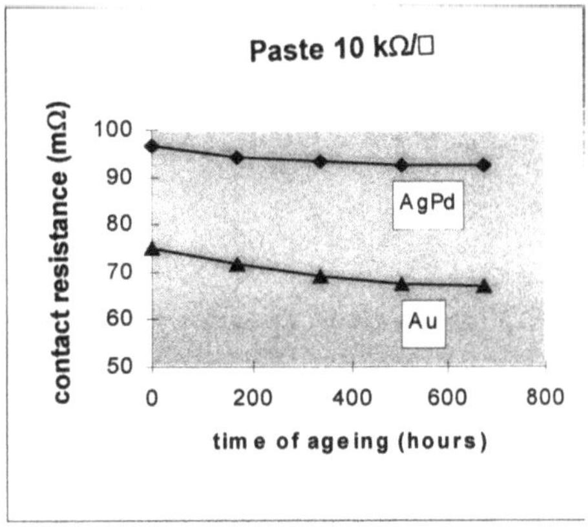

Figure 6. Contact resistance between resistive and conductive layer

The characteristics of the contact resistance and resistance are very similar. That is to say that the area of the mixed material between the conductive and resistive layer does not influence the properties of the resistor substantially.

7. Conclusions

Measurement of a big group of thick film resistors shown that the nonlinearity of resistors is in good correlation with their noise. Therefore it can be assumed that the relationship between component reliability and nonlinearity corresponds very closely with the already accepted relationship between resistor current noise and reliability. It was also found that the course of changes of the contact resistance and resistance of resistors are very similar.

The use of nonlinearity for reliability assessment has also some disadvantages. It is focused only on linear electronic components, such as resistors and capacitors. This method is not the absolute one. It is not possible to determine reliability of a component according to its nonlinearity when correlation between reliability and level of nelinearity is not known. However, when this relationship is known, it is very easy to assess reliability of the components and classify them as less reliable or more reliable on the basis of the measurement of their nonlinearity.

moisture-induced failures (see, for instance, Altimari et al., 1992, Kimura et al., 1992, Ilyas and Poborets, 1993, Ganssan and Berg, 1993, Liu, 1993, van Doorselaer, 1993). Such failures (Fig. 1) typically occur during surface-mounting the packages onto printed circuit boards (PCB) by means of high temperature (220-240°C) reflow soldering and are usually attributed to (although might not be limited by) high internal water vapor pressure caused by a sudden evaporation of the absorbed moisture.

Although the phenomenon and various consequences of moisture-induced damages in plastic packages are not yet completely understood, it has been established that the following factors and their unfavorable combinations play an important role, as far as such damages are concerned:
- High moisture content that not only leads to high water vapor pressure, but might result also in substantial decrease in the ultimate, fatigue, and brittle strength of the molding compound;
- Interfacial delaminations that lead to high stress concentrations, high water vapor pressure (due to the moisture accumulated in the gap between the delaminated surfaces) and, most importantly, result in considerable structural weakening of the package because of its inability to perform as a composite structure: the delaminated portion of the molding compound becomes isolated from the rest of the package and bends independently of the package as a plate;
- Low fracture toughness of the molding compound that makes the material unable to effectively withstand the initiation and propagation of cracks;
- Insufficient thickness (low section modulus) of the molding compound, especially if delamination takes place;
- Elevated thermally induced stresses caused by the thermal expansion (contraction) mismatch of the dissimilar materials in the package (mainly the chip and the molding compound), as well as by the temperature gradients in the compound (especially in the through-thickness direction)
- Initial stresses and initial bowing (Kelly, 1992) of the package resulting from the thermal expansion (contraction) mismatch of the materials in the package when it is cooled down from the manufacturing (curing) temperature to room temperature.

Clearly, each of these factors can play, in a particular package design, a greater or lesser role depending on the package's geometry, materials, and loading. It is clear also that the results of experimental investigations alone, valuable as they might be, inevitably reflect the combined effect of a variety of factors on the package failure, whereas what is needed for the understanding of the mechanical behavior, life prediction, and structural and materials optimization, is the knowledge of the role of each particular parameter affecting the package reliability. Therefore theoretical modeling can be very helpful for understanding, prediction, and optimization of the mechanical behavior of a plastic package subjected to reflow soldering operation during its surface mounting on a printed circuit board.

In the analysis which follows we describe two analytical stress models aimed at the assessment of the plastic package propensity to moisture-induced structural failure. The first model (Suhir, 1995, 1997) is based on the constitutive equations obtained as a generalization of von-Karman's equations for large deflections of plates. This generalization considers the combined effect of the lateral water vapor pressure and thermal loading. In the second model (Suhir and Ilyas, 1996) we suggest a simple and easy-to-use geometric parameter which reflects the role of the major geometric factors affecting the package

Moisture Induced Failures in a Typical Plastic Package

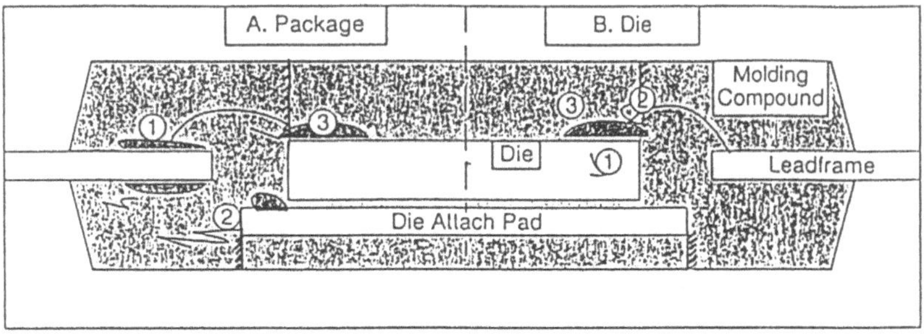

A. Package

1. Cracks between leads
2. Cracks in plastic around the die-attach pad (can be internal and/or extend to the surface)
3. Delamination of plastic from chip surface leading to cracks in the plastic

B. Die

1. Cracks in the silicon
2. Bond-wire failures
3. Passivation cracking
   Interlevel dielectric cracks
   Metal interconnect deformation
   Electrical parameter shifts

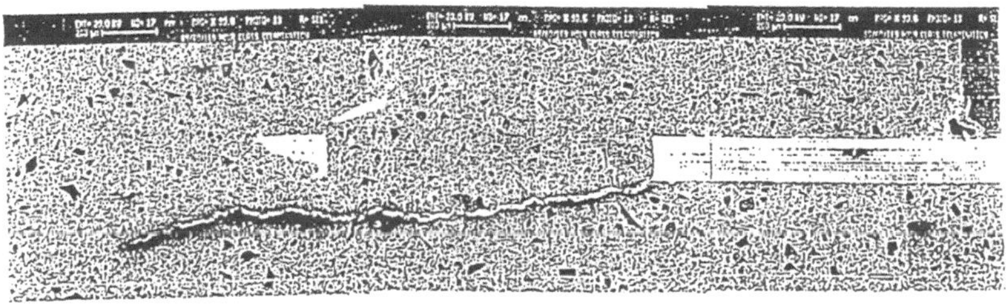

Fig. 1

$$\left.\begin{aligned} & w=0, \quad \text{for} \quad x=\pm\frac{a}{2}, \quad y=\pm\frac{b}{2} \\ & \frac{\partial w}{\partial x}=0, \quad \text{for} \quad x=\pm\frac{a}{2} \\ & \frac{\partial w}{\partial y}=0, \quad \text{for} \quad y=\pm\frac{b}{2} \end{aligned}\right\} \tag{10}$$

Here a is the dimension of the underchip layer of the molding compound along the x axis, and b is its dimension along the y axis. The origin O of the rectangular coordinates x, y, z, is in the center of the main plane.

The equations (1) are not different from the "conventional" von-Karman's equations for large deflections of plates subjected to a distributed lateral load p. The only difference is in how the flexural rigidity D and the in-plane compliance C are computed. These, when the thermally induced strains are accounted for, should be evaluated by the formulas (2) and (4), i.e. with consideration of the change in Young's modulus in the through-thickness direction as a function of temperature.

In an approximate analysis, the equations (1) can be further simplified by assuming that the aspect ratio of the underchip portion of the molding compound, treated as a clamped plate, is larger than 2.5, even if it might not be the case in an actual situation. This conservative assumption enables one to examine the bending of an elongated plate, rather than a plate of finite aspect ratio.

3. Calculation Procedure

In a practical calculation procedure the input data are: chip width, c; paddle width, a; underchip thickness, h; Young's modulus, E, and the coefficient of thermal expansion, α, of the molding compound as a function of temperature, T; Poisson's ratio, ν, of the compound; water vapor pressure, p.

The calculations can be carried out in the following sequence:

1) For the given temperature gradient, calculate the distribution of Young's modulus in the through-thickness direction and compute effective Young's moduli, E_D and E_C, either by the formulas (3) and (9), or by the simplified formulas (Suhir, 1997)

$$E_D = E_0 \frac{8(e_+ + e_- + e_+ e_- + 1) - (e_+^2 + e_-^2)}{5(e_+ + e_- + 4)}, \quad E_C = E_0 \frac{e_+ + e_- + 4}{6} \tag{11}$$

In the formulas (11), $e_+ = E_+/E_0$ and $e_- = E_-/E_0$ are the ratios of Young's moduli E_+ and E_- on the "hot" (outer) and the "cold" (inner) surfaces of the compound, respectively, to the midplane Young modulus E_0. The formulas (11) are obtained under an assumption that the distribution of Young's modulus of the molding compound in the through-thickness direction can be approximated by a parabola.

2) For the given thickness-to-width ratio h/*a* and the computed p/E_D ratio of the lateral pressure, p, to the effective Young modulus, E_D, in bending, determine the pressure factor

$$\chi_p = \frac{1-\nu^2}{32}\sqrt{\frac{254}{35}\frac{E_C}{E_D}\left(\frac{a}{h}\right)^4}\frac{p}{E_D} \tag{12}$$

3) For the given c/*a* ratio of the width, c, of the loaded area (chip's width) to the width, *a*, of the underchip portion of the compound, and the calculated pressure factor, χ_p, determine, from the plot in Fig. 2, the parameter u, which characterizes the level of the in-plane ("membrane") stress.

4) For the given c/*a* ratio and the computed u value, calculate the factors χ_f and χ_M, which consider the effects of the c/*a* ratio and the u value on the maximum deflection, f, of the plate and the maximum bending moment M. These factors can be calculated by the formulas

$$\chi_f = \frac{24}{u^4 \sinh u}\left\{\left(u\frac{c}{a}\right)(1-\cosh u) + \sinh\left(u\frac{c}{a}\right) + \right.$$
$$\left. + \sinh\left[u\left(\frac{c}{a}\right)\right] + \left[\frac{u^2}{2}\frac{c}{a}\left(2-\frac{c}{a}\right) - 1\right]\sinh u\right\}, \tag{13}$$

$$\chi_M = \frac{3}{u^2}\frac{\left(u\frac{c}{a}\right)\cosh u - \sinh\left(u\frac{c}{a}\right)}{\sinh u}, \tag{14}$$

or can be taken from the plots in Figs. 3 and 4, respectively.

5) Determine the maximum bending stress by the formula

$$\sigma_b = \sigma_0 \chi_M\left(u, \frac{c}{a}\right), \tag{15}$$

where

$$\sigma_0 = \frac{pa^2}{2h^2} \tag{16}$$

is the bending stress in the case of zero in-plane force (u = 0) and a lateral load distributed over the entire width of the plate (c = *a*).

6) Determine the maximum "mechanical" in-plane stress (i.e. the stress due to the pressure p only) by the formula

$$\tilde{\sigma}_x^0 = 0.002362(1-\nu^2)E_C\left(\frac{p}{E_D}\right)^2\left(\frac{a}{h}\right)^6\chi_f^2 \tag{17}$$

As evident from this formula, the in-plane ("membrane") stress can be rather high, if the lateral (water vapor) pressure p is appreciable and the ratio *a*/h is large. This is thought to be the case in thin small outline packages (TSOP's).

7) Calculate the ratio, f/h, of the deflection, f, to the underchip thickness, h, of the compound by the formula:

u / c/a	0	1	2	3	4	6
0.2	0.296	0.274	0.226	0.180	0.144	0.0994
0.3	0.436	0.404	0.335	0.267	0.215	0.149
0.4	0.568	0.527	0.439	0.352	0.284	0.198
0.5	0.688	0.639	0.535	0.432	0.350	0.246
0.8	0.944	0.884	0.753	0.622	0.516	0.375
1.0	1.000	0.949	0.806	0.667	0.562	0.417

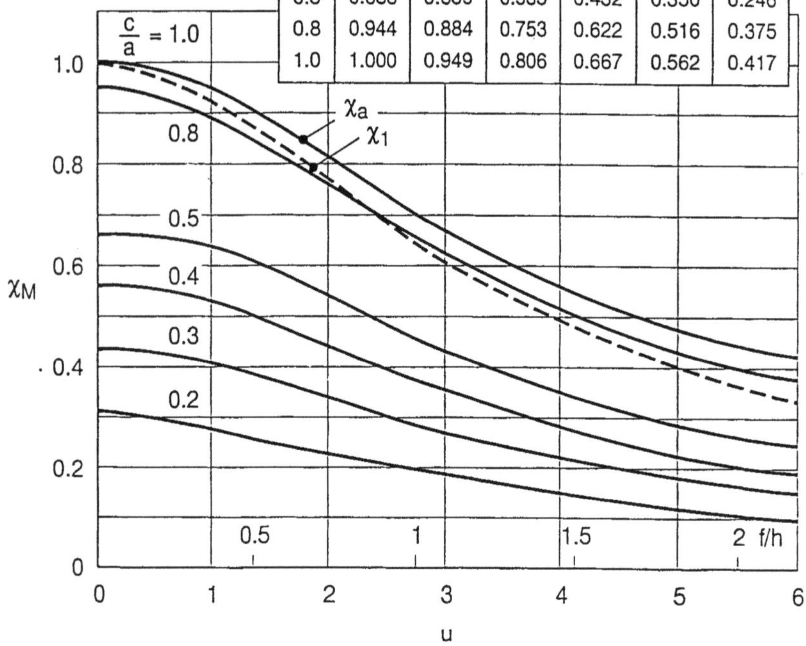

Fig. 4

$$\frac{f}{h} = \sqrt{\frac{35}{254}\frac{E_D}{E_C}}\,u = 0.3712\sqrt{\frac{E_D}{E_C}}\,u \qquad (18)$$

8) Determine the parameter

$$\eta = \frac{E}{E_C}\frac{\tilde{\sigma}_x^0}{\sigma_b} \qquad (19)$$

which considers the role of the membrane stresses (E is the specified value of the Young modulus of the molding compound) and compute the von-Mises stress

$$\sigma_M = \sigma_b\sqrt{1+(2-\nu)\eta^2+(1-\nu+\nu^2)\eta+[1+(1+\nu)\eta\frac{\alpha\Delta T}{1-\nu}\frac{E-E_C}{\tilde{\sigma}_x^0}} \qquad (20)$$

Numerical example

Let the temperature gradient in the through-thickness direction for the underchip portion of the molding compound be such that Young's modulus of the material is $E_+ = 0.10\times10^6$ psi = 70.3 kg/mm^2 on the "hot" (external) side of the package, $E_- = 0.15\times10^6$ psi = 105.5 kg/mm^2 on the "cold" (internal) side, and $E_0 = 0.20\times10^6$ psi = 140.6 kg/mm^2 in the midplane area of the compound. Then the formulas (11) yield: $E_D = 0.1538\times10^6$ psi = 108.1 kg/mm^2, E_C=0.1750×10^6 psi = 123.1 kg/mm^2. Let the water vapor pressure be p = 360 psi = 0.253 kg/mm^2 (which corresponds to the pressure of a saturated water vapor at 220°C), the width of the underchip portion of the molding compound be a = 20 mm, and its thickness be h = 1 mm. Assuming for Poisson's ratio ν = 0.4, the formula (12) yields: $\chi_p = 28.250$. Let the chip be c = 10 mm wide, so that the chip-to-paddle with ratio is c/a = 0.5. The plot in Fig. 2 yields: u = 5.65. From the formula (18) we find: $\frac{f}{h} = 1.966$. Then from the plot in Fig. 3, for c/a = 0.5 and u = 5.65, we obtain: $\chi_M = 0.26$. If the in-plane stresses were not accounted for and the lateral water vapor load were assumed distributed over the entire width of the delaminated underchip portion of the compound, then the bending stress, as predicted by the formula (16), would be σ_0 = 72000 psi = 50.6 kg/mm^2. Thus, the fact that the tensile force is not zero and the lateral load acts only in the midportion of the package, leads to a substantial, almost by a factor of four, reduction in the bending stress: $\sigma_b = \chi_M\sigma_0 = 0.26 \times 72000 = 8720$ psi = 6.13 kg/mm^2. The "mechanical" transverse in-plane stress (acting in the x direction) can be determined by the formula (17) and is: $\tilde{\sigma}_x^0 = 4870$ psi = 3.42 kg/mm^2. From (19) we have: η = 0.2230. Let the coefficient of thermal expansion of the molding compound be $\alpha = 60 \times 10^{-6}$ 1/°C, and the change in temperature from the curing temperature to the reflow soldering temperature be ΔT = 80°C. Then the formula (20) results in the following value of the von-Mises stress: σ_M = 22016 psi = 15.5 kg/mm^2. This stress is quite high and can possibly result in the failure of the molding material.

Let us estimate now, whether an increase in the thickness of the underchip portion of the molding compound can lead to an appreciable decrease in the level of the predicted von-Mises stress. Let the thickness of this portion be, say, h = 1.5 mm, instead of h = 1.0 mm in the previous example. Then the formula (12) yields: $\chi_p = 5.580$. From the plot in Fig. 2 we find: u = 2.65, and the formula (18) results in the following value of the deflection-to-thickness ratio: f/h = 0.9221. The corresponding deflection f = 1.383 mm is appreciably smaller than the deflection f = 1.966 mm in the case of a 1 mm thick compound. The bending stress, calculated in the absence of the in-plane forces for a lateral load distributed over the entire width of the plate, is σ_0 = 32,000 psi. The factor χ_M, which considers the effect of the finite c/a ratio and the finite value of the tensile stress on the maximum bending moment, is $\chi_M = 0.20$, and the corresponding bending stress is σ_b = 6,400 psi. This value is significantly lower than the bending stress in a 1 mm thick compound. The "mechanical" in-plane stress, predicted by the formula (17), is $\tilde{\sigma}_x^0$ =2,410 psi. From (19) we have: $\eta = 0.3228$, so that the von-Mises stress is σ_M= 8,023 psi = 5.64 kg/mm^2. Thus, an increase in the thickness of the underchip portion of molding compound by a factor of 1.5 led to a 2.74-fold reduction in the von-Mises stress.

Calculated von-Mises and Bending Stresses

The maximum von-Mises, σ_M, and bending, σ_b, stresses calculated for the molding compound, assuming its complete delamination from the pad, are shown in Table 1. The systematic calculations were carried out for different underchip thicknesses, h, different widths, a, of the paddle, different widths, c, of the chip, different water vapor pressures, p (these were assumed uniformly distributed over the underchip area), and different Young's moduli, E, of the molding compound. The bending stresses were calculated from the deflections computed with consideration of the membrane stresses, and therefore the data for the bending stress in this table reflect the contribution of the bending stress to the maximum value of the von-Mises stress evaluated from the nonlinear equations. Clearly, the bending stresses were substantially higher if they would be calculated from the linear theory.

As evident from the calculated data, the underchip thickness is the most important geometrical characteristic which has the strongest effect on the stress level. When this thickness changes from h = 0.600 mm in a "thin" package to h = 1.200 mm in a relatively "thick" package, i.e. by a factor of two, the calculated nonlinear maximum bending stress decreases approximately by a factor of four (in a linear analysis this factor would be exactly "four"). This can be seen from the following formula, obtained for the maximum bending stress in the delaminated underchip portion of the molding compound treated as an elongated rectangular plate clamped at the ends and partially loaded in its midportion by a uniformity distributed load (Suhir, 1995):

$$\sigma_b = p\frac{ac}{4h^2}\left(3-\frac{c^2}{a^2}\right) \tag{21}$$

In this formula, σ_b is the maximum bending stress in the molding compound at the chip corner in the middle of the longer side of the chip.

The maximum von-Mises stress, calculated with consideration of the finite aspect ratio of the underchip layer of the molding compound (i.e. with consideration of the increase in the total stress due to bending in the longitudinal direction), as well as with consideration of the effect of the in-

TABLE 1

		Calculated Von-Mises and Bending Stresses, kg/mm.sq																		
		h,mm	0.120						0.600						1.200					
p,ps	E × 10⁻⁴, psi	a,mm	4	8			12		4	8			12		4	8			12	
		c,mm	4	4	8	4	8	12	4	4	8	4	8	12	4	4	8	4	8	12
200	5	σ_M	16.6	16.3	26.9	16.0	26.2	35.5	3.14	5.70	8.24	5.67	9.18	11.45	0.78	2.17	3.14	3.15	5.21	6.11
		σ_b	14.2	12.9	22.9	12.4	21.1	30.1	3.07	4.97	7.35	4.65	7.74	9.93	0.78	2.12	3.07	2.93	4.79	5.65
	10	σ_M	20.8	20.5	33.8	20.2	33.0	44.6	3.15	6.92	9.85	7.19	11.56	14.15	0.78	2.17	3.15	3.37	5.77	6.75
		σ_b	17.8	16.3	28.8	15.6	26.7	37.9	3.11	6.22	8.97	6.04	9.89	12.39	0.78	2.14	3.11	3.22	5.47	6.41
	15	σ_M	23.7	23.5	38.7	23.2	37.8	51.0	3.14	7.57	10.75	8.23	13.18	15.97	0.78	2.16	3.14	3.41	5.94	6.95
		σ_b	20.3	18.7	32.9	17.9	30.6	43.3	3.12	6.94	9.93	7.02	11.39	14.07	0.78	2.15	3.12	3.31	5.71	6.70
360	5	σ_M	24.8	24.0	40.0	23.7	38.7	52.6	5.54	8.50	12.53	8.32	13.54	17.16	1.41	3.85	5.54	5.07	8.18	9.67
		σ_b	21.1	19.0	33.9	18.3	31.2	44.6	5.33	7.25	11.03	6.72	11.31	14.80	1.41	3.70	5.33	4.57	7.34	8.76
	10	σ_M	31.0	30.3	50.2	29.9	48.8	66.2	5.66	10.59	15.26	10.58	17.11	21.30	1.41	3.90	5.66	5.76	9.57	11.21
		σ_b	26.5	24.0	42.7	23	39.4	56.1	5.54	9.28	13.65	8.70	14.46	18.49	1.41	3.82	5.54	5.37	8.84	10.41
	15	σ_M	35.3	34.8	57.4	34.3	55.9	75.7	5.67	11.88	16.97	12.16	19.59	24.11	1.41	3.90	5.67	5.99	10.18	11.91
		σ_b	30.3	27.5	48.8	26.4	45.2	64.2	5.59	10.60	15.36	10.14	16.60	21.04	1.41	3.85	5.59	5.69	9.57	11.24
500	5	σ_M	30.9	29.9	49.8	29.5	48.2	65.5	7.51	10.58	15.79	10.32	16.82	21.49	1.96	5.24	7.51	6.48	10.38	12.35
		σ_b	26.4	23.6	42.3	22.7	38.8	55.5	7.14	8.94	13.82	8.27	13.99	18.49	1.95	4.98	7.14	5.73	9.20	11.08
	10	σ_M	38.7	37.7	62.6	37.2	60.7	82.4	7.82	13.29	19.35	13.11	21.28	26.72	1.96	5.40	7.82	7.60	12.43	14.60
		σ_b	33.1	29.8	53.2	28.7	49.0	69.9	7.59	11.49	17.16	10.68	17.88	23.11	1.95	5.25	7.59	6.98	11.31	13.40
	15	σ_M	44.1	43.2	71.6	42.6	69.5	94.3	7.86	15.05	21.65	15.09	24.39	30.29	1.96	5.42	7.86	8.08	13.47	15.78
		σ_b	37.8	34.2	60.8	32.8	56.2	79.9	7.71	13.22	19.40	12.44	20.64	26.32	1.95	5.32	7.71	7.56	12.49	14.70

h = underchip thickness of the molding compound
a = pad width
c = chip width
E = Young's modulus of the molding compound
p = water vapor pressure

plane ("membrane") stresses, is somewhat higher than the maximum bending stress. The difference in the calculated von-Mises stress and in the bending stress is due primarily to the difference in the level of the membrane stresses and is higher for thinner packages, larger chips and smaller the chip-to-paddle widths ratios.

The difference in the role of the in-plane ("membrane") stresses, as far as bending of the underchip portions of the molding compound is concerned, can be taken as a suitable structural criterion, which enables one to distinguish between "thick" and "thin" packages. In thick packages (i.e. in packages with relatively thick underchip layers) with small chips, and/or low water vapor pressures, bending stresses only could be considered. In a situation, when the ratio of the underchip thickness of the molding compound to the chip (paddle) width is larger than, say, 10%, the shearing stress should be also considered (we do not account for this stress in our analysis). Thin packages, however, are characterized by appreciable in-plane ("membrane") stresses. These should be considered, in addition to bending stresses, when evaluating the total stress in the molding compound of a thin package. The role of the in-plane stresses increases with a decrease in the underchip thickness of the compound, and with an increase in the chip and the paddle size. One can use the ratio of the von-Mises stress to the bending stress to judge on whether the package should be considered a "thick" or a "thin" one: in a "thick" package this ratio is close to one, while in a "thin" package this ratio is substantially larger than one.

In the case of a package with an underchip thickness of h = 0.600 mm, pressure level of p = 200 psi and Young's modulus of the molding compound of E = 5×10^4 psi (above glass transition temperature), the calculated maximum von-Mises stress, in the case of a 8mm wide chip, whose widths coincides with the width of the pad (σ_M = 8.24 kg/mm^2), is by a factor of 1.121 greater than the maximum bending stress (σ_b = 7.35 kg/mm^2). This factor increases to 1.186 when the paddle width increases to 12mm. When both the chip and the paddle are 12mm wide, the von-Mises stress exceeds the bending stress by a factor of 1.153. In the case of an underchip thickness of h = 1200 mm, the maximum von-Mises stress, for a 8mm wide chip and 8mm wide paddle, exceeds the bending stress by a factor of only 1.023. This factor increases to 1.088 for a 12mm wide pad. When both the chip and the substrates are 12mm wide, the ratio of the von-Mises stress to the bending stress is just a little smaller: 1.081.

When the applied pressure increases, both the bending and the membrane stresses increase. While the bending stress, calculated for the given deflection, whether linear or nonlinear, is proportional to the applied pressure, the membrane stress is proportional to the applied pressure squared. Therefore, the ratio of the maximum von-Mises stress to the maximum bending stress increases with an increase in the pressure level. For the pressure level of p = 360 psi (which is the pressure of a saturated water vapor at the reflow soldering temperature of 220°C), and Young's modulus of the material of E = 5×10^4 psi, the calculated maximum von-Mises stress, in the case of an underchip thickness of h = 0.600 mm, chip width c = 8 mm and the paddle width a = 8mm, is about σ_M = 12.53 kg/mm^2, and the maximum bending stress is σ_b = 11.03 kg/mm^2, so that the stress ratio is 1.136. When the paddle width is a = 12 mm, these stresses are σ_M = 13.54 kg/mm^2 and σ_b = 11.31 kg/mm^2, respectively, and their ratio is σ_M / σ_b = 1.197. When both the chip and the paddle are 12mm wide, the von-Mises stress is σ_M = 17.11 kg/mm^2, and the bending stress is σ_b = 14.46 kg/mm^2, so that their ratio is 1.183. For a thick package with an underchip thickness of h = 1.200 mm, the von-Mises and the bending stresses are σ_M = 9.67 kg/mm^2 and σ_b= 8.76 kg/mm^2, respectively, and their ratio is only 1.103.

The increase in Young's modulus results in the increase in the von-Mises stress. This is due primarily to the fact that larger Young's moduli lead to higher membrane stresses. Since the membrane stresses are especially high in thin packages, there is an incentive for the application of low modulus molding compounds in thin packages, experiencing appreciable membrane stresses, especially if the expected water vapor pressure is high. In thick packages, the application of low modulus materials is not expected to result in a substantial reduction in the stress level. If, for instance, in a package, experiencing water vapor pressure of p = 200 psi and characterized by the geometric characteristics h = 0.600 mm, c= 8 mm, and a = 8 mm, the Young modulus decreases from E = 15×10^4 psi to E = 5×10^4 psi, the calculated von-Mises stress decreases from 16.97 kg/mm^2 to 12.53 kg/mm^2, i.e. by a factor of 1.354. In a h = 0.120 mm thick underchip layer, this factor is appreciably high: 1.423. In a h = 1.200 mm thick layer it is only 1.023.

The von-Mises and bending stresses were calculated also for a hypothetical case of a very thin package, in which the underchip thickness is only h = 0.120 mm. Such a case might be of interest from the viewpoint of how far can one "push" the existing technology to make plastic packages thinner-and-thinner and the chips larger and larger. The calculated data indicate that for such a package both the von-Mises and the bending stress are very high. Indeed, for a package with the geometric characteristics c = a = 8mm, pressure level of p = 200 psi and Young's modulus of the material of E = 5×10^4 psi, the calculated von-Mises stress is as high as σ_M = 26.9 kg/mm^2, i.e. by a factor of 3.264 higher than this stress in a package with the underchip thickness of of h = 0.600mm. When the applied pressure becomes p = 360 psi, the von-Mises stress increases to σ_M = 40.0 kg/mm^2. This stress is very high and, as evident from the Table 2, exceeds significantly the ultimate strength of a typical molding compound. The actual situation could be even worse, since this data were obtained for room temperature condition and for materials properties corresponding to a relatively dry material.

The von-Mises stresses calculated for several actual plastic packages, assuming constant pressure of p = 360 psi, are shown in Table 3. As evident from this table, the von-Mises stress varies in a very wide range, from 0.300 kg/mm^2 in the case of a thick, 68 PLCC, package (h = 1.513 mm) to 20.070 kg/mm^2 for a very thin, 100 TQFP, package (h = 0.600mm), i.e. by a factor of 66.9! Although in reality the difference will be less significant (thinner packages are expected to experience lower water vapor pressure at failure), it is still expected to be large. Indeed, no failures were observed in the 68 PLCC package, while the 100 TQFP package failed at a relatively low moisture level (Ilyas and Poborets, 1993). Note that since in very thin packages the "membrane" stress plays an important role and, at the same time, is affected substantially by the Young modulus value, it is imperative that this modulus is known with sufficient accuracy at the reflow soldering temperatures (i.e. in the region above glass transition temperature). Since this information is typically unavailable (Table 2 gives Young's moduli values at room temperature), in Table 1 it is simply assumed that Young's moduli of the molding compound at the reflow soldering temperatures are as low as E = (5 $\rightarrow$ 15) 10^4 psi = (35 $\rightarrow$ 105) kg/mm^2. This is by an order of magnitude lower than the Young moduli in Table 2 whose values are taken from the vendor specifications.

TABLE 2

MATERIAL / PROPERTIES	YOUNG'S MODULUS.	COEFFICIENT OF THERMAL EXPANSION, α, 1/°C		GLASS TRANSITION T-RE.	MOLDING T-RE.	MOLDING PRESSURE	ULTIMATE STRESS	THERMAL FACTOR $\sigma_e = E\Delta\alpha\Delta t$,	STRESS RATIO	SPIRAL FLOW LENGTH AT 180°C.	THERMAL CONDUCTIVITY cal/cm×sec×°C
	E, kg/mm^2	ABOVE T_g	BELOW T_g	T_g, °C	t_m °C	p, kg/mm^2	σ_f, kg/mm^2	$kg/mm^{2*)}$	$\eta = \frac{\sigma_f}{\sigma_e}$	l, cm	
NITTO KS44	1200	58×10^{-6}	13×10^{-6}	150	165-185	0.70	16	1.47	10.88	95	15×10^{-4}
HITACHI CEL-9000	2100		12×10^{-6}	120			18	1.76	10.25	76	
HITACHI 4650-SX4	1820	55×10^{-6}	10×10^{-6}	165	175-180	0.56-0.84	15	1.73	8.86	84	16×10^{-4}
NITTO X-7000	1650		16×10^{-6}	120			17	2.01	8.47	100	
NITTO X-7000H	1650		16×10^{-6}	120			17	2.01	8.47	70	
TOSHIBA KE 1300	1550		12×10^{-6}	155			15	1.77	8.46	85	
SUMITOMO 7302	1850		12×10^{-6}	155	175	0.70	17	2.12	8.03	90	16×10^{-4}
SHIN-ETSU KMC-180-V-B	1350	55×10^{-6}	11×10^{-6}	180	160-180	0.49-0.98	12.5	1.63	7.66	57-78	16×10^{-4}
SHIN-ETSU KMC-180-V-A	1250	55×10^{-6}	13×10^{-6}	165	160-180	0.49-0.98	12.5	1.71	7.29	51-76	15×10^{-4}
PLASKON 3400	1400	69×10^{-6}	17×10^{-6}	155	170-185	0.52-0.70	16	2.52	6.87	90	16×10^{-4}
PLASKON ULS-12	1440		12.5×10^{-6}	155			11	1.74	6.32	89	
NITTO MP-180	1100	65×10^{-6}	15×10^{-6}	169	165-185	0.70	11	1.87	5.88	90	15×10^{-4}
SUMITOMO	1100	72×10^{-6}	15×10^{-6}	185	175	0.70	12.2	2.08	5.87	75	16×10^{-4}
SHIN-ETSU KMC-165-4	1200	65×10^{-6}	16×10^{-6}	165	160-180	0.49-0.98	12.5	2.15	5.81	102	15×10^{-4}
SUMITOMO EME 6300HD	1200		17×10^{-6}	160	175	0.70	12.5	2.24	5.59	85-115	16×10^{-4}
SUMITOMO EME 6300HD-86	1200	72×10^{-6}	17×10^{-6}	162	175	0.70	12.5	2.27	5.51	86	16×10^{-4}
TOSHIBA KE 1000	1180		15×10^{-6}	200			13	2.44	5.33	85	
SUMITOMO EME 6300HD-102	1200	73×10^{-6}	17×10^{-6}	161	175	0.70	11.6	2.25	5.15	102	16×10^{-4}
SUMITOMO 7700	1200		15×10^{-6}	190	175	0.70	12	2.33	5.14	80	16×10^{-4}
HITACHI CEL 4620SX	1650	62×10^{-6}	16×10^{-6}	158	175-180	0.56-0.84	14	2.81	4.98	71	14×10^{-4}
HITACHI CEL-X-7600	1270	65×10^{-6}	17×10^{-6}	165	180	0.56-0.84	12	2.45	4.89	89	18×10^{-4}
NITTO MP 119	1250	65×10^{-6}	18×10^{-6}	170	165-185	0.70	13	2.68	4.85	90	15×10^{-4}
SUMITOMO EME 6300HD-115	1200	73×10^{-6}	18×10^{-6}	167	175	0.70	12.2	2.52	4.84	115	16×10^{-4}
PLASKON S-7	1300	39×10^{-6}	16×10^{-6}	155	170-185	0.52-0.70	10	2.16	4.62	90	16×10^{-4}

*) $\Delta t = T_g - 25°C$, $\Delta\alpha = (\alpha - 3.2 \times 10^{-6})$ 1/°C

TABLE 3

Package type	Die size (c) (mm)	Paddle size (a) (mm)	Underchip Thickness (h) (mm)	Cracked (*)	Calculated von-Mises stress (kg/mm^2)
Predicted von-Mises stress for Surface Mount Plastic Packages					
132 PQFP	8.910 × 8.700	10.160 × 10.160	1.422	no	6.311
164 PQFP	8.520 × 8.310	10.668 × 10.668	1.422	no	6.668
100 TQFP	6.850 × 6.950	8.001 × 8.001	0.457	yes	20.070
28 SOJ	3.549 × 4.679	7.620 × 4.826	1.041	no	2.469
44 PLCC	9.000 × 8.580	9.398 × 9.398	1.498	no	4.964
68 PLCC	6.700 × 6.760	8.890 × 8.890	1.651	no	3.383
84 PLCC	8.130 × 8.130	9.144 × 9.144	1.651	no	3.836
160 EIAJ	6.810 × 10.520	7.391 × 11.709	1.346	no	3.804
68 PLCC	4.200 × 4.200	6.800 × 6.800	1.513	no	0.300
32 PLCC	6.800 × 5.600	8.900 × 7.380	1.090	no	2.799
44 PLCC	8.100 × 6.500	9.000 × 9.000	1.579	no	1.614
80 PQFP	8.132 × 8.078	9.205 × 9.201	0.600	yes	15.888
240 PQFP	9.650 × 9.410	12.790 × 12.780	1.286	no	5.339

* 30°C/60% RH/168 hrs test conditions

Measured Weight Gain in a Plastic Package Due to Absorbed Moisture

Measured weight gains in several plastic packages due to the absorbed moisture are shown in Fig. 5 (Ilyas and Poborets, 1993). The tested packages are listed also in Table 4 (phase I of moisture study). As evident from Fig. 5 and Table 4, the underchip thickness of the molding compound has the major effect on the weight gain at the point of the package failure. This conclusion is in good agreement with the calculated data shown in Tables 1, 3, and 4, as well as with the formula (21). The formula (21) indicates that for not very thin packages (all the packages in Table 4 are, in effect, not very thin from the structural analysis standpoint), the bending stress, calculated on the basis of a linear theory, adequately reflects the stress level in the package (see the three last columns in Table 4). The plot in Fig. 5 indicates also, that "thick" packages are less sensitive to the change in the amount of the absorbed moisture than "thin"packages. This plot enables one to determine the "safe zone," in which the absorbed moisture does not lead to a structural failure, as well as the "crack" zone, in which moisture induced failures were observed.

It should be pointed out, however, that the plot in Fig. 5 considers the role of the underchip thickness of the molding compound, but does not consider the roles of other geometric characteristics of the package, and, first of all, the effect of the chip size. In order to account for this effect, the percent of the weight gain in plastic packages due to moisture absorption was plotted in Fig. 6 as a function of the ratio h/c of the underchip thickness, h, to the chip width, c. The area below the solid line is the "safe zone," and the area above the broken line is the "crack zone." The area between these lines is a "neutral zone," in which no definite statement can be made, as far as the effect of the moisture content on the compound cracking is concerned. As evident from the Fig. 6, in which the results of additional (phase II) tests are also included, ``thick" packages with relatively small chips (such as PLCC, SOJ, EIAJ, PQFP and MQFP) fail at the moisture level higher than 0.18%, while ``thin" packages (such as TQFP) with relatively large chips fail at the moisture level below 0.15%. Fig. 6 indicates also that moisture content at failure rapidly decreases with the decrease in the h/c ratio, if this ratio is below 10%.

As has been pointed out, the paddle size plays also an important role, as far as the propensity of a plastic package to moisture induced cracking is concerned. The effect of the paddle size can be accounted for by introducing, based on the formula (21), a geometric factor

$$f = \frac{ac}{4h^2}\left(3 - \frac{c^2}{a^2}\right), \quad \frac{c}{a} \leq 1. \tag{22}$$

This factor changes from zero to

$$f_{max} = \frac{c^2}{2h^2} = \frac{a^2}{2h^2} \tag{23}$$

when the chip's width, c, changes from zero to the paddle width, a.

In Fig. 7 the measured moisture content is plotted versus the inversed geometric factor

Moisture Absorption Weight Gain (Wt.%) vs. Thickness of the Mold Compound Under the Die

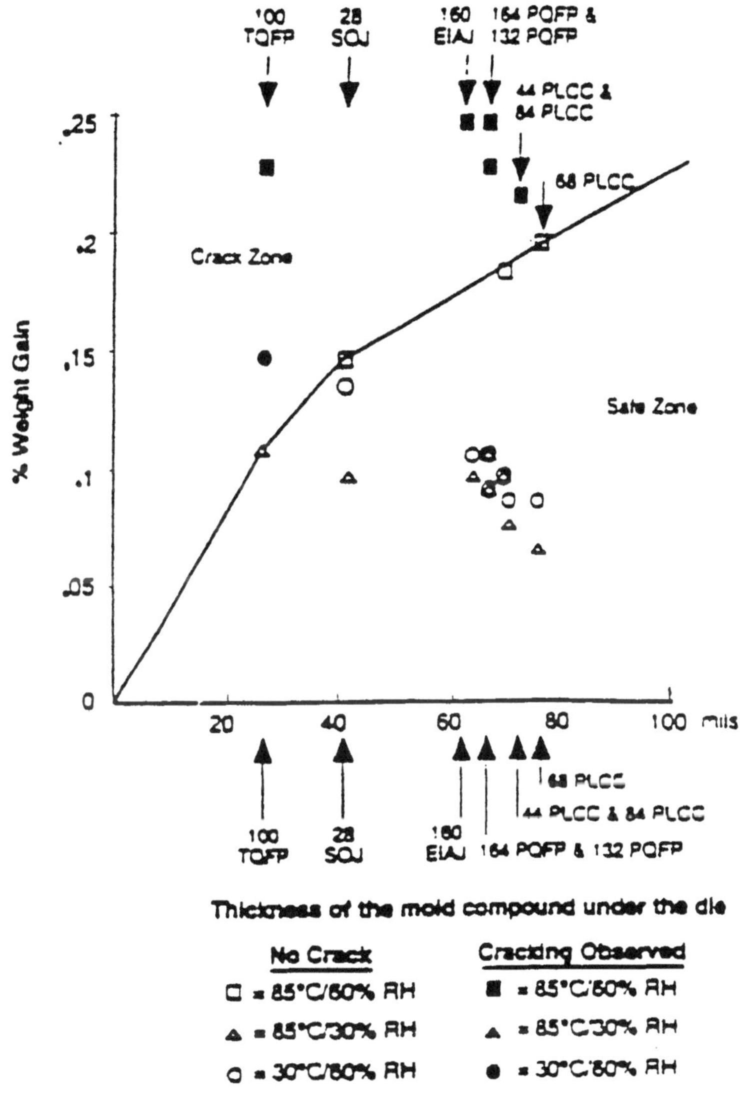

Fig. 5

TABLE 4

PHASE OF STUDY	PACKAGE TYPE	UNDERCHIP THICKNESS, h,mm	PAD WIDTH, a,mm	CHIP WIDTH, c,mm	MOISTURE CONTENT, %		GEOMETRIC FACTOR[2]		CALCULATED STRESS FOR p = 360psi, kg/mm^2		
					SURVIVED	FAILED	f	1/f	von-MISES[1]	BENDING[1]	BENDING[2]
I[3]	100TQFP	0.457	8.001	6.850	0.11	0.15;0.23	148.730	0.00672	20.070	17.537	37.653
	164PQFP	1.422	10.668	8.310	0.11	0.23	26.230	0.03812	6.668	6.496	6.640
	132PQFP	1.422	10.160	8.700	0.09	0.25	24.772	0.04037	6.311	6.176	6.271
	44PLCC	1.498	9.398	8.580	0.10;0.19	—	19.462	0.05138	4.964	4.900	4.927
	84PLCC	1.651	9.144	8.130	0.08;0.09	0.22	15.065	0.06638	3.836	3.806	3.814
	160EIAJ	1.346	7.391	6.810	0.10;0.11	0.25	14.940	0.06693	3.804	3.775	3.782
	68PLCC	1.651	8.890	6.700	0.07;0.09	0.20	13.286	0.07527	3.382	3.358	3.363
	28SOJ	1.041	4.826	3.549	0.10;0.14;0.15	—	9.709	0.10300	2.469	2.459	2.458
II[5]	100TQFP	0.457	7.010	6.350	0.16;0.18	—	116.130	0.00861	18.450	16.357	29.400
	80TQFP	0.457	6.604	4.521	0.10;0.17	—	90.469	0.01110	15.187	13.394	22.903
	208SQFP	1.346	10.490	8.128	0.12	0.18	28.233	0.03542	7.155	6.942	7.148
	160MQFP	1.346	9.601	7.112	0.09	0.22[4]	23.097	0.04330	5.888	5.760	5.847
	68PLCC	1.651	6.604	4.572	0.12	0.19	6.980	0.1433	1.771	1.767	1.767

[1] Calculated based on the general theory (Suhir, ECTC, 1995), assuming E = 135,000psi

[2] Calculated by the formulas (Suhir, ECTC, 1995)

$$f = \frac{ac}{4h^2}\left(3 - \frac{c^2}{a^2}\right), \quad \sigma_b = pf$$

[3] Q. Ilyas and B. Poborets experimental data (''Structural Analysis in Microelectronics and Fiber Optics'', E. Suhir, ed., ASME, 1994)

[4] No cracks observed. Failed electrically because of delamination

[5] Experimental data (Ilyas and Potter, ECTC, 1996)

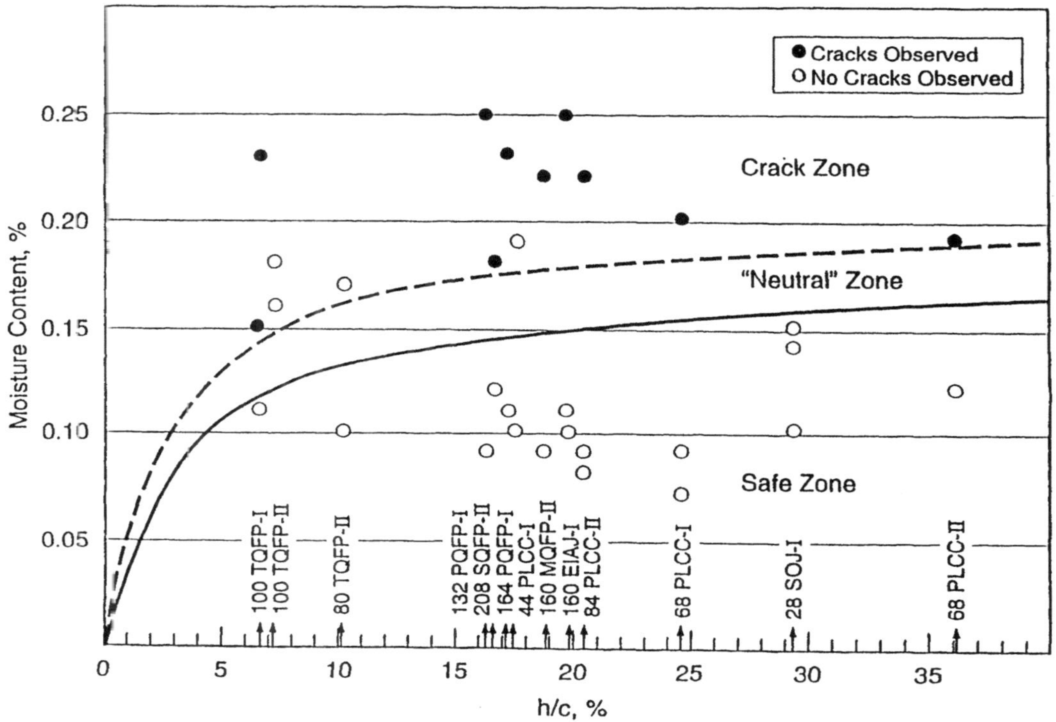

Cracks Observed
No Cracks Observed
Crack Zone
"Neutral" Zone
Safe Zone
0.25
0.20
0.15
0.10
0.05
Moisture Content, %
100 TQFP-I
100 TQFP-II
80 TQFP-II
132 PQFP-I
208 SQFP-II
164 PQFP-I
44 PLCC-I
160 MQFP-II
160 EIAJ-I
84 PLCC-II
68 PLCC-I
28 SOJ-I
68 PLCC-II
0
5
10
15
20
25
30
35
h/c, %

Fig. 6

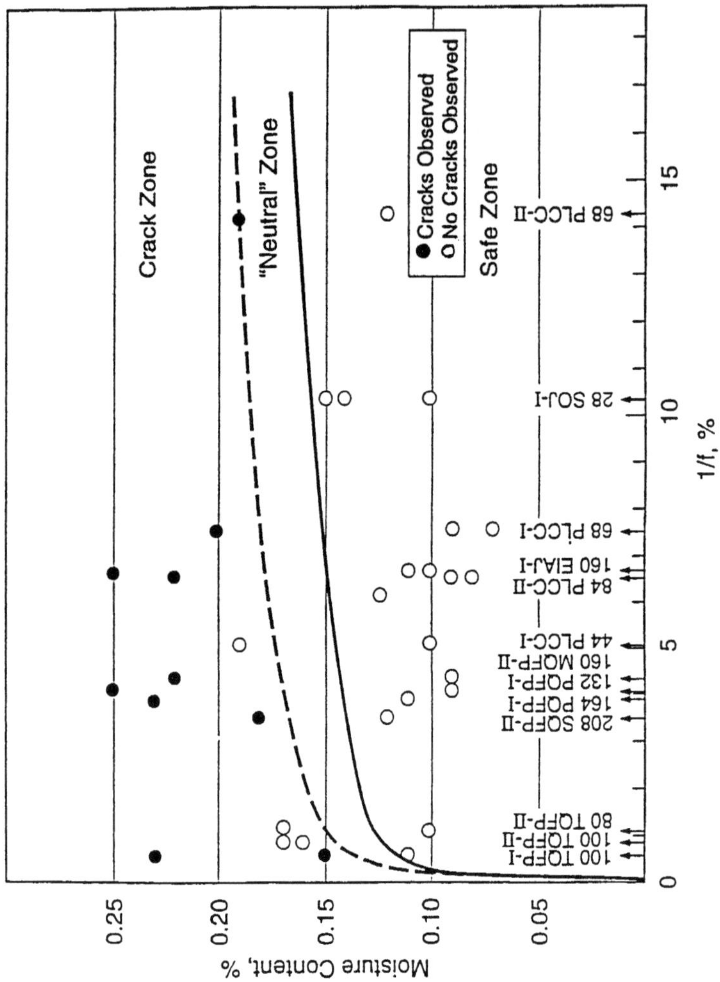

Crack Zone
"Neutral" Zone
Safe Zone
● Cracks Observed
○ No Cracks Observed
Moisture Content, %
0.25
0.20
0.15
0.10
0.05
1/f, %
0
5
10
15
100 TQFP-I
100 TQFP-II
80 TQFP-II
208 SQFP-II
164 PQFP-I
132 PQFP-I
160 MQFP-II
44 PLCC-I
84 PLCC-II
160 EIAJ-I
68 PLCC-I
28 SOJ-I
68 PLCC-II

Fig. 7

$$\frac{1}{f}=\frac{4h^2}{ac}\frac{1}{3-\frac{c^2}{a^2}} \tag{24}$$

This inversed factor reflects the geometric "robustness" of the package and considers the effect of the major geometric characteristics of the package on its propensity to moisture induced failure. The formula (24) can be written as

$$\frac{1}{f}=2\left(\frac{h}{c}\right)^2\chi\left(\frac{c}{a}\right), \tag{25}$$

where the function

$$\chi\left(\frac{c}{a}\right)=2\left(\frac{\frac{c}{a}}{3-\frac{c^2}{a^2}}\right) \tag{26}$$

considers the effect of the pad size on the magnitude of the inversed geometric factor. This function increases from zero to one, when the chip-to-pad width ratio $\frac{c}{a}$ increases from zero to one. Based on the Fig. 6 and Fig. 8 data, we conclude that thin packages of the TQFP type are significantly more moisture sensitive than all the other packages. One can conclude also that no moisture induced failure is likely to occur, if the moisture content in a "thick" package is kept below 0.12%.

It should be pointed out that the calculated geometric factors and the induced von-Mises and bending stresses can be effectively used as adequate criteria of the plastic package propensity to moisture-induced failure. As evident from the Table 8, the theoretically evaluated increase in the inversed geometric factors (these reflect the "robustness" of the package structure) and the corresponding decrease in the predicted stresses reflect very well the experimentally determined increase in the moisture content at the point of the package failure.

The plot in Fig. 7 can be effectively used to assess the propensity of a new plastic package design (i.e. a design in which one or more geometric parameters are different from those that were tested) to moisture induced failures. This can be done assuming that different packages, characterized by the same $\frac{1}{f}$ value are equally prone to moisture induced failure, as long as the molding material is the same.

Effective Water Vapor Pressure

In the calculations, whose results are shown in Table 1, it has been assumed that the level of the water vapor pressure is the same for all the packages and is as high as p = 360 psi. This corresponds to the pressure of a saturated water vapor at the reflow soldering temperature of 240°C. The results of the theoretical analysis and the experimental evaluations enable one to tentatively assess the effective water vapor pressure which results in the plastic package

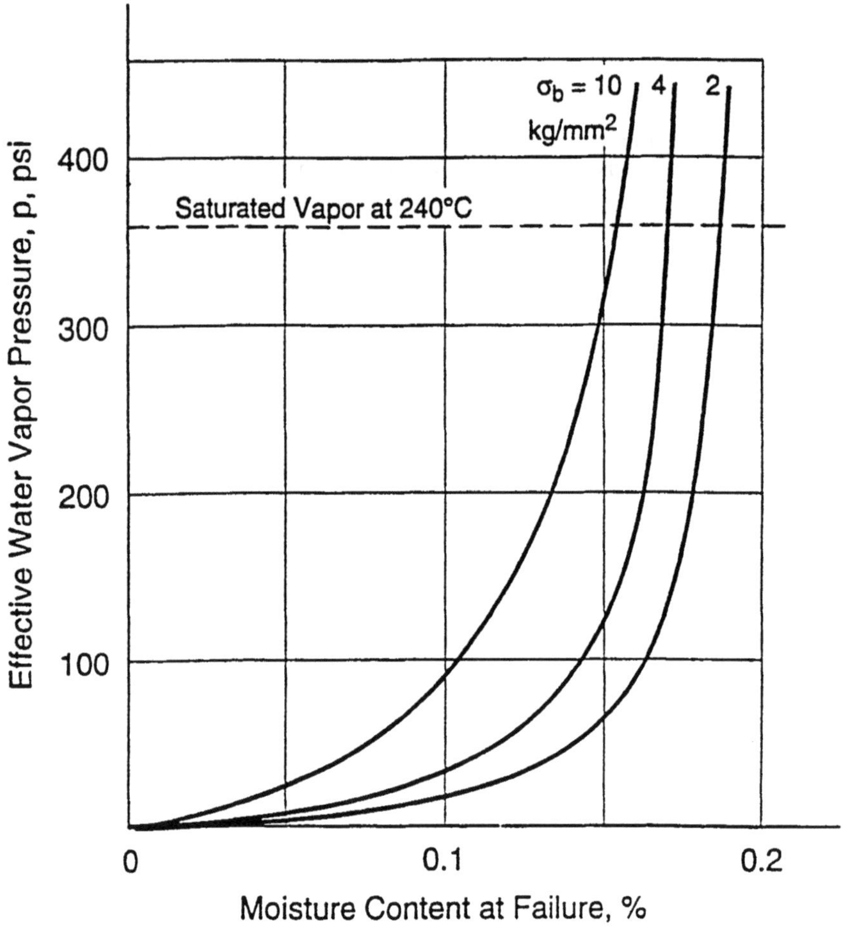
σ_b = 10
kg/mm²
4
2
Saturated Vapor at 240°C
Effective Water Vapor Pressure, p, psi
400
300
200
100
0
0.1
0.2
Moisture Content at Failure, %

Fig. 8

failure during reflow soldering. Indeed, if the bending stress at failure, σ_b, is determined experimentally for the specimens made of the given molding compound material for different moisture levels (say, by using three- or four point bending), then the lateral pressure, which results in the stress σ_b, can be evaluated, based on the formulas (21) and (22), as follows:

$$p = \sigma_b \left(\frac{1}{f}\right) . \qquad (27)$$

Then, using the experimental relationship between the moisture level and the $\frac{1}{f}$ values from the Fig. 3, one can obtain the relationship between the moisture content at the point of the package failure and the level of the effective water vapor pressure at this point. This relationship is plotted in Fig. 4 for different σ_b values. These values are selected from the region below the specified ultimate stress (Table 2) which is an overestimation of the actual (expected) strength of the molding compound at reflow soldering conditions. The obtained data indicate that for moisture contents typical for the existing plastic packages, the pressure at failure can be significantly lower than p = 360 psi, especially for ``thin" packages. Clearly, this tentative conclusion is based on an assumption that the calculated stress at failure is a strength characteristic of the material (not the package), and can be considered the same for different package geometries.

Conclusion

The following major conclusions can be drawn from the performed analysis.

1. As is known, the current practices of "baking" and "bagging" of "thick" and "thin" packages are not different. Application of an approach, based on the combined information, obtained from the theoretical modeling and experimental evaluations of the plastic package propensity to moisture induced failures, enables one to quantitatively assess the effect of the plastic package geometry on the allowable moisture level for the given package design. As is known, the current practices of "baking" and "bagging" of "thick" and "thin" packages are not different. The obtained data indicate that there is an obvious incentive to try to develop guidelines ("Figures of Merit") that would enable one to separate packages that need to be "baked" and "bagged" (or "rebaked" and "rebagged") from those that supposedly do not. Application of such guidelines might result in significant savings of expense and time.

2. The inversed geometric factor

$$\frac{1}{f} = \frac{4h^2}{ac} \frac{1}{3 - \frac{c^2}{a^2}}$$

can be used as an adequate criterion which characterizes the "robustness" of a plastic package design with respect to its propensity to failure during reflow soldering. In this formula, h is the underchip thickness of the compound, c is the chip's width, and a is the pad's width. This criterion indicates that the underchip thickness plays the most crucial role, as far as the effect of the package geometry is concerned. Such a conclusion is in good agreement with experimental observations. The application of this parameter enables one to evaluate the propensity of new package designs to moisture induced failures, at least for sufficiently "thick" packages, in which only the bending stresses could be considered. In the case of "thin" packages, the general theory (Suhir, 1995, 1997) should be applied for this purpose.

3. "Thick" packages (practically all the tested packages, except, probably, the TQFP ones, belong, from the Structural Analysis point of view, to this category) are not expected to fail during high temperature reflow soldering operation, if the moisture content in these packages is below 0.12%. Thin TQFP packages can fail at the moisture content level as low as 0.10%.

4. The assessed effective water vapor pressure at the point of the package failure depends on the package geometry, and can be well below the level of of 360psi (saturated water vapor at 240°C), especially for thin packages. Detailed theoretical and experimental investigations are necessary, however, to determine the effective water vapor pressure, depending on the molding material, package geometry, and reflow soldering conditions.

Acknowledgement

The author acknowledges, with thanks, the help of J. Weld who carried out the stress calculations.

REFERENCES

1. Altimari, S., et al., (1992). "Role of design factors for improving moisture performance of plastic packages," *42-nd ECTC*, pp. 945-950.
2. EIA/JEDEC (1995). Standard, Test Method A112-A, ``Moisture-induced stress sensitivity for plastic surface mount devices," *Electronic Industries Association, Engineering Department*, Arlington, VA.
3. Ganssan, G. S., and Berg, H. (1993). "Model and analyses for reflow cracking phenomenon in SMT plastic packages," *43-rd ECTC*, pp. 653-660.
4. Ilyas, Q. and Poborets, B. (1993). "Evaluation of moisture sensitivity of surface mount plastic packages," EEP-vol. 7, "*Structural Analysis in Microelectronics and Fiber Optics*," E. Suhir, ed., ASME, pp. 145-156.
5. Kelly, G., et al. (1992). "Investigation of thermo-mechanically induced stress in a PQFP 160 using finite element techniques," *42-nd ECTC*, pp. 467-472.
6. Kimura, H., et al. (1992). "New analytical study for popcorn phenomenon (package cracking)," *IEEE CHMT Trans.*, *CHMT 15(6)*, pp. 1117-1122.
7. Liu, S. (1993). "Debonding and cracking of microlaminates due to mechanical and hydro-thermal leads for plastic packaging," in EEP-vol. 7, "*Structural Analysis in Microelectronics and Fiber-Optics*," E. Suhir, ASME, 1993, pp. 1-11.
8. Suhir, E. (1991). Structural Analysis in Microelectronic and Fiber-Optic Systems, Van Nostrand Reinhold, New York.
9. Suhir, E. (1995). "Failure Criterion for moisture sensitive plastic packages of IC devices," *45-th ECTC*.
10. Suhir, E., (1997). Failure criterion for moisture-sensitive plastic packages of integrated circuit (IC) devices: Application of von-Karman's equations with consideration of thermoelastic strains, *Int. Journal of Solids and Structures*, to be published.
11. Suhir, E. and Ilyas, Q. S. M., (1996). Thick" plastic packages with "small" chips vs "thin" packages with "large" chips: how different is their propensity to moisture-induced failures?", EEP-vol. 6, E. Suhir, ed., "*Structural Analysis in Microelectronics and Fiber Optics*", ASME, pp. 101-125.
12. van Doorselaer, K., et al., (1993). Thin type packaging: an effective way to improve the popcorn resistance of plastic packaged IC's, 31-st *Reliability Physics Symp.*, pp. 244-249.

THE IMPACT OF STRESS ON THE NOISE, QUALITY AND RELIABILITY OF PASSIVE AND ACTIVE DEVICES

Josef SIKULA and Petr VASINA
Department of Physics, Technical University
Zizkova 17, 602 00 Brno, Czech Republic
Tel. +425 7261650 Fax +425 7261 666

ABSTRACT

In the framework of our research of applicability of noise measurement to the quality testing and reliability prediction of electronic devices we studied noise characteristics of conducting thin resistors, capacitors, p-n junction and MOS devices.

Low frequency noise and third harmonic index (THI) measurements are used as reliability indicators for description of ageing stability of thin film resistors. A method for testing of tantalum and aluminium capacitors is based on relation between noise spectral density and DC leakage current. We have investigated noise characteristics in diodes, LED's, transistors, power diodes and thyristors. Similar investigation on RTS noise have been made for MOS structures. It is supposed that the low frequency noise can be used as a quality and reliability indicators. Furthermore, it has been demonstrated on several occasions that the low frequency noise is much more sensitive to hot-carrier (HC) related degradation of the interface/oxide properties of a MOSFET.

1. Thin Film Resistors

1.1. INTRODUCTION

The objective of our research consisted in finding a correlation between the spectral density of the low frequency noise, THI measurements, temperature dependencies and the frequency exponent on one hand and the technology standard, device time stability and reliability on the other.

The layers are not perfectly continuous from the microscopic point of view, consisting of conducting metallic grains separated by a dielectric. The predominant

R. Tummala et al. (eds.), Electronic Packaging for High Reliability, Low Cost Electronics, 53–60.

transport mechanism in thin films consists in charge carrier tunnelling between the grains. The conducting paths between the grains are time variable, which results in an 1/f noise. Traps in the contact barriers and tunnelling via substrate traps give rise also to 1/f noise. The noise spectral density is than related to the sample technology [1].

1.2. NOISE SPECTRAL DENSITY

Excess noise spectral density as a function of frequency for two different samples is shown in Fig.1. Curve 1 is noise spectral density of a low noise resistor prepared by "good" technology and curve 2 is noise spectral density of a resistor with burst noise source.

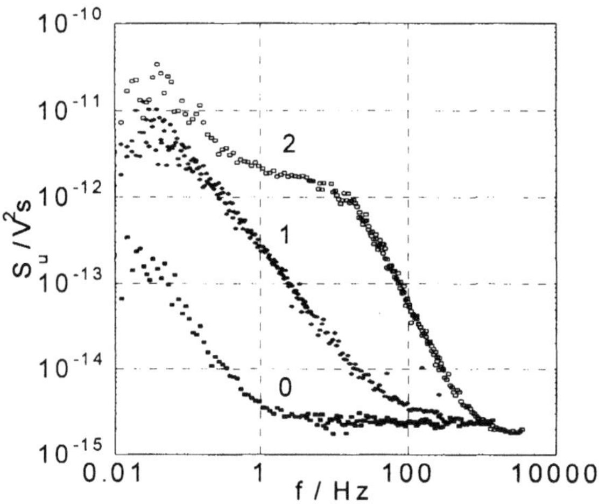

Fig. 1. Noise spectral density of two 100 kΩ resistors. Curve 0 is the apparatus noise, , curve 1 is noise spectral density of a low noise resistor and curve 2 is noise spectral density of a resistor with burst noise source.

There are two different frequency dependencies Lorentzian and 1/f noise. Lorentzian component or burst noise is generated in resistor contacts or in thin film cracks (see Fig.1, curve 2). For such processes it holds that the fluctuation is directly proportional to the mean value. It means that the noise voltage is proportional to the DC voltage applied across the sample.

1.3. NOISE RELIABILITY INDICATORS

Qualitative characterisation of the excess noise is possible with the use of noise spectral density at given frequency. In our investigation, the measurable quantity is an indicator of sample quality C_Q given as

$$C_Q = \frac{S_U}{U^2} f \qquad (1)$$

Some conclusion as to the technology can be drawn from these measurements. The lower value of C_Q and its dispersion, the better is the technology. This hypothesis was tested by classical ageing methods.

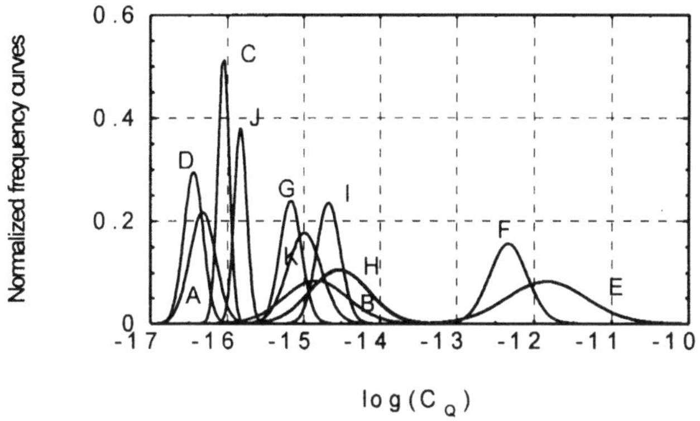

Fig. 2. Normalised frequency curves of C_Q for different technologies.

Some technologies, for example E, see Fig. 2, has high dispersion and high value of C_Q. Generally, we can conclude that low dispersion of C_Q indicator have ensembles with low average value of C_Q. This can be illustrated by Fig. 2, where the normalised frequency curves of $\log C_Q$ for selected Japanese, US and European resistors production are plotted.

1.4. THI MEASUREMENT

Important non ohmic regions are at contacts and in vicinity of defects of the thin film resistor structure and defects of the thin film layer. These non-ohmic regions are screened by the third harmonic voltage measurement.

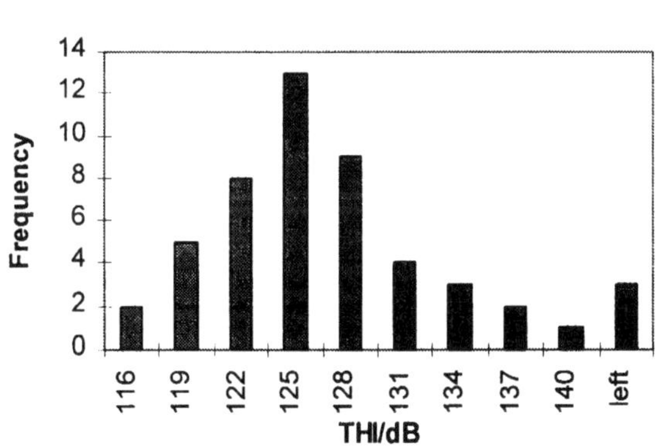

Fig. 3. Bar chart diagram of THI for resistor ensemble No. 1.

It can be seen in Fig. 3., where 6 samples, out of 50 samples, are screened by this method. The samples with higher value of nonlinearity exhibit also extreme value of TCR, low stability, etc.

2. Electrolytic Capacitors

2.1. INTRODUCTION

A new method for testing of tantalum and aluminium capacitors is described. This is based on relation between noise spectral density and DC leakage current. The most important sources of fluctuations consist in regenerative microbreaks, the insulation layers width variation, which are due to random dissolution of the oxide and gas bubbles generation. Other important sources are macroscopic capacitance fluctuations, which are due to non-uniform insulator surface at the electrodes. Fluctuation of polarisation and fluctuations of mechanical strain may cause still another kind of noise which may be of importance. Last is the contact resistance noise component, which also makes some capacitors be noisy. Two kinds of noise may be distinguished. Partial discharges and regenerative microbreaks cause two-state impulse-like noise, whereas the polarisation and charge transport fluctuation bring about a multi-state continuous spectrum noise. These irreversible processes are responsible for capacitance structure degradation.

2.2. NOISE SPECTRAL DENSITY

The current flowing through defects fluctuates. Noise depends on structure perfection, number of grain boundaries, homogeneity, bubble creation, etc. Frequency dependence of noise spectral density in low frequency region is in Fig. 4. In mHz frequency region noise spectral density is 1/f type. This gives important information on slow irreversible processes due to crystallisation, oxide reduction and creation of electrical field inhomogenities.

It was found, for the same value of DC component in different samples, have different value of dispersion or the noise spectral density. This was used as a reliability indicator. The ratio of noise spectral density and leakage current was used as the reliability indicator. When a high intensity current flows through the layer, irreversible changes take place in the layer which results in a degradation of the capacitor quality.

2.3. AGEING

To get more information on the microscopic process in the capacitors two ageing procedures were applied : (i) stressing at a temperature of 70° C and a DC voltage of

250 volts and (ii) thermal oxidation only at 85° C. The capacitor ensemble consisted of 50 samples. The samples were characterised in the initial state and subsequently after intervals of 100, 250, 500 and 1 000 hours. The capacitors have been assessed by their capacitance, impedance, tanδ, leakage current and noise spectral density. These characteristics were evaluated at these time intervals. The current noise spectral density have been measured by the sampling of the noise voltage in time domain and its transformation by the FFT.

In the initial state the ensemble can be described by a normal distribution. After 100 hours of stressing the average DC leakage current goes down which can be attributed to the oxide layer stabilisation. Samples with higher values of the leakage current are more clearly distinguished from the rest of the ensemble.

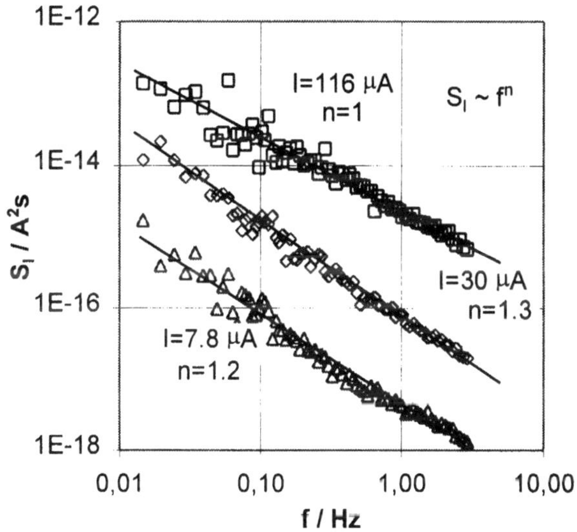

Fig. 4. Current noise spectral density vs. frequency of tantalum capacitor

For all ensembles no pre-selection of the capacitance magnitude was done, thus in their initial state their distribution has large dispersion. After 1000 hours of ageing the average value of capacitance decreased from 1.18 to 1.10 μF and the ensemble in Fig. 5. can be considered as a superposition of at least three clean ensembles. In general, samples exhibiting high noise in the initial state, show a large capacitance drop in the course of ageing. In eight samples the capacitance decreased more than by 10 per cent. Comparison of the four mentioned ensembles shows that the lower the standard deviation of the capacitance, leakage current and noise voltage the better the reproducibility of the manufacturing process and the service life of the capacitors.

By selecting in the course of the manufacturing, the reliability can be improved to certain extent. However, it is to be taken into account that screening does not make these processes deterministic.

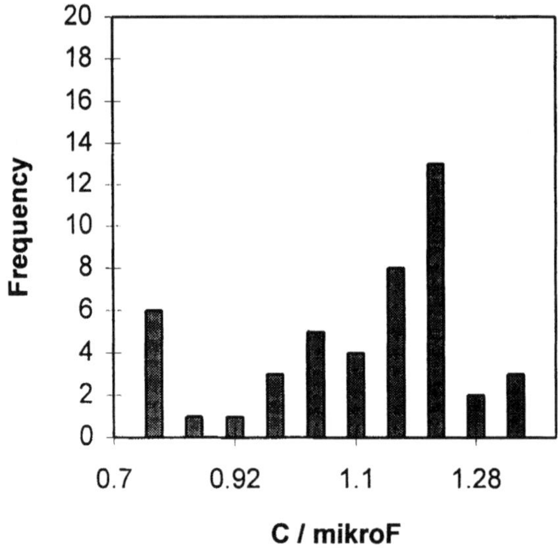

Fig. 5. Bar chart diagram of capacitance for 1 mF capacitors ensemble after 1000 hours of ageing.

Ageing process revealed that samples with higher noise spectral density in the initial state exhibit higher capacitance, impedance and leakage current change. Degradation process in samples with stronger noise sources is more intense.

3. Active Devices

3.1. INTRODUCTION

We have investigated noise characteristics in diodes, LED's, transistors, power diodes and thyristors. Similar investigation on RTS noise have been made for MOS structures. We suppose that 1/f , burst and RTS noise can be used as a quality and reliability indicators. The amplitude of burst or RTS pulses observed in these samples is ranging from nA (for MOS and LED's) to mA (for power devices).

3.2. DEFECTS IN BIPOLAR DEVICES

Occurrence of these noise sources is related to inhomogeneities in bipolar devices [2]. In structures with p-n junction a correlation between these sources and the excess current was observed. In reverse direction mode burst noise is observed when U-I characteristics have differential negative resistance region. In all cases existence of burst noise is related to irreversible processes and degradation.

More or less all of these devices have two levels systems which in frequency domain give the noise spectral density of g-r noise and in time domain can be ascribed to one dimensional renewal process. A bistable noise is generated by channel modulation and in the p-n junction devices the channel appears nonintentionally.

The actual reliability of electronic devices is usually less than the maxim theoretical value of the reliability depending on the attained manufacturing level. It may be due to irregularities in the used manufacturing process.

3.3. IMPACT OF HOT CARRIER DEGRADATION ON NOISE IN MOSFETS

Recent evidence has been provided that the down-scaling of CMOS technologies results in a drastic increase of the sample to sample dispersion in LF noise which is an issue of concern for mixed analog/digital circuits. Furthermore, it has been demonstrated on several occasions that the LF noise is much more sensitive to hot-carrier (HC) related degradation of the interface-oxide. Noise components which are due to the interface or surface states observed in this systems are 1/f and RTS noise.

The stressed p-MOSFETs, twin-well, single-poly 0.7 μm CMOS technology, having LDD's and 0.25 μm spacers were studied. The device width is 20 μm and the DL = 0.08 μm. Stress is applied for half an hour and for a gate voltage V_{gs} = - 3 V and a drain voltage V_{ds} = - 6 V. The impact of HC stress on the LF noise in linear operation and in saturation were studied as a function of the effective device length L_{eff}. The observed noise increase is compared with the degradation of the static parameters.

The normalised changes (in absolute value) of the static device parameters are small for longer device studied and considerable degradation was observed for the shorter devices. The largest change is found for the maximum transconductance, which shows a HC induced increase of up to 15 % for the L=0.65 μm p-MOSFET. This is stronger than the $1/L_{eff}$ dependence reported by Mistry and Doyle [3] for short-channel n-MOSFETs. It appears that exponent n of the HC related change is in the range 2-4, depending on whether the transconductance or the threshold voltage change is used as the lifetime criterion.

This stronger dependence for p-MOSFETs can be understood qualitatively by considering the fact that the HC degradation in p-channel devices reduces in a shortening of the channel length, giving rise to an increase in the measured transconductance. The current noise spectral density S_I at f = 10 Hz in linear operation shows in general a pronounced increase up to 30 % for the L=0.65 μm p-MOSFET, while negligible change, if any, is found for L=2 μm. F and R mode yield very similar noise levels in linear operation, in agreement results obtained on stressed n-MOSFET.

In saturation on the other hand, there is a tendency for higher noise in R mode, which is in agreement with the fact that the HC damage is closer to the drain. When pinching off the channel, the effect of the damaged region on the carrier transport will be negligible for the F mode. For the R mode, the damaged region is close to the source and will have full impact on carrier capture and scattering.

4. Conclusion

It is commen experience that there are good and poor specimens. Therefore, the noise level can vary considerably between nominally identical devices made with he same manufacturing techniques and even belonging to the same processing batch. The sensitivity of the noise magnitude to small deviations is typical of the impact of impurities, defect and damage on the noise behaviour. Since this effect is due to a variable which cannot be controlled, the identification of the exact source of noise is not easy. Therefore, extensive, detailed and time consuming experiments are required to give a positive identification. Generally accepted conslusions state that the noise level depends on: (i) the chemical purity of the sample, (ii) the perfection of the crystal structure, number of point defects, grain boundaries, point and linear defects, (iii) the condition of the surface and the associated parameters.

Noise generating defects ususally are manifested by also changing some other measurable quantities, such as a change of the resistance in passive components, an excess current component in p-n junction devices and a change in the threshold voltage and the transconductance of MOS transistors. It has also been observed that the chemical conditions of the surface can also affect the magnitude of the noise spectral density. The effects of various devices are illustrated by means of the humidity factor, i.e., a wet or a dry ambient. The 1-f noise increases in a wet ambient and show slow changes with time.

The changes of these parameters have been reported to be by orders of magnitude smaller than those of the excess noise. Therefore, noise studies are a sensitive tool for quality characterization and reliability prediction.

Acknowledgement

This research vas supported by the CEC within the framework of the European Laboratory for Electronic Noise No. ERBCHRXCT920047 and by the Grant Agency of the Czech Republic No 103/97/0899.

5. References

1. Moeser, R. (1995) Chip Resistors: Pulse load capability and current noise. *Proc. of CARTS-EUROPE'95 Symposium*, Amsterdam Holland.
2. Jones, B.K.(1993) Electrical noise as a measure of quality and reliability in electronic devices, *Advanced in Electronics and Electron Physics* **87**, 201-57.
3. Mistry, K., and Doyle, B.S. (1989)An empirical model for the L_{eff} dependence of hot-carrier lifetimes of n-channel MOSFETs, *IEEE E-D Lett.*, **10**, 500 - 502.

SMART UNIT FOR TEHNOLOGICAL MONITORING IN ELECTROCHEMICAL INDUSTRY

RADU IONESCU, SORIN GANCIULESCU, DAN CARAVASILE
RAMELECTRO Ltd.
Sos. Iancului, 101A, sector 2, Ap. 16, Bucharest-73376, Romania

OVIDIU TOL, PAUL SVASTA, DANIEL LEONESCU
University Politehnica of Bucharest
Splaiul Independentei 313 Bucharest Romania

1. Electrochemical Processes. Characteristic Parameters

Electrochemical processes can be defined through the complex phenomena which succeeds on electrode / electrolyte interface.

At the industrial scale this processes find a series of applications in various activity fields, from which we can mention:

- metals extraction and refine;
- chlorine and sodium hydroxide fabrication;
- hydrogen and oxygen obtaining;
- metal galvanic plating and metallic alloys in functionally or decorative-protective purposes;
- battery and accumulator fabrication;
-technical applications of electrokinetic phenomena's (electrophoretic and electro-osmotic procedures) [1].

From technological point of view, these processes can be controlled through a series of parameters which has an influence on process dynamic for realization of functional characteristics of the finished product.

This parameters are included in one of the groups [2]:

a - for composition of electrolyte (specifically concentration of metallic ions for complex agents, pH, addition agents or conductive salts concentration).

b - for operating (current density, temperature, electrolyte moving).

c - various (cathodic efficiency, current wave form, basic metal type)

Automation of this dynamic kind of process has the purpose to monitor and to control the composition and some of the operating parameters. This action presumes the processing of a large data volume and taking real time control decisions, and like results:

R. Tummala et al. (eds.), Electronic Packaging for High Reliability, Low Cost Electronics, 61–67.

- insuring the quality of finished product;
- increasing installation efficiency;
- reducing rejects;
- minimizing of specific consumption of energy and raw materials.

The authors of this work propose the achievement of a smart electronic unit for electrochemical processes automation. It is made in a configurable architecture which can be adapted to various technological applications and can fulfill above mentioned functions.

Automation unit development was made around Intel 80C51 type microcontroller, chosen under the following reasons:

- taking over capacity of a big amount of data;
- possibility of I/O direct processing;
- existence of internal resources: timers, serial interface, interruption system, I/O ports.

Unit achievement is exemplified for electrochemical metal plating case, when parameters from Table 1 are controlled.

TABLE 1. Parameters controlled for electrochemical metal plating case

No	Parameters	Type	Value	Sense
1	Temperature	analog	-5° ÷ 100°C	input
2	voltage on current shunt	analog	0 ÷ 75 mV	input
3	pH	analog	0 ÷ 14	input
4	Ionic concentration	analog	1 ÷ 500 g/l	input
5	Electrolyte level	digital	min.÷ max.	input
6	voltage command of current supply	analog	0 ÷ 5 V	output
7	Thermal control	digital	on - off	output
8	Bright ag. adjustment	digital	on - off	output
9	pH correction	digital	on - off	output
10	Electrolyte refill	digital	on - off	output
11	Probes self-drive	digital	on - off	output

As a sequel there are described few aspects of the way in which each one of these parameters (in pre-established ranges) are direct influencing the metal electroplating (in the same time the cathodic layer ions concentration) and the interdependence between them.

a. Current density [3] is the most important operating parameter, directly determines (in give interval) the layer microcristaline structure.

b. Electrolyte temperature is the unique factor which promotes macrocristaline depositions - temperature increasing has a positive global effect on process speed and current efficiency.

c. Electrolyte pH (concentration of hydrogen ions) has and important part in metallic ions deposition, for obtaining an corresponding layer from the point of view of aspect, adherence and porosity.

d. Ionic concentration of the covering metal is a composition variable, characteristic to each electrolyte.

2. Automation Unit

From table 1 we can see that we have two distinguished functions:

- parameters monitoring (positions 1-5), and
- their adjustment command (positions 6-11).

Each entering signal is tacked, compared with memorized values and depending of the result of this operation his correction is commanded.

The signals overrated through specialized probes, directly from electroplating bath, need previous conditioning by which it is minimized hardware processing.

Output command signals are passed through buffers (digital way) and the analog one by A/D conversion and they are specific conditioned for execution elements command.

Through software package it is insured the measurement-operating correlation, also values of controlled parameters.

Because of the numerous internal resources of the controller, the unit functions can be valorized by the connection to local terminals (display, keyboard, printing unit - for which there are insured hardware connections) to a central unit (PC type - through included serial interface) or by the change of controlled parameters limits with "remote control".

3. Block diagram description

In the schematic design the authors followed an unitary concept of the hardware and software structure, so that the designing costs to be allocated rationally between those two segments. There were chosen electronic components to insure an optimal performances/price report.

According to figure no. 1, the unit is composed from two modules which are connected between them by an back sliding valve. First of them (marked with interrupted line in fig. 1), is the developing system around the 8051 microcontroller, with an architecture adapted to the application [4] , [5]. The second one is the I/O module.

The microcontroller is working with a clock signal of 12 MHz which gives resources for real time function. We mention that it has an incorporated oscillator, beside being attached only the Quartz crystal.

As the microcontroller is multiplexing the inferior address octet with the data octet it is necessary their separation, using a HCT 373 buffer type commanded by ALE signal. During instruction cycle, in buffer we can find the inferior address octet, within at the microcontroller outputs we will find the data octet.

The system has two memory blocks: program memory (Flash type) and data memory (RAM type). Both of them have 32 Kbytes, but depending of necessities, their dimension can be changed.

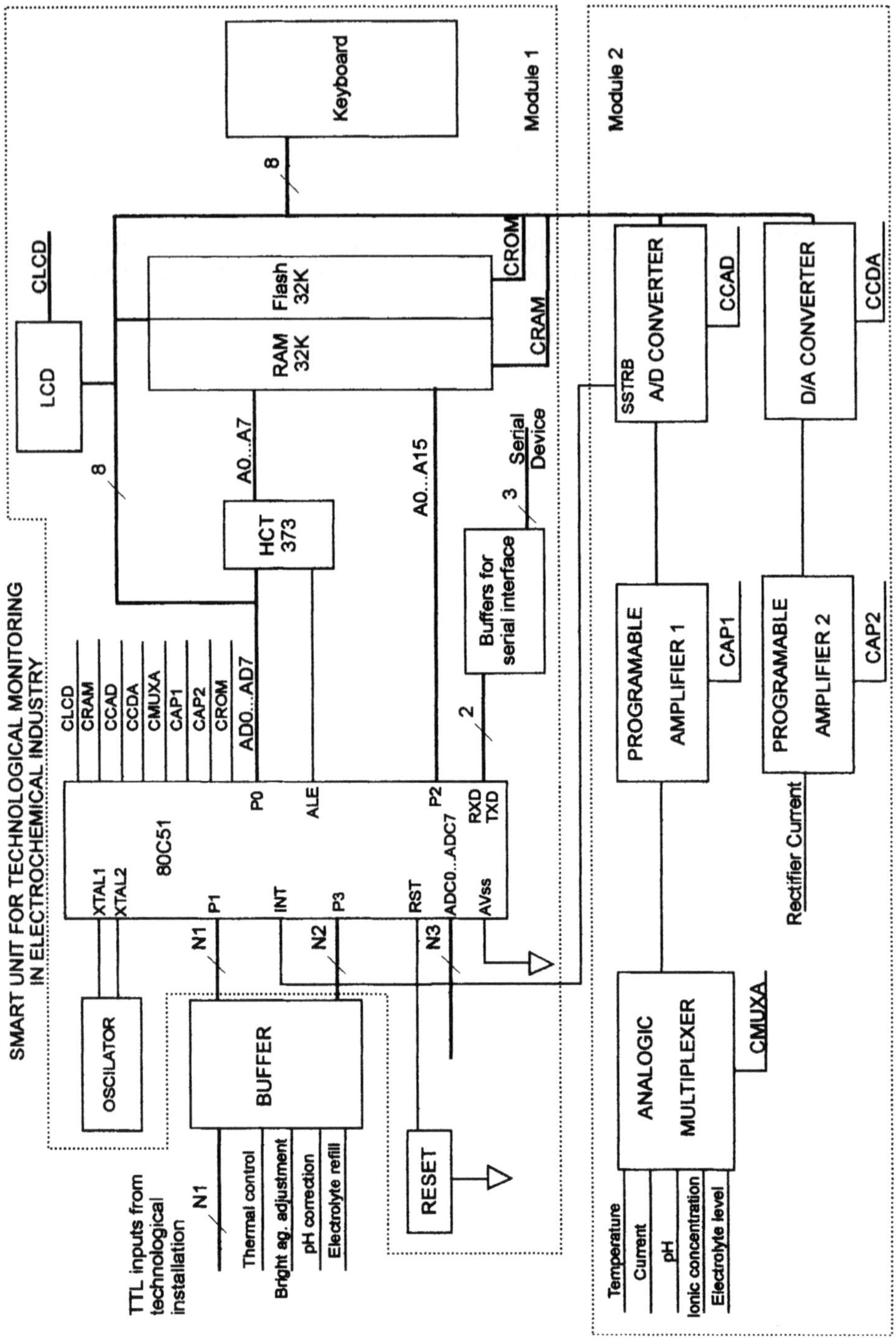

Figure 1. Schematic design

Choosing the Flash memory was determinate by the process necessities: modification of the measurement channels characteristic coefficients (at selfcalibration), of the measured parameters limits or the commands to be executed. This memory keeps also the monitor program which supervise all module function.

During a work session, RAM memory keeps data proceed from technological equipment, also being used for various calculations. We mention that this memory selection is made through RD and PSEN signals conjunction, so that the possibility for RAM to work as program memory was provided too (at putting in function or testing).

The microcontroller has a serial interface UART type. Connection with exterior it is made through two RxD and TxD lines. Because the signals provided by microcontroller are not compatible with line signals (they have TTL level), it was provided an interface buffer MAX 232 or MAX 233 type. Through serial interface the system can be connected to an central computer which has to take over data, when the monitoring of full technological line is followed and to execute complex work. Through this interface it is an existing possibility to transfer some commands from central computer.

Although it is a willing that the presence of the human operator to be reduced, there are moments when his intervention is necessary (for example at installation starting). In order to make local programming of the unit (depending on the necessities) was provided a dedicated keyboard. This allows data introduction or selection of some commands. We attached also an smart LCD, LTN 211 type (a possible local monitoring of the equipment), for displaying some parameters or some error messages.

The I/O module contains like principal components the analog/digital and digital/analog converters. The analog/digital converter is MAX 186 type having 8 channels Single Ended , has a conversion rate of 133 KHz and it is provided with internal reference and sample and hold circuit [6]. The circuit was chosen because of its good performances and low price, also for its very "economical" microcontroller connection (through only 4 wires). We considered to be the principal peripheral, so that the SSTRB signal of finished conversion is connected with interruption line of the microcontroller. Towards the converter the signals are getting through an analog multiplexor, MMC 4051 type, being proper calibrated with a programmable amplifier AD526 type.

Because the system has to supply analog signals too, the module has a digital/analog converter MAX 532 type. This works at 6MHz frequency, being connected with microcontroller through only 3 wires. The signal resulted is amplified through a programmable amplifier like above mentioned.

The LCD, CRAM, CORM, CAD, CODA, CRUX, CAP1, CAP2 signals are the selection signals and command for LCD, RAM memory, ROM memory, A/D converter, D/A converter, analog multiplexed, programmable amplifiers number 1 and 2.

4. Soft description

The monitor program which is coordinating the system functionality is in the Flash memory. The diagram of module function is give in figure 2.

According to figure 2, all peripherals (excepting the analog to digital converter) are interrogated in polling mode. The only interruption validated in the system is the one which comes from the analog to digital converter, considered priority. Each peripheral has a flag which shows the treatment necessity. When it has to realize a transfer, the serial interface is in a peculiar situation, setting the RT0 bit from the attached register and waiting to be accessed.

Because the microcontroller has the possibility to work on bit, the clocks for the converters are simply created by setting/resetting one bit from one port.

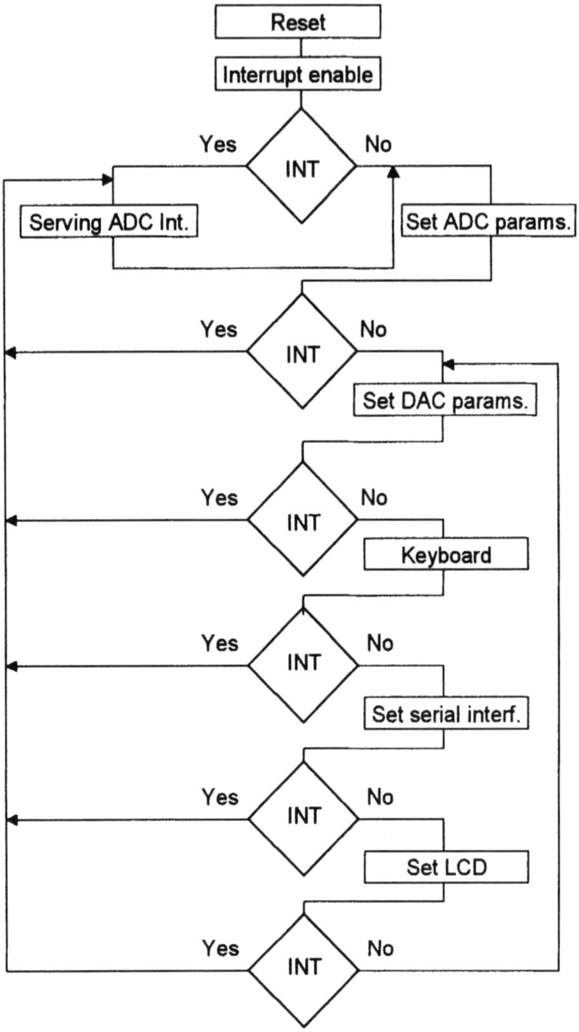

Figure 2. Block diagram

5. Technological achievement.

Technological design has to respect the concrete working conditions, in other way said, it has to be insured the conditions of electromagnetic compatibility. We mentioned before that the unit is made from two modules, one of them being the development system (where we have only digital logic), and the other one being dedicated to I/O interface in which we have digital and analog signals. For sure, the most exigent from constructive point of view, is the I/O module, where beside the general design conditions regarding signal lines has to be taken care of the protection at the eventual electric and magnetic fields which can appear. The working temperature is also an important element: the digital circuits functionality could be wrong if some limits are passed.

6. Conclusions

The realization of this smart unit allows:

- A client - server structure.
- The local processing.
- A high speed serial interface .
- A "remote control" configuration.
- A power management.
- The modular structure.
- The connection possibility to a block using signal processor

The compact construction, as well as the large necessary number of smart units used in equipment with more technological lines, provides the manufectoring costs at lower level.

The designed structure of this unit allows the posibility to take in count development of MCM packaging too .This would creat efficiency on large product series at low costs.

References

1. Firoiu, C. (1983) *Tehnologia proceselor electrochimice*, Ed. Didactica si Pedagogica, Bucuresti.
2. Brenner, A. (1963) *Electrodeposition of Alloys, vol. I (Principle and practice)*, Academic Press, New York and London.
3. Gruenwald, E (1978) *Galvanotehnica*, Ed. Stiintifica, Bucuresti.
4. *** (1995) *8 Bit Microcontroller Family*.
5. *** (1995) *Intel Embedded Applications Vol I + Vol II*.
6. *** (1995) *MAXIM - New releases data book*.

LASER PATTERNING OF PRINTED WIRING BOARDS FOR MCM-LS

ZS. ILLYEFALVI-VITÉZ, J. PINKOLA, M. RUSZINKÓ, P.J. SZABÓ
Department of Electronics Technology, Technical University of Budapest
Goldman t. 3., Budapest, H-1111, Hungary

1. Introduction

The current state of the art process for manufacturing the interconnect boards of laminated multichip modules (MCM-Ls) is essentially identical with that used for highly dense multilayer printed boards, but includes higher requirements of fine-line pattern definition, etching, selective plating, additive plating, special materials etc. The five standard types of processes, which are characteristic of MCM-L technology, are as follows: photoresist application for imaging the different layers; chemical etching of clad and/or plated copper foils/layers; lamination and curing of precured adhesive dielectric layers to insulate and fix the conductive patterns of the multilayer structure; drilling of holes through single layers for buried and blind vias and through the entire laminate for plated through holes; and, plating of drilled holes to accomplish electrical connections between the conductive patterns.

The need for fast and cheap prototyping becomes more and more important in the production of MCMs. A prototyping process should fulfill the following requirements:

- rely basically on the same processing and materials that are applied in the production,
- fast turnaround.
- flexibility in modifying the interconnection layout,
- apply relatively low cost special methods,
- possibly avoid the high cost artwork and photo-mask production,
- enable as low batch numbers as one piece.

A new approach is being developed for prototyping MCM-Ls that fulfills the former requirements. It combines the advantages of laser surface patterning and wet chemical etching [1,2]. The steps of processing of the sole interconnection layers are as follows:

- etch-resistant coating on copper clad laminates,
- direct cutting the interconnection pattern into the coating using laser processing,
- copper etching,
- coating removal by chemical method.

For surface imaging a laser micromachining system is applied, which can also be used for structuring surface layers, engraving decorative pictures, marking mechanical or electronic parts, trimming film type components, soldering, drilling etc. The pattern is engraved on the surface applying focused laser beam delivered by computer controlled rotating mirror galvanomotor scanning system. The results of surface profile and material analyses are demonstrated in the paper.

R. Tummala et al. (eds.), Electronic Packaging for High Reliability, Low Cost Electronics, 69–78.

2. Facilities of MCM-Ls at the Department

The laboratories devoted for the technology of printed wiring boards where the laminates are fabricated and the processes are investigated contain the basic equipment elements as follows:

- a laser photo-plotter for photo-mask direct exposition (max. 480x570 mm, 0.25 μm accuracy);
- automated developing system for photo-mask preparation;
- fabrication line for dry film photo-resist technology;
- wet chemical etching apparatus for printed circuit boards;
- CNC drilling machine;
- electroplating line for the fabrication of through-hole-plated printed circuit boards;
- electroplating line for gold plating of direct connectors;
- laminating press for the fabrication of multilayer boards.

The equipment of the PWB laboratories provides good possibilities for the fabrication and technological development of multilayer plated-through-hole printed wiring boards with max. 8 layers, min. 0.3 mm hole diameter, 0.2 mm line width and spacing. The fabrication of various other components made by chemical and/or electrochemical processing, such as chemically etched metal parts, is also possible. Considering the higher requirements of the laminates of MCM-Ls, these processing technologies need improvements.

As it was mentioned., there are five standard types of processes, which are characteristic of the technology of multilayer printed circuit laminates. Accordingly, the development strategies of these processes and the processing sequences are as follows:

1. Dry film photoresist application will remain the characteristic process for imaging the different layers, although liquid photoresists offer better resolution and, as a consequence, it is also within the scope. Dry film photo-resist imaging applied in an additive or a semi-additive processing sequence, however, provides fine resolution, and a proper solution for MCM-Ls. As part of the continuing reconstruction, aqueous dry film photo-resist technology, including the suitable stripping equipment and process, were recently introduced at the Department, primarily in order to meet the requirements of environmental protection.
2. Wet chemical etching of copper layers after a due reconstruction of the equipment will meet the requirements. Switching to the more advanced dry plasma etching is beyond the possibilities, since its cost and incompatibility with the other processes.
3. The equipment for the lamination and curing of precured adhesive dielectric layers of multilayer structures is compatible with the applied materials, and although presently fulfils the requirements, it needs improvement in the near future.
4. Mechanical drilling of holes through single layers for buried and blind vias and through the entire laminate for plated through holes will certainly remain the main process in this field. Since the possibility of laser drilling is steadily developing at the Department, its application is also in scope.
5. Plating of drilled holes to accomplish electrical connections between the conductive patterns is a key process in any MCM-L processing sequence. During the last two years

the main parts of the equipment of the electroplating line for panel and pattern plating, and the gold plating line of contact fingers have been changed, developed or modified, and recently direct plating technology has been introduced [3], thus the plating process can be considered as the most advanced and reliable part of the sequence.

In the last few years, one of the most significant research activity of the Department has been the laser image transfer development. As the main result, a laser micro-machining system has been developed [4-5] for structuring surfaces, films, engraving patterns, marking mechanical or electronic parts, trimming thin and thick films type, soldering, drilling etc. The pattern is engraved on the surface applying focused laser beam delivered by computer controlled rotating mirror galvanomotor scanning system, described elsewhere [6-9]. The system is capable to accept computer generated figures, especially the imaging and/or trimming patterns of computer simulated and designed hybrid circuits. Moreover, this laser system can be a successful tool in a novel technology, in the combined processing sequence of tin electroplating (or any other etch-resistant coating), laser surface imaging and wet chemical etching, which offers a promising solution of fine resolution laminate fabrication (Fig. 1.), providing flexible fast prototyping possibilities for multichip modules.

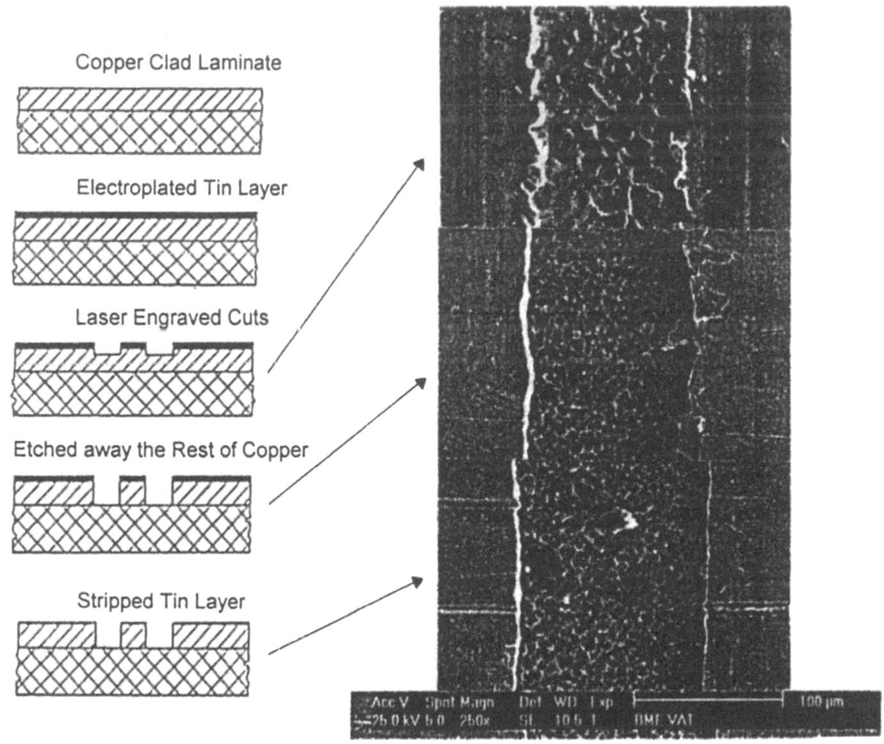

Figure 1. The combination of electroplating, laser direct image transfer, and wet chemical etching for the fabrication of fine resolution printed wiring laminates

3. Initial Investigation on Laser Image Transfer

Utilising the existing possibilities of the Department, initial investigations have been carried out. Applying the standard electroplating technology, single-sided copper clad epoxy laminates with nominally 17 μm thick Cu layer were covered by approximately 2.5 μm thick tin (Sn) layer.

This structures were imaged by laser. The Nd:YAG laser worked in the Q-switched multi-mode operation. Depending on the different parameters of the laser operation, mainly depending on the pumping current ranging 16-28 A, the peak power of the pulses varied 10-200 kW (Figure 2.), while the beam intensity in the focal spot was in the 0.05 to 1 GW/m^2 range. The samples were moved under the laser spot with a velocity of 10 mm/s. The image drawing was carried out with different pumping currents, and multiplied laser processing was also tried. In Figure 3. the width vs. pumping current function is plotted. The optimum was found at about 22-24 A pumping current and single processing.

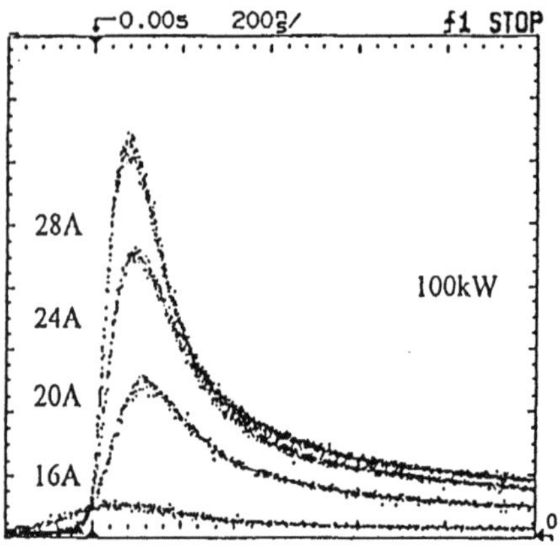

Figure 2. The power of processing laser pulses vs. time and pumping current

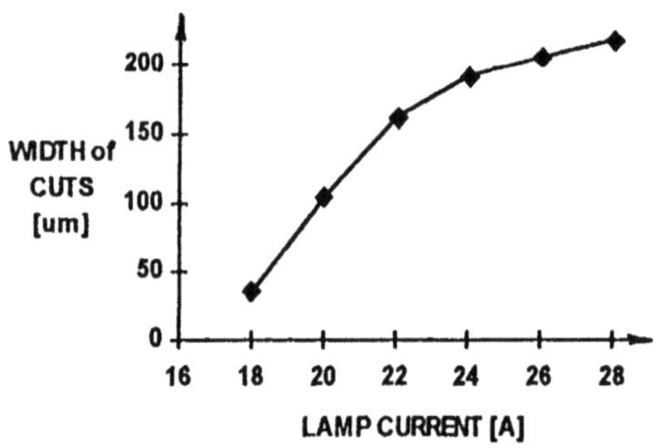

Figure 3. The final widths of cut vs. pumping current

The laser pulses opened line-shaped windows in the tin layer on the surface, while they had only a slight effect on the copper bottom layer. Figure 4.a. (upper left, next page) shows the surface profile of the sample across the line, processed by higher pumping current (24 A) than the optimum conditions.

The profiles were plotted by a Tencor Alpha Step 500 type stylus profiler with 12.5 μm radius. The depth of the cut after laser processing is about 2.5 μm, which means that the 2.5 μm thick tin layer is just cut through. The width of the cut is about 50 μm. Large amount of material, mainly tin, can be observed along the edges.

Etching was carried out in a conveyor spraying equipment, using a solution of copper-tetramin-complex. The temperature of the solution was 50 °C, and the etching time was 1 min. The resulted cut profile is presented in Figure 4.b. (lower left). The width of the cut is much larger of about 180 μm, and the etch process went through the approximately 15 μm thick copper layer. The undercut of the etching is significant, in the range of 1 μm at both sides.

Figure 4.c. (upper right) shows the cross-section of the cut after removing the tin mask. The profile in the copper layer practically did not changed, the width of the cut is 195 μm, the accumulated materials of tin, however, disappeared from the edges.

For better understanding, in Figure 4.d. (lower right) the curves of Figure 4.a., b. and c. are fitted to each other. In this way the thickness of the tin layer and the cut depth just reaching the copper surface are well illustrated.

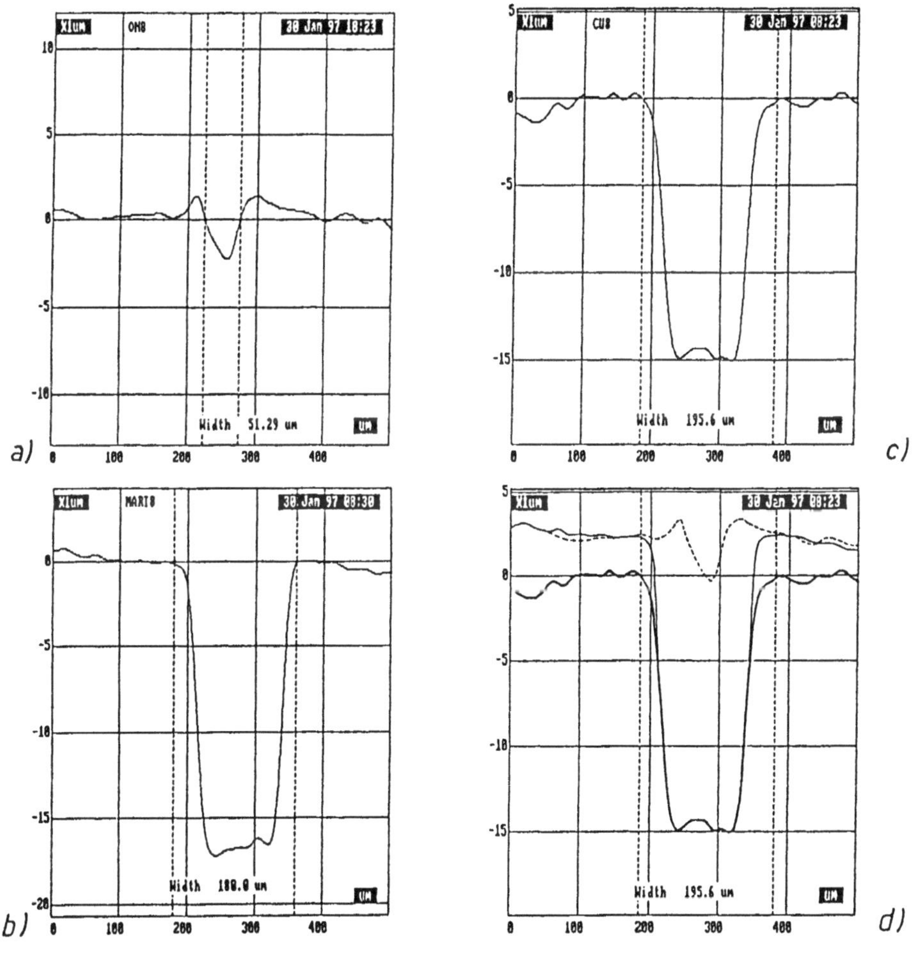

Figure 4. Cut profiles at subsequent processing steps:
a) laser cut processed into a tin layer on copper; b) after copper etching;
c) after tin etching; d) the fitted curves of Fig. a., b. and c. illustrate
the effect of the processing steps on the cut depths

It is to be noted, that with the application of the multi-mode laser beam and a relatively obsolete etching equipment, the possibilities of the technology were not exploited, thus the initial stage of the experimental results should be emphasised. On the basis of the wavy edge contour of the lines and the rough surface of the bottom after laser processing (see the photos in Figure 1.), the necessity of further investigations to find a more suitable etch-resistant coating is also raised.

4. Materials Analysis of Samples with Different Etch-Resistant Coatings

Materials analysis was carried out to find the reason of the rough surface in the laser processed cuts. The results of the analysis on two samples processed with 24 and 27 A pumping currents are shown in Figure 5. It is seen that strikingly large amount of tin (Sn) remained in the cuts. At higher pumping currents the amount of tin is larger. Fortunately the resulting tin-copper solid solution can be etched through by the usual etching process and etchant of the copper layers.

Although the cuts seem to be clear, and the measured insulation resistance across the cuts was normal, ion migration failure analyses [12] and long life tests were necessary to estimate the reliability of the processing method.

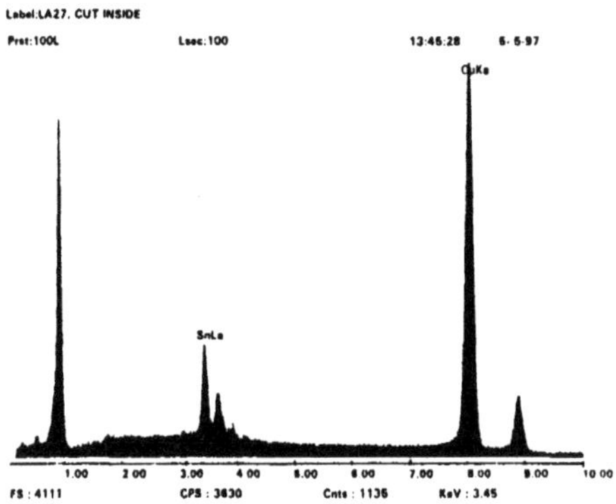

Element	K Ratio	Weight %
SnL	0.2981	32.328
CuK	0.7019	67.672
Total		100.000

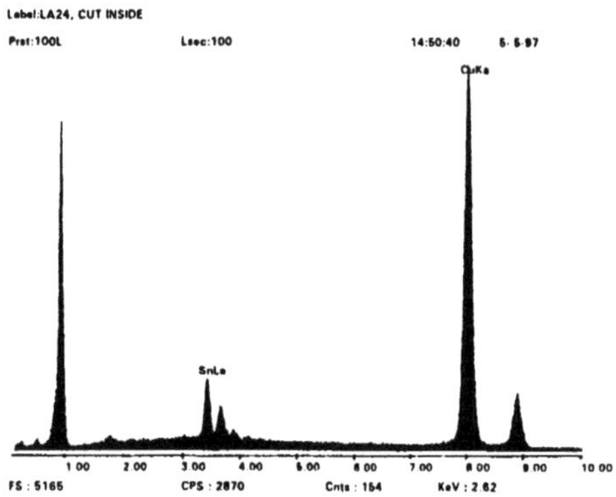

Element	K Ratio	Weight %
SnL	0.2226	24.649
CuK	0.7774	75.351
Total		100.000

Figure 5. Comparison of materials analysis results of cuts in tin layer on copper, processed with different pumping currents: 27A (left) and 24A (right)

In the course of searching for more suitable etch-resistant coatings, electroplated gold (Au) and thin polymer layers were tested. Although the laser processed cut in the gold layer apparently was very similar to the tin layered samples, the bottom of the cut remained etch-resistant. Materials analysis results of these cuts are shown in Figure 6.

Since very high percentage of gold remained in the cut, the insolubility of the layer can be reasoned by a special Cu-Au solid solution phase. It is very interesting that the composition of the layer in the cut has a good correspondence to that of a well-ordered solid phase (see the phase diagram in Figure 6.). As a consequence of their well-ordered structures, these phases usually show higher etching and corrosion resistivity than the other compositions.

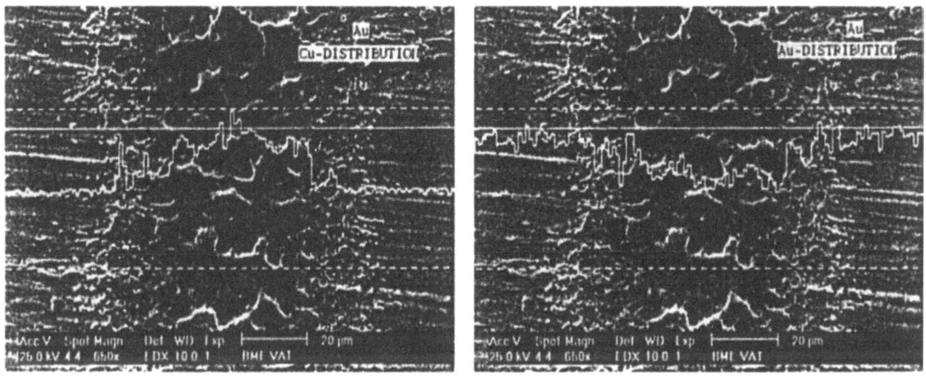

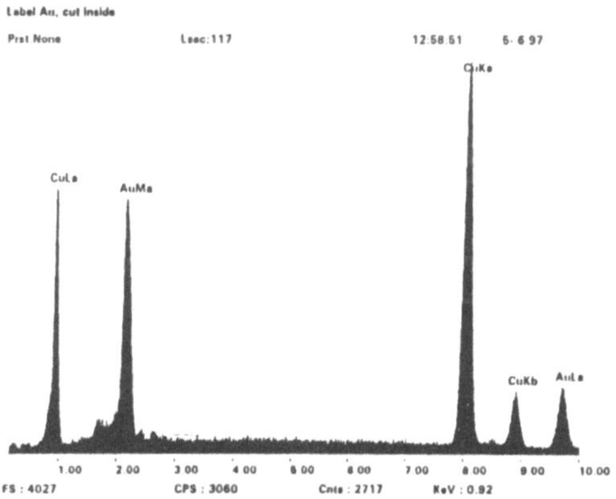

Element	K Ratio	Weight %
CuK	0.5203	46.767
AuL	0.4797	53.233
Total		100.000

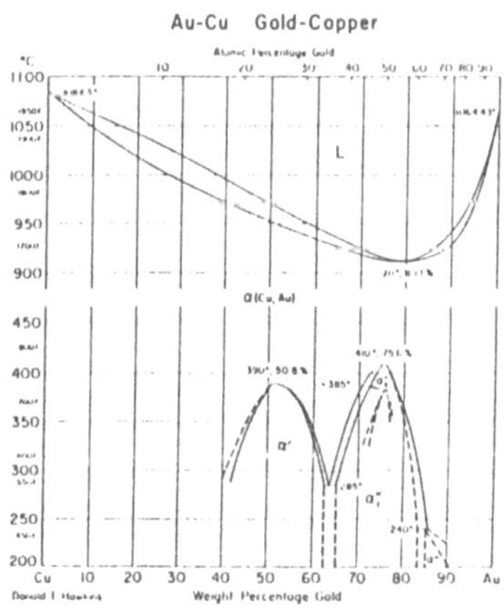

Figure 6. Analyses of laser processed electroplated gold layer on copper. Upper left and right: copper and gold line-distribution analyses across the cut. Lower left: material spectrum in the cut. Lower right: phase diagram of Au-Cu alloy.

The polymer layers were easily removable by laser processing, thus smaller modus aperture and lower laser intensity were used. Material analysis showed neither carbon nor any other contamination in the cut, and the etchability was also perfect. Unfortunately the waviness of the contour along the edges of the lines is relatively high (see Figure 7.), that can be reasoned by the irregular removal of weakly adhesive layer parts influenced by the intensive laser beam.

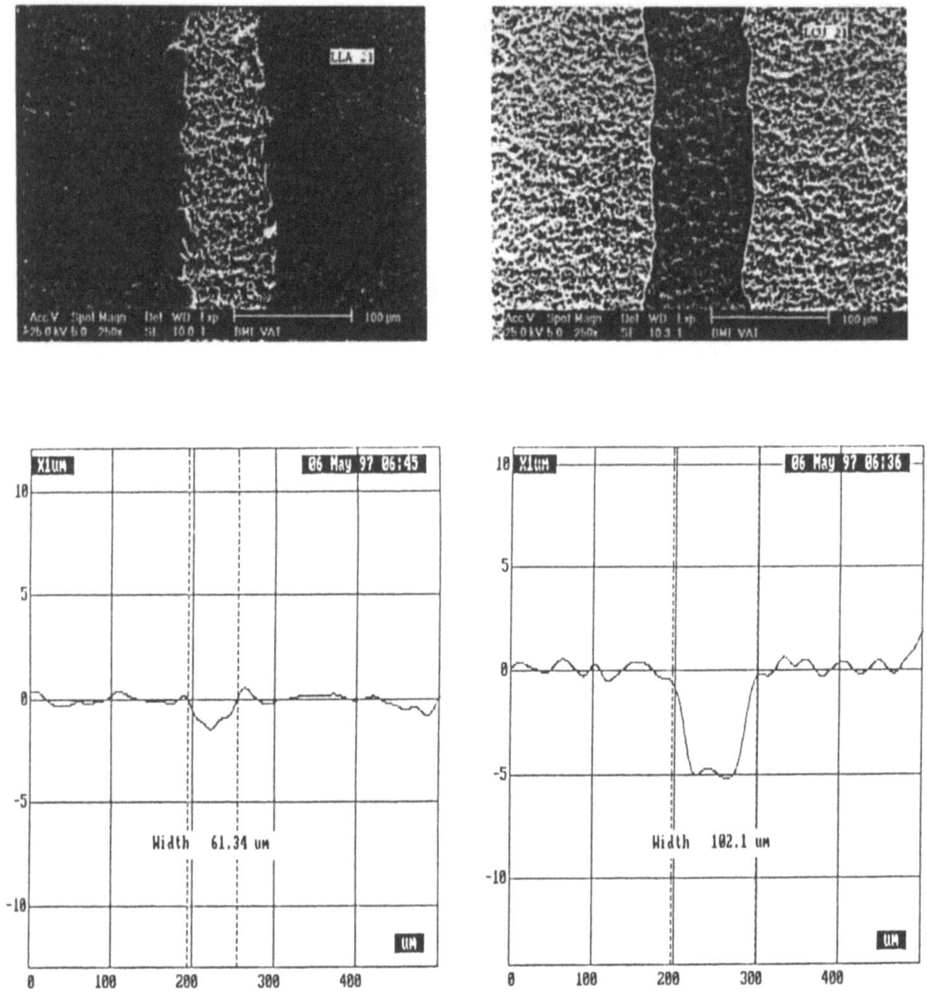

Figure 7. SEM photos (upper figures) and Alpha-step surface profiles(lower figures) of laser processed polymer film on copper. The left figure shows the cut after laser processing, while the right figure shows it after etching.

Finally, for the demonstration of the applicability of the combined laser imaging process, a chip carrier pattern was cut out, its microscopic view is shown in Figure 8.

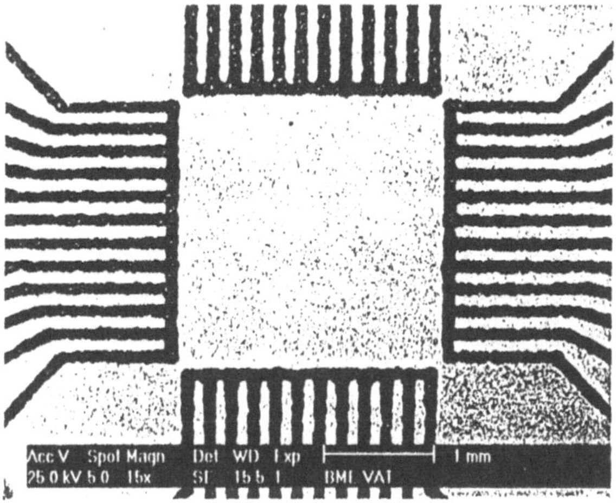

Fig. 8. Microscopic view of a laser processed and etched chip carrier pattern

5. Conclusions

A new approach was established in fast and cheap prototyping of MCM-Ls. The method combines the advantages of the application of etch-resistant coatings (e.g. electroplated tin layers or polymer films), laser direct pattern transfer and wet chemical etching. The first results indicate that the resolution is on an appropriate level, and the processing provides flexible possibility for prototyping in low batch numbers. Since the waviness along the line edges is high, further investigations are necessary to make the processing competitive with photoresist imaging for larger series.

6. Acknowledgement

This research is supported by the European Commission in the frame of INCO Copernicus Project No. ERBIC 15CT 960743.

7. References

1. Illyefalvi-Vitéz,Zs.; Pinkola,J.: *Application of Laser Engraving for the Fabrication of Fine Resolution Printed Wiring Laminates for MCM-Ls.* Proceedings of the 47th Electronic Components and Technology Conference, May 18-21, 1997, San Jose, California.

2. Harsányi,G.; Pinkola,J.; Tóth,E.: *CM^2-Cheap Multichip Module: a New Approach for Fast and Cheap Prototyping of MCM-Ls.* Second Pan Pacific Microelectronics Symposium and Tabletop Exhibition, Jan. 28-31, 1997, Maui, Hawaii.
3. Gál,L.; Hajdu,I.; Pinkola,J.; Tóth,E.: *New Metal Plating Technology of Nonconductive Surfaces.* Proceedings of the 18th International Spring Seminar on Electronics Technology, June 26-30, 1995, Temesvar (Czech Republic), pp. 254-258.
4. Illyefalvi-Vitéz,Zs.: *Laser Processing in MCM-C Technologies,* NATO-ARW on MCM-C/Mixed Technologies, 1994, Islamorada, USA, p.26. also in NATO-ASI Series 3: High Technology - Vol. 2., "MCM-C/Mixed Technologies and Thick-Film Sensors" (ed. by Jones,K.W., Kurzweil,K., Harsányi,G., Mergui,S.) Kluwer Academic Publishers, Dordrecht, The Netherlands 1995., pp. 69-76.
5. Illyefalvi-Vitéz,Zs.: *Analysis of Laser Induced Surface Effects on AlN Substrates,* NATO-ARW on MCM and Sensor Technologies, 1995, Budapest, Hungary, p.7. and also in NATO-ASI Series 3: High Technology, Vol.16, "Multichip Modules with Integrated Sensors" (ed. by Jones,K.W. and Harsányi,G.) Kluwer Academic Publishers, Dordrecht, The Netherlands, 1996., pp. 283-288.
6. Enyedi,L.; Illyefalvi-Vitéz,Zs.; Ruszinkó,M.: *Analysis of a Laser Beam Deflecting Mechatronics System.* Mechatronics: The Basis for New Industrial Development, Computational Mechanics Publications, Southampton (ISBN: 1853123676), Boston (ISBN: 1562522914), (Proceedings of the Joint Hungarian-British International Mechatronics Conference, September 21-23, 1994, Budapest), pp. 247-252.
7. Illyefalvi-Vitéz,Zs.: *Mechatronics Technology Improves the Reliability of Motion Systems.* Proceedings of the 18th International Spring Seminar on Electronics Technology, June 26-30, 1995, Temesvar (Czech Republic), pp. 47-53.
8. Enyedi,L.; Illyefalvi-Vitéz,Zs.; Ruszinkó,M.: *Sensor Applications in a Laser Micromachining System.* Proceedings of the Workshop on Thick and Thin Film Sensors and their Application in Ecology, Szklarska Poreba (Poland), June 12-15, 1994, pp. 162-165.
9. Enyedi,L.; Fülöp,S.; Illyefalvi-Vitéz,Zs.; Ruszinkó,M.: *Simulations and Measurements in a Laser Beam Delivery Mechatronics System.* Mechatronics'96, Sep 18-20, 1996, Guimaraes (Portugal), Proceedings of The 5th UK Mechatronics Forum International Conference (ISBN 972-8063-08-3), pp. 2-335-2-340.
10. Harsányi,G.; Pércsi,L.: *Failure Analysis of Migrated Short Circuits in MCMs Using Soldering.* NATO ARW on MCM and Sensor Technologies, May 18-20, 1995, Budapest (Hungary) in Jones,W.K.; Kurzweil,K.; Harsányi,G.; Mergui,S. (editors): MCM and Sensor Technologies, NATO ASI Series, Kluwer Academic Publishers, Dordrecht (The Netherlands), 1996.

SILVER FILLED ADHESIVES - CONTROLLING RHEOLOGY AND ELECTRICAL CONDUCTIVITY

P. SEXTON
K. MCNEILLY
Chemet Corporation
52 Gardner St.
Attleboro, MA 02703 USA
S. SCHLAG
Metaux Precieux SA Metalor
Av. du Vignoble
CH-2009 Neuchatel, Switzerland

ABSTRACT

As packaging criteria become more demanding, new silver filled conductive adhesives must be developed to meet the changing requirements. Epoxy and polyamide materials are being supplemented by new compositions based on silicones, bis-malimide, cyanate esters, etc. Development of suitable conductive adhesives requires understanding of rheological and chemical considerations. Rheology can be predicted by models using the base resin viscosity and the physical properties of the metal filler. Electrical conductivity in some cases is dependent upon both physical properties of the metal filler and the residual lubricant coating. Rheological behavior is also a function of the relative strengths of the networks. These networks strengths are dependent on the organic coatings on the metal powders.

INTRODUCTION

Silver filled adhesives continue to be major materials in semiconductor die attachment. As devices progress in technology new resins are supplementing the originally used epoxies and polyimides. Among these are bis-malimide, cyanate esters, silicone based materials, and other proprietary compounds. These new materials bring with them challenges for the formulator to develop the desired rheology and electrical conductivity.

Many of the resins new to these applications have unloaded viscosities substantially higher than the former resins, e.g. 2,000 cp versus 400 cp.

R. Tummala et al. (eds.), Electronic Packaging for High Reliability, Low Cost Electronics, 79–90.

Because of this, the producers of the Ag flake fillers are required to develop materials that have a lower contribution to the system viscosity.

Since silver flakes have a chemisorbed layer resulting from the lubricant and solvent used in production, the resultant electrical conductivity requires chemical compatibility of the Ag flake lubricant and the resin to develop adequate electrical conductivity. For a discussion of the production techniques for manufacturing silver flake refer to E. M. Jost, et al., "Silver Flake Production and Optimization for Use in Conductive Adhesives"[1]. The following summarizes the research done to date in the Chemet Corporation laboratory to predict rheology and electrical conductivity of silver filled resins used as conductive adhesives.

RHEOLOGY AS A FUNCTION OF PHYSICAL PROPERTIES

Figure 1 shows an envelope in XY space where X is specific surface area and Y is median particle size. Current technology allows silver flake to be produced with properties X, Y provided X, Y are both within the envelope. At any given median particle size the corresponding maximum specific surface area is limited by the fragility of the particles. Beyond this, a particle disintegrates into multiple particles and therefore a lower mean particle size.

To characterize a powder the mean particle size and particle size distribution are of great use. In the case of platelets, such as the silver flake discussed herein, mean particle size, particle size distribution, and a measure of thickness are required to characterize these materials. Many of the measurements used have some difficulty associated with them. For example, particle size distribution is only accurate to the degree that agglomerates can be disassociated into their fundamental particles and Fisher subsieve size makes the assumption that all particles are perfect monosize spheres. Jost, et al. have shown mathematically that for flakes that are totally platelets surface area, to a first approximation, is an inverse function of particle thickness[2]. Refer to figure 2.

Viscosity of a series of experimental adhesive formulations was measured. The test resin was 60% Shell Epon 1002 and 40% butyl carbitol acetate by weight. Loading was 80% silver flake by weight. The viscosity of the unloaded resin was approximately 400cp. Viscosity was measured using a Brookfield Model HBT with a CP51 spindle.

We filled resins with a wide range of silver flakes at a constant weight per cent loading. Figure 3 shows viscosity versus median particle with surface area variable. No functional relationship is evident. Similarly viscosity is not a function of specific surface area if particle size is variable, figure 4.

If particle size is held constant viscosity is a function of surface area, figure 5.

Using 3 dimensional plotting of data it can be seen that there is a minimum viscosity and that physical property changes away from this increase viscosity, refer to figure 6. The practical significance of this is that there is not a unique pair of physical properties to result in a given viscosity. The authors have also noted that near the viscosity minimum flakes tend to align. This has been observed in formulations used for through hole inks for printed circuit boards and internal electrode inks for multilayer ceramic capacitors[3]. Deviations from the viscosity minimum create arrays that are not closely packed which result in undesirable properties of the precious metal loaded film.

Plotting viscosity versus mean particle size and specific surface area was done in a range of resins with varying viscosities and at a range of shear rates. Increases in unloaded resin viscosity increases the resultant viscosity, however, the general shape of the viscosity plots are relatively unchanged. Our work was limited to silver flakes with normal particle size distributions. In many actual conductive adhesive systems, bimodal particle size distributions are used to enhance shear thinning. Measurement of this shear thinning is done by measuring viscosity at 2 different shear rates, typically 0.5 rpm and 5.0 rpm. The ratio of the viscosities is commonly referred to as the "thixotropic index".

ELECTRICAL CONDUCTIVITY AS A FUNCTION OF LUBRICANT SOLVENT SYSTEM AND FLAKE SIZE

Prior to our experimental program, we were aware of a variety of anecdotal information, empirical results, and theoretical physical models predicting electrical conductivity. The most widely accepted were:

- Certain lubricant types (and specifically oleic acid) maximize electrical conductivity.
- Electrical conductivity increases with increasing particle size due to minimizing the number of point contacts.

Production of Experimental Silver Flakes

Silver flakes were milled in a model 1S attritor using 7 different lubricant/solvent pairs. These are listed in Table 1. A pair of silver flakes were produced with each type of lubricant. Pairs were comprised of Type A silver flakes and Type B silver flakes. Class A flakes had a median particle size by SEM of 8.5μ ± 0.6 and a specific surface area of 0.75 m^2/g. Refer to tables 2A and 2B.

MEASUREMENT OF ELECTRICAL RESISTIVITY

Three resin/hardener systems were used. The compostions of Types I and II are in the public domain and Type III is a proprietary system. These are given in Table 3.

Resin Type I and II were loaded with 77.5% silver flake by weight and resin Type III was loaded to 75.0%. This was done to result in similar viscosities of each. All 14 flakes were put in each resin system. All conductive adhesives were degassed. Slides were made by spreading each filled adhesive between parallel 3M® tape 810. After removing the tape a strip of silver filled epoxy 2.54mm x 63.5mm remained. Each was cured for 1 hour at 175°C. After cooling to room temperature resistance was measured using a Fluke 8012A digital multimeter. Resistivity was calculated from resistance measurements and physical dimensions.

RESULTS

The volume resistivity as a function of lubricant type acid flake class is shown graphically in figures 7-9 .

Results can be summarized as follows:

Some resin systems are not sensitive to lubricant type and exhibit low resistivity over a wide range of lubricants (An exception is the lubricant/solvent type 2 which is a thiol system. This is not suitable for silver but is generally used with gold). Resin systems exhibiting this behavior are Type III and to a slightly lesser extent Type I, figures 7 and 9.

Some resin types are very sensitive to lubricant/solvent type and have widely varying resistivity results. An example of this is resin type II, figure 8.

Some resins are insensitive to the size of the silver flake and exhibit low resistivity for both large (Class A) flakes and small (Class B) flakes. Examples of this are resin Types I and III, figures 7 and 9.

Some resin systems are very sensitive to the size of the silver flake where large flakes (Class A) give lower resistivity than small flakes (Class B). An example of this is resin Type II, figure 8.

For the limited number of systems studied here, the systems sensitive to flake size were also sensitive to lubricant/solvent type, figure 8.

CONCLUSIONS

Viscosity of silver loaded resins can be predicted by physical properties of the particles when mean particle size and specific surface area are known. A viscosity minimum exists and there is not a unique solution for a given viscosity.

In silver filled adhesives the dependence of resistivity on flake size and lubricant/solvent coatings is system dependent. Some adhesive resin systems are very insensitive to the size and lubricant/solvent coating. In these systems a wide range of silver flake sizes produced with varying lubricant/solvents give similar resistivities when loaded to a constant weight of silver.

Lubricant/Solvent Type	Lubricant Type	Solvent Type
1	Unsaturated Fatty Acid	Alcohol
2	Thiol	Proprietary
3	Proprietary 1	Proprietary
4	Proprietary 2	Proprietary
5	Saturated Fatty Acid	Kerosene
6	Salt of a Fatty Acid	Alcohol
7	Soap	Water

Table 1. Lubricants used for producing the silver flakes.

Lubricant Type	1	2	3	4	5	6	7
Apparent Density	2.69	2.55	3.04	2.43	2.87	1.95	2.06
Tap Density	4.13	3.85	4.38	3.97	4.7	3.5	3.58
Surface Area	0.4	0.32	0.29	0.38	0.43	0.4	0.45
Weight Loss @ 110°C	0.02	0.02	0.01	0.04	0.02	0.06	0.03
Weight Loss @ 538°C	0.21	0.21	0.15	0.23	0.21	0.58	0.74
M.V	8.9	9.04	7.86	8.16	7.3	9.1	9.4

Table 2A. Summary of physical properties for class A silver flakes.

Lubricant Type	1	2	3	4	5	6	7
Apparent Density	2.09	1.89	1.76	2.27	2.08	1.44	1.4
Tap Density	3.72	3.24	3.27	3.52	3.8	2.63	2.61
Surface Area	0.75	0.67	0.67	0.68	0.7	0.82	0.81
Weight Loss @ 110°C	0.02	0.06	0.05	0.06	0.01	0.04	0.04
Weight Loss @ 538°C	0.27	0.45	0.35	0.22	0.31	0.78	0.76
M.V.	4.6	4.21	4.61	5.87	3.83	4.23	3.75

Table 2B. Summary of physical properties for class B silver flakes.

Resin Type	Epoxy	Diluents	Curing Agent
Type I	51% Epon® 1002F & 15% Bisphenol F	30% Butyl Glycidyl Ether	4% BF_3 Monoethylamine
Type II	82% Epon® 828	15% Butyl Glycidyl Ether	3% Dicyandiamide
Type III	Proprietary system	N/A	N/A

Table 3. Summary of resin types.

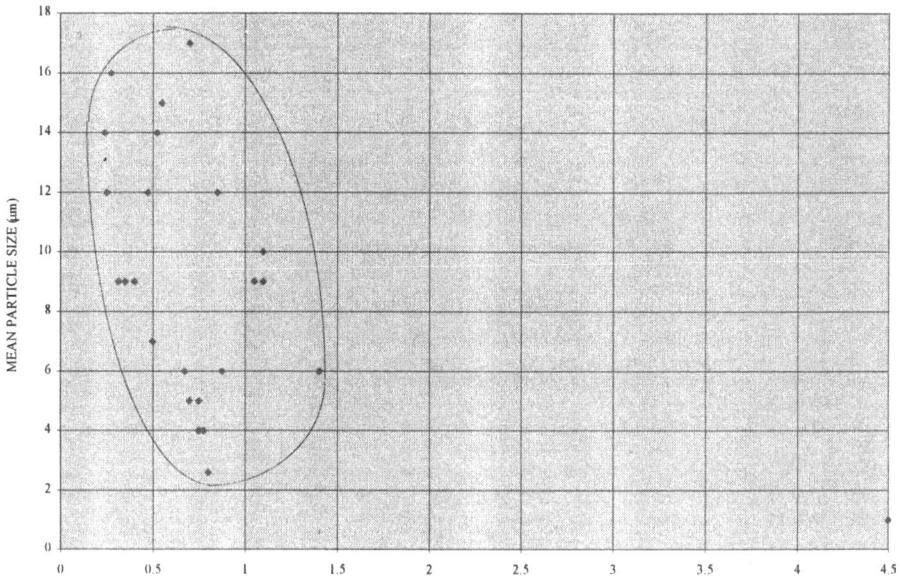

Figure 1. Mean particle size and specific surface areas of silver flakes.

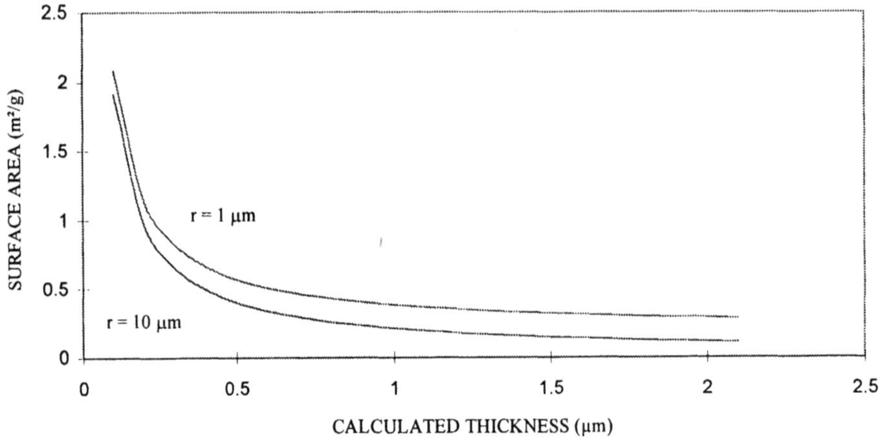

Figure 2. Specific surface area vs. calculated average thickness of platelets.

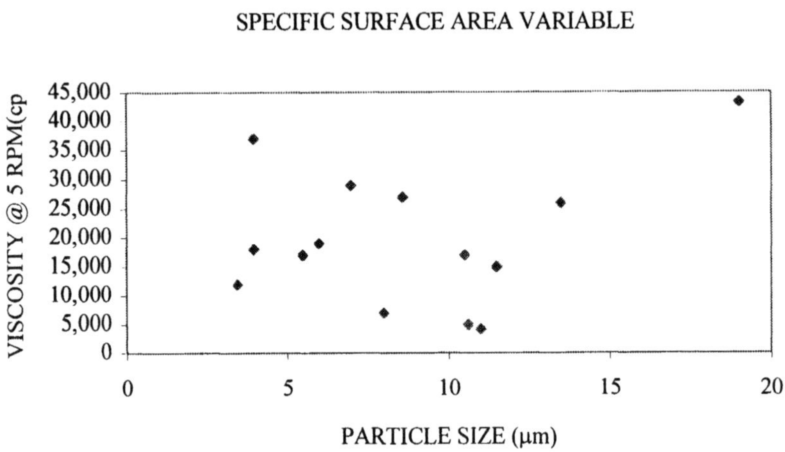

Figure 3. Viscosity vs. particle size (specific surface area variable).

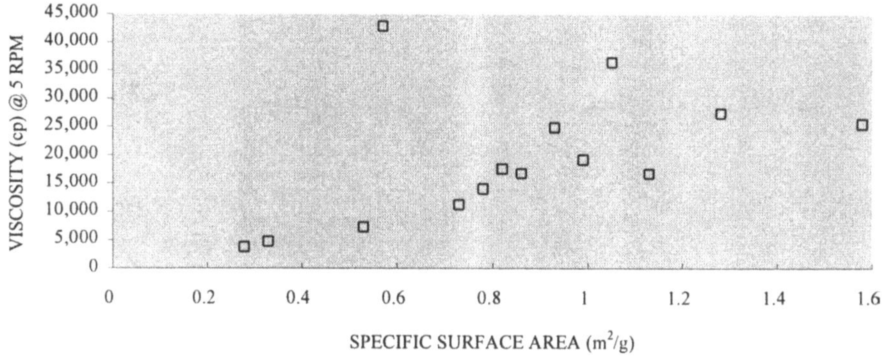

Figure 4. Viscosity vs. specific surface area (particle size variable).

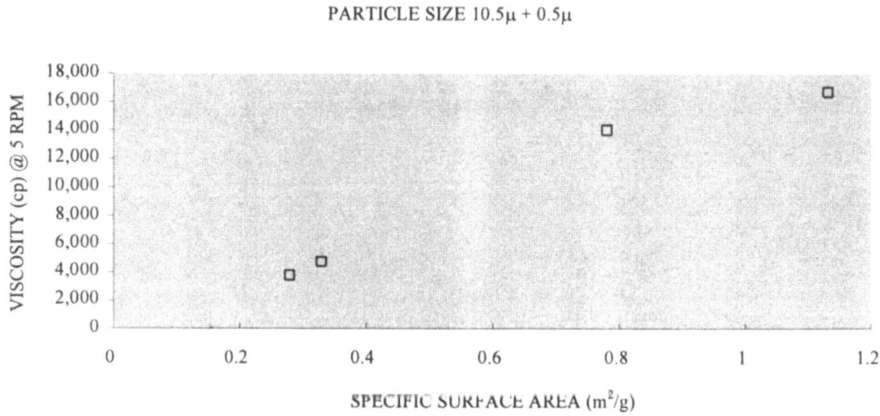

Figure 5. Viscosity vs. specific surface area (particle size constant).

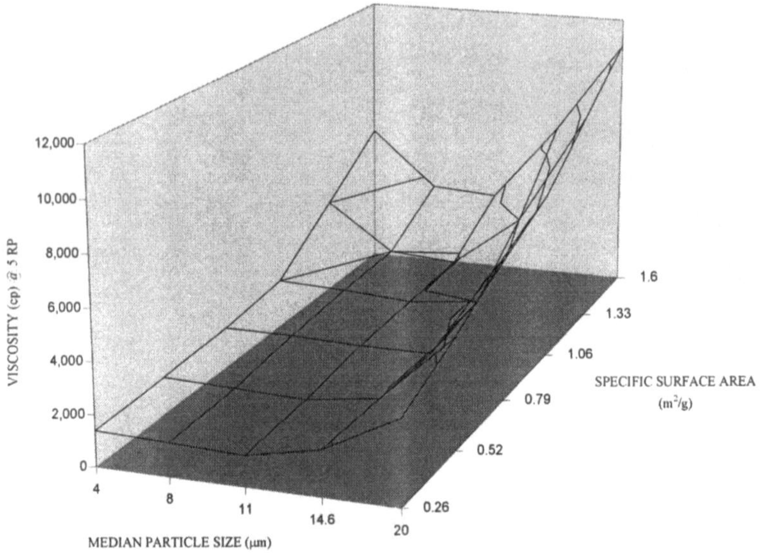

Figure 6. Viscosity vs. specific surface area and particle size.

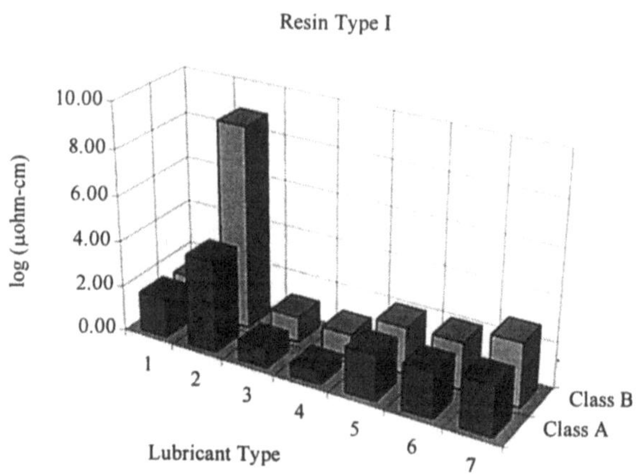

Figure 7. Resisitivity vs. lubricant type and flake size, resin Type I.

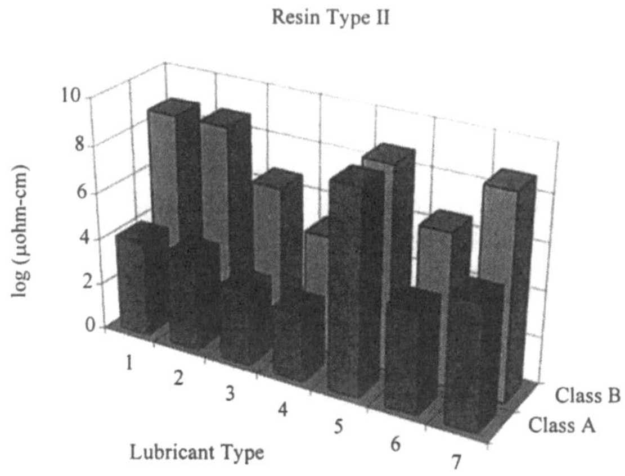

Figure 8. Resisitivity vs. lubricant type and flake size, resin Type II.

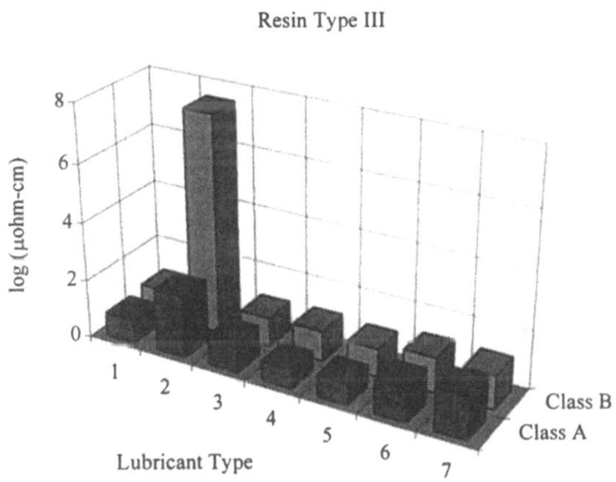

Figure 9. Resisitivity vs. lubricant type and flake size, resin Type III.

BIBLIOGRAPHY

1. E. M. Jost, K. McNeilly, T. Santala, "Silver Flake Production and Optimization for Use in Conductive Polymers", Proceedings of the 1987 International Symposium on Microelectronics, ISHM, Reston, VA pp 548-553.

2. E. M. Jost, K. McNeilly, P. Sexton, "Silver Flakes for State of the Art Conductive Adhesive Applications, Proceedings of the 7th International Microelectronix Conference, SHM, Yokohama, Japan, pp 422-428.

3. Multiple confidential private communications.

REFERENCES

1. Ashdown, C.P.; Domalavage, P.K.: "Ultra Fine Alloy Powder for Injection Molding", Advances in Powder Metallurgy 1989, Metal Powders Industry Federation, Princeton, NJ, pp. 45-ff.

2. Casson, N.; "A Flow Equation For Pigment-Oil Suspensions Of The Printing Ink Type", Rheology of Disperse systems, ed, C.C. Mills, Pergamon Press, London, pp. 85-103, 1959.

3. Evenson, G.F.; "Rheology of Dispersed Systems", ed. C.C. Mills, Pergamon Press, London 1959.

4. Jost, E.M.; McNeilly, K.; Santala, T.: "Silver Flake Production and Optimization for Use in Conductive Polymers", Proceedings of the 1987 International Symposium on Microelectronics, ISHM, Reston, VA pp 548-553.

5. Kirkpatrick, Scott: "Percolation and Conduction", Reviews of Modern Physics, vol. 45, num. 4, pp. 574-588 (1973).

6. Kitano, T.; Kataoka, T.; Shirota, T.: Rheol. Acta, 20, pp. 207-209, 1981.

7. Lovinger, Andres J.: "Development of Electrical Conduction in Silver-filled Epoxy Adhesives," J. Adhesion, vol. 10, pp. 1-15, 1979.

8. McNeilly, K.; Jost, E.; Sexton, P.: "Silver Flakes For State of The Art Conductive Adhesive Applications", Proceedings of the 7th International Microelectronics Conference, SHM, Yokohama, Japan, pp. 422-428, June 1992.

9. McNeilly, K.; Jost, E.; Sexton, P.: "Effect of Lubricant Solvent System on the Electrical Properties of Silver Filled Polymers", Proceedings of the 8th International Microelectronics Conference, SHM, Omiya, Japan, June 1994.

10. Metzner, A.B.: "Rheology of Suspensions in Polymeric Liquids", Journal of Rheology, vol. 29, pp 739-775, 1985.

11. Mills, P.: "Non-Newtonian behavior of flocculated suspensions:, Journal DePhysique, vol 7, pp. 301-309, April 1985.

12. Roos-Kozel, Barbara L.; Casavecchia, Frances M.: "Parameters Affecting Silver Flake and Powder Performance in Silver Filled Polymers," Solid State technology, Aug 1984.

13. Ruschau, G.R.; Yoshikawa, S.; Newnham, R.E.: "Effects of Filler Particle Size on the Resistivity of Conductive Composites," Intl. J. Hybrid Microelectronics, vol. 13, num. 4, (1990).

SILICIDES FORMATION IN METAL — SILICON FILM SYSTEMS: THE ROLE OF IMPURITIES

Yu.N. Makogon, S.I. Sidorenko

National Technical University "Kyiv Politechnical Institute"
NTUU "KPI", 37 Prospect Peremogy, Kyiv, 252056, Ukraine

ABSTRACT

Silicide thin films structure and physical properties formation laws were investigated by physical metallography methods during the different methods of their making: one-, two- and three layer systems on the silicon alternate deposition and homogeneous silicide substrates sputtering, synthesized by powder metallurgy method. Transition metals silicide films composition, structure and properties formation are defined by following factors: sputtered material chemical composition, film deposit conditions, heat treatment conditions, structure-phase condition of the substrate for metal on silicon deposition case.

INTRODUCTION

The contact systems to silicon designing is based on the knowledge of silicides thin films structure, phase composition and properties formation processes patterns [1, 2]. Complexities of this patterns determination are generally connected with hard controlled impurities influence on the silicide formation processes. Initial evaporated material (its purity), residual atmosphere or gas medium of evaporating setup are impurities sources [3].

Impurities influence on the crystalline structure formation and properties of silicide films during the solid body reactions in the "transition metal — silicon wafer" system, and also during the silicide phases crystallization into amorphous films is investigated in this work.

EXPERIMENT

Titanium, molybdenum, tungsten, platinum and nickel disilicides films were obtained by electron beam metal-silicon co-sputtering or by metal deposition on the silicon substrates with following vacuum or inert gas annealing.

Magnetron deposition of targets (with corresponding composition) produced by powder metallurgy method was used also.

Amorphous silicide films crystallization was carried out by hydrogen or vacuum annealing under the pressure of 10^{-2} — 10^{-4} Pa in the 773 — 1273 K temperature range during 1 h. Annealing by noncoherent luminous radiation pulses was used also for the films transition from amorphous condition into crystalline one. Pulse energy was approximatly 30 J /sm^2, exposure time 0,1 — 0,35 sec.

Silicon single crystal wafers of (100), (111) orientation exposed to annealing by boron or phosphorus were used as a substrates. Investigates were conducted by four-probe measurement, X-ray diffraction, SEE and Auger analyses.

R. Tummala et al. (eds.), Electronic Packaging for High Reliability, Low Cost Electronics, 91–98.

RESULTS AND DISCUSSION

The sequence of forming phases and phase composition of silicide films can be substantially different from predicted by binary system diagrams for the corresponding bulk materials depending on the impurities content.

It was established that impurity atoms located in the silicon substrate slow down silicide formation solid body reaction rate. It was determined that a number of intermediate compounds like Mo_5Si_3, Ni_3Si, Pt_2Si weren't observed during solid state reactions in the Mo, Ni, Pt films on single crystal silicon substrate with inconsiderable O_2 content. This phases formation was established in films sputtered on amorphous silicon layer containing 2 at.% O. Oxygen impurity atoms can blockade reaction development on the intermediate silicide formation stage in dependence on concentration [4].

Carbon impurity atoms within Mo films decrease growth rate of Mo_3Si phase, however reaction type is not changed in this. Silicide layer thickness grows according to parabolic low.

$MoSi_2$, WSi_2, $TiSi_2$ films crystallization processes are developed through intermediate modifications formation under the 10^{-2} — 10^{-5} Pa vacuum annealing in the 873 — 1173 K temperature range. There are C40 hexagonal modifications in $MoSi_2$, WSi_2 disilicides and orthorhombic structure C49 in $TiSi_2$.

Mo+Si — $MoSi_2$(hex)C40 — $MoSi_2$(tet)C11b,
W+Si — WSi_2(hex)C40 — WSi_2(tet)C11b,
Ti+Si — $TiSi_2$C49(orthorhom) — $TiSi_2$(orthorhom)C54.

Oxidative residual atmosphere of annealing camera prevents the phase transition from amorphous state into crystalline one. Films decontamination — nitrogen and hydrogen formation - proceeds during vacuum heating. However, oxygen of residual atmosphere is absorbed by silicide film (practically under the any vacuum), because vapours pressure during oxygen dissociation are less than discharges achieved by modern equipment [5].

Oxygen impurity atoms which form compounds of film's components slow down the intermediate phase transition into the finite one and can blockade crystallization process on the intermediate modification formation stage. Some amount of amorphous phase is saved. It shows itself in higher values of films resistance (Fig.1). More developed crystalline structure and lower resistance values can be obtained by hydrogen annealing, where any traces of the oxidative atmosphere are absent.

It was shown in the [6] that increased (>3at%) carbon content is a very important factor during $TiSi_2$ C49 structure formation.

As it was established by the Auger and SEE analyses directed diffusion of O_2, C_2 impurity atoms (and other one located in the metal film) to the surface layers proceeds during the silicide layer growth. It can be demonstrated by example of $TiSi_2$ crystalline structure formation in the "titanium film - silicon" system (Fig.2,3).

Impurity atoms directed diffusion to the film surface and to the film-silicon substrate interface is observed also during transition metal silicides amorphous films crystallization (Fig.4, 5). In the pulse heating by noncoherent luminous radiation conditions impurity atoms redistribution takes the masstransfer character with the anomalous high rate, which by the 4 order of magnitude greater than the diffusion rate

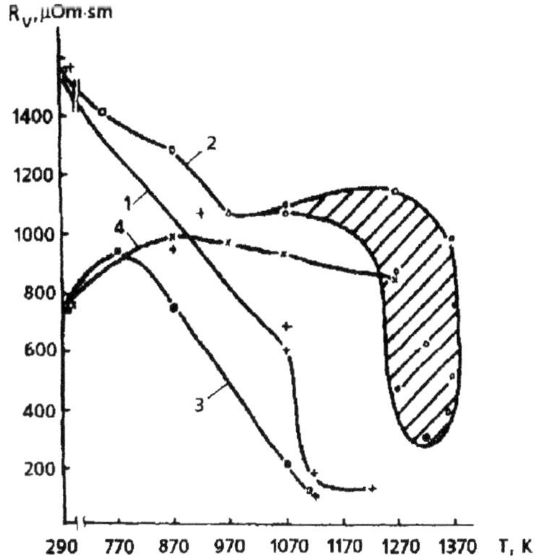

Figure 1. Resistance changes of $MoSi_2$ (1, 2 curves) and WSi_2 (3, 4 curves) films in dependence on the vacuum annealing temperature (exposure time was 1h) (2, 4 curves) and on the annealing temperature in the hydrogen (exposure time was 0.5 h) (1, 3 curves); film thickness — 250 nm, silicon orientation — (100).

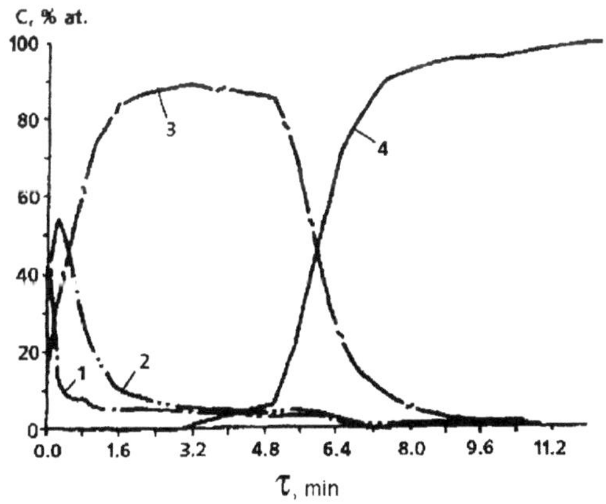

Figure 2. Elements concentration change at the titanium film thickness and the transitional zone in Ti/Si structure after the electron beam sputtering; Ti film thickness — 100 nm, silicon orientation — (111): 1 — C, 2 — O, 3 — Ti, 4 — Si

during vacuum annealing in the disilicide titanium films. Diffusion effective coefficient for film $TiSi_2$ increases from 10^{-14} to 10^{-10} sm^2/s.

Redistribution effect is shown non-identically for the different impurities in the same "metal — silicon" system. The largest masstransfer is observed during the phase transition for the oxygen atoms.

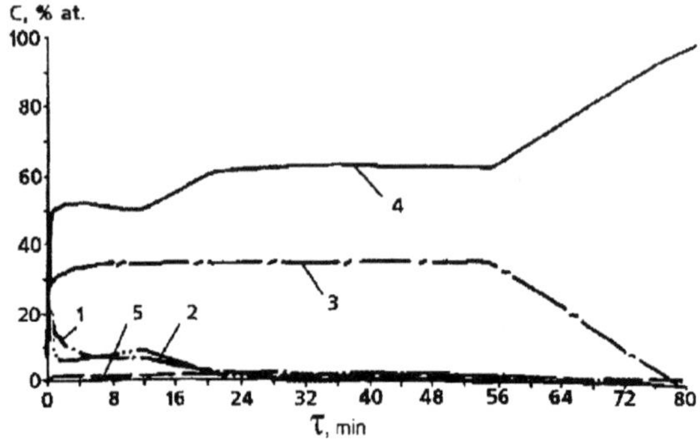

Figure 3. Distribution of concentration of elements in the Ti/Si structure after the vacuum annealing at 1070 K (exposure time was 0.5 h); Ti thickness — 100 nm, silicon orientation — (111): 1 — C, 2 — O, 3 — Ti, 4 — Si, 5 — Ar

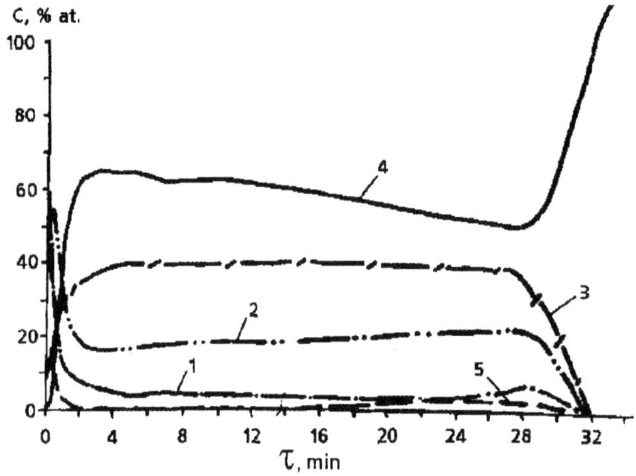

Figure 4. Distribution of concentration of Ti, C, O, Si elements at the thickness of the produced by magnetron deposition Ti+Si/SiO_2/Si film after the sputtering: 1 — C, 2 — O, 3 — Ti, 4 — Si, 5 — Si by SiO_2

It was determined that silicide film resistance, stress level change are affected byoxygen atoms more effectively than by carbon one. During Ti_5Si_3 film formation by electron beam co-sputtering method under the partial oxygen pressure of 10^{-3} — 10^{-6} Pa absorbed by silicide film oxygen atoms content is quickly increased with vacuum degree decrease [7]. It is in agreement with quick change of film's stress level and resistance (Fig.6, 7). Oxygen amount absorbed by film is constant under the different oxygen partial pressures sputtering. Oxygen content increase to ~3 at % practically doesn't affect operating performances. Carbon atoms have influence on the films electric and physical properties and microdevices operating performances, in which film-silicon substrate interface condition is very important factor in the formation of its parameters, for

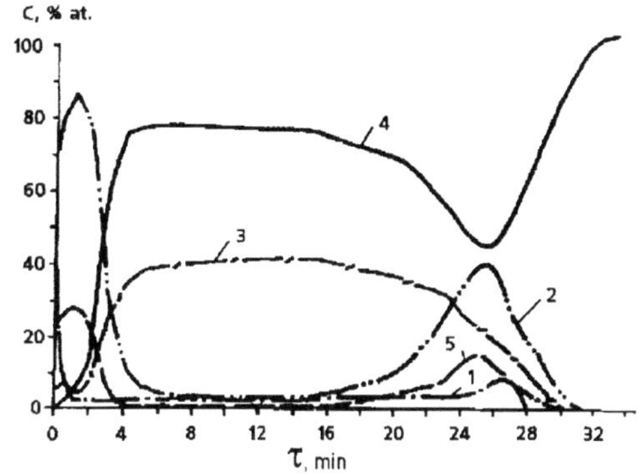

Figure 5. Ti, C, O, Si elements concentration change at the Ti+Si/SiO_2/Si film thickness after the pulse annealing: 1 — C, 2 — O, 3 — Ti, 4 — Si, 5 — Si by SiO_2

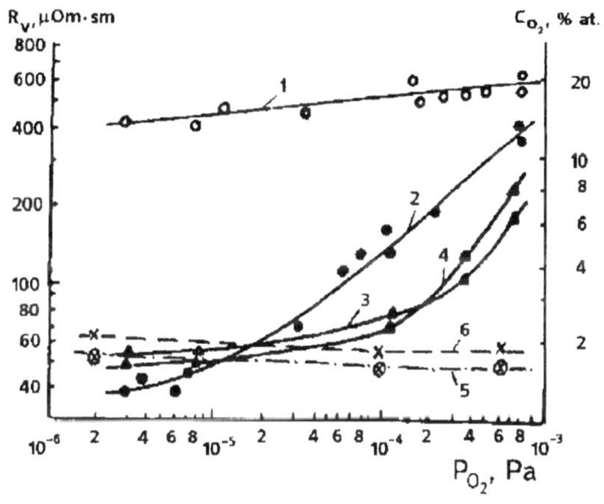

Figure 6. Resistance (1, 2) and content of oxygen (3, 4) dependence on the oxygen partial pressure in deposition camera before the annealing (1, 3) and after the annealing (2, 4). Content of carbon after the sputtering (5), after the crystallization (6)

example, voltage-currents of Shottky diodes. It can be demonstrated on the example of binary Pt/Ni compositions on the silicon substrate. Carbon atoms segregations are formed on the silicide structures interface during the $Pt_xNi_{1-x}Si$ silicide formation at annealing optimal temperatures rise (>773 K) (Fig.8). It makes voltage-current of diodes on the $Pt_xNi_{1-x}Si$ base worse (breakdown voltage decreases U<25 V at 1 mA current, creeping current increases).

O, C impurity atoms redistribution to the film surface layers caurses oxides and carbides of metal, a silicon formation during the sliced films crystalline structure formation. Silicide formation processes developing is slowed down by this, however, saturated by impurity atoms layers can be barriers for undesirable silicon — metal of microdevice metallization following layer (Al, Au) interdiffusion. Changes in the

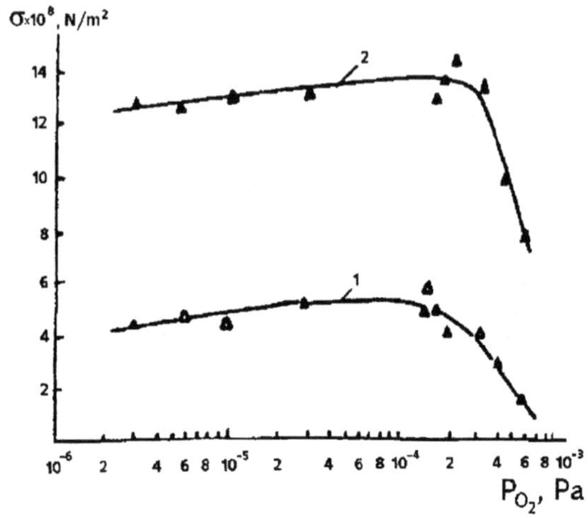

Figure 7. Stress dependence on the oxygen partial pressure in the deposition camera before annealing (1) and after annealing (2) in the Ti+Si/SiO_2/Si films

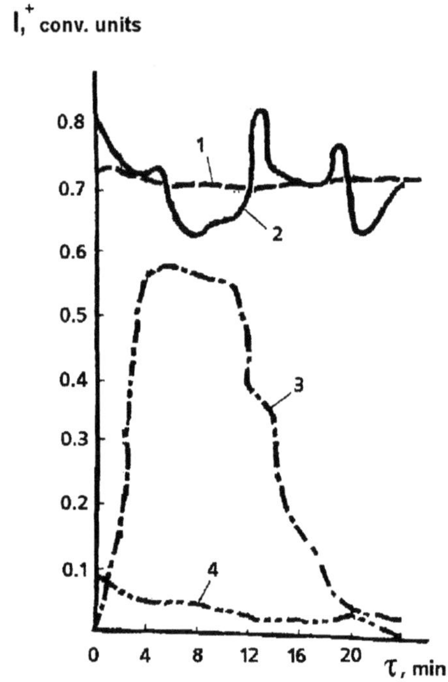

Figure 8. Secondary ions emission intensity dependence on the deposition time of the Ni/Pt/Si system vacuum condensates, what it are spoilage at the electrical parameters (thickness: Ni — 75 nm, Pt — 15 nm); 1 — $^{28}Si^+$, 2 — $^{12}C^+$(x100), 3 — $^{58}Ni^+$, 4 — $^{16}O^+$(x100)

technology of Shottky diodes on the $MoSi_2$ base production and this layer partial employment make it possible to improve the voltage-currents to reduce the current, to increase the breakdown voltage and temperature-time stability of their operating performances [8] (Fig.9).

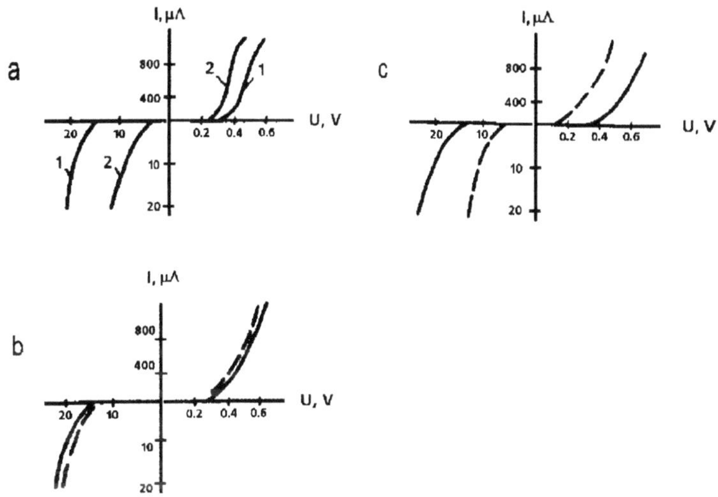

Figure 9. Diodes Shottky voltage-currents of systems (a — Al-Mo-$MoSi_2$-nSi (1), Al-$MoSi_2$-nSi (2); b — Al-Mo-$MoSi_2$-nSi after the additional annealing at 673 K (exposure time was 1 h); c — Al-$MoSi_2$-nSi after the additional annealing at 823 K (exposure time was 0.5 h); dash lines are initial voltage-currents

CONCLUSIONS

1. Formation of the phase composition, structure, and properties of transition metals silicides thin (about 600 nm thickness) films on silicon substrate which are formed in vacuum up to 10^{-5} Pa is affected by: chemical composition of output material which is evaporated, deposition and heat treatment conditions, structural and phase state of the substrate, impurities; in this case the role of impurities (content of which is about 10 at % within films) is predominate.

Pollution impurities (dependent on concentration) can reduce “metal — silicon” solid state reaction speed, blockade its development on the stage of formation of interval silicide or phase metastable modification — the final product of transformations in the present system; as the result phases' consistency, and phase composition of films can be essentialy different from phase balance predicted by diagrams for corresponding massive materials.

2. Beside impurities kinetics of crystallization processes of amorphous layers formed by metal — silicon co-sputtering, diffusion phase formation in "metal film — silicon" system is essentialy affected by structural and phase state of silicon substrate.

Silicide phases crystalline structure formation within metal films on silicon substrate is accompanied by impurity atoms directed diffusion up to "film-substrate" interface and film surface; during amorphous silicide films crystallization in conditions of pulse heating by uncoherensive luminous radiation impurity atoms redistribution proceeds with anomalous speed by 4 orders of magnitude greater (diffusion's effective coefficient reaches $\cong 10^{-10}$ sm^2/s), than the speed of processes during traditional heat treatment.

3. Utilization of films annealing within hydrogen, heat treatment impulsive methods allows to form crystalline structures of stable disilicides of molybdenum, titanium, tungsten, tantalum with the low level of stresses (10-70 MPa) and specific electric resistance values within the interval of 15 — 120 $\mu\Omega\cdot$sm from amorphous silicide films.

Mechanical stresses level, silicide films electric resistance value are more dependent on oxygen atoms content, than on carbon ones.

REFERENCE

[1] Marc-A. Nicolet, S.S. Lan, VLSI Electronics Microstructure Science (Academic Press, New York London, 1983).

[2] C.Y.Ting, F.M. d'Heurle, S.S. Lyer, J. Electrochem. Soc. (1986) 12.

[3] S.P. Murarka, Siliside for VLSI Applications (Academic Press, New York London, 1985).

[4] M.V. Belous, V.M. Koptenko, Yu.N. Makogon, Fizika i Himiya Obrabotki Materialov (1987) 3.

[5] V.M. Amonenko, L.N. Pyabchikov, Voprosy Atomnoy Nauki i Tehniki, (1978) 2.

[6] Yu.N Makogon, Neorganicheskiye Materialy, 28 (1992) 8.

[7] A.A. Krutko, Yu.N. Makogon, S.I. Sidorenko, Fizika i Himiya Obrabotki Materialov, (1989) 3.

[8] Yu.N. Makogon, V.N. Nazarenko, A.I. Ostapchuk, S.I. Sidorenko, Neorganicheskiye Materialy, 29 (1993) 4.

MECHANICAL PROPERTIES OF Pb-Sn-In-Ag SOLDERS FROM -100°C To 150°C

W. KINZY JONES ,YANQING LIU and MILIND SHAH
Florida International University
Department of Mechanical Engineering
Miami, FL, 33199, USA

1.Abstract

The decrease in soldered interconnect dimensions with the advent of flip chip and chip scale packaging, coupled with new substrate technologies and operating ranges, have placed new demands on understanding the mechanical responses of solders. With an estimated 10^{13} joints per year, the mechanical properties of solders must be known over the temperature ranges in which they must operate to insure continued reliability and performance with the changes in assembly technologies found in high performance, low cost electronics. In this paper, the mechanical response of five solder alloys from the Pb-Sn-In-Ag alloy system were determined over the temperature range of -100°C to 150°C. All alloys exhibited increases in strength ($\sigma_{0.2}$ and UTS) and elastic modulus and decreases in elongation with decreasing temperature. The eutectic and indium based alloys exhibited superplastic behavior between 100°C and 150°C. Properties analyzed included elastic modulus, yield strength, ultimate tensile strength (UTS), and elongation.

2. Introduction

There are many factors which effect the application of solders to low cost, high performance applications, including environmental pressures to remove lead (Pb) and the competing pressures of more interconnects with decreased interconnect pitch. This emphasis on new lead free alloys and greater demands on alloys with a history of use since the introduction of printed wiring technology over 50 years ago, has pushed for more research in solders during the last 5 years. Even with this expanded research, the understanding of the mechanical response of solder alloys is complex, at best. Because the operating temperature is in the range of 0.5 to0.9 of the melting point (T_m) for all of these alloys, thermally assisted deformation, such as creep, becomes significant . The resulting properties are then a mixture if properties, i.e., fatigue response is now a combined response of fatigue and creep and classical elastic response is a combination of elastic and inelastic responses, making the elastic properties strain rate dependent. In this

R. Tummala et al. (eds.), Electronic Packaging for High Reliability, Low Cost Electronics, 99–106.

study, the mechanical properties were evaluated using uniaxial tension testing for the mechanical response under load at a constant strain rate and acoustic pulse testing for evaluation of the strain rate independent elastic modulus, although the mechanically determined values are appropriate for use in performance predication. Advances in finite element method (FEM) have allowed for the incorporation of time dependent (creep, inelastic response) and non linear material responses.. However, accurate modeling is only possible when the materials properties are available over the entire desired temperature range. Most modeling is performed with room temperature values, resulting in model predictions, which offer little more than trends in the response. However, as the accuracy of the materials data increases, the ability to model processes, including time dependent processes and environmental stress will follow in increases accuracy of prediction.

In this study, the mechanical properties of eight Pb-Sn, In-Sn, In-Pb solders were evaluated over the temperature range from -100°C to 150°C. The deformation and fracture process was investigated, and their fracture mechanism is proposed. According to the matrix type of the alloys, the solder alloys can be broken into two groups: 1) eutectics alloys (63Sn/37Pb, 62Sn/36Pb/2Ag, 96Sn/4Ag, 49In/48.5Sn/2.5Ag, and 50In/50Sn); and 2) solid solutions alloys (50In/50Pb, 5Sn/95Pb and 10Sn/90Pb).

3. Experimental Procedure

Commercially available solders (95Pb/5Sn, 90Pb/10Sn, 63Sn/37Pb, 62Sn/36Pb/2Ag, 96Sn/4Ag, 50In/50Pb and 50In/50Sn) were used in the experiment. One alloy solder (49In/48.5Sn/2.5Ag) was melted and cast into ingot by mixing the In-Sn eutectic solder with pure Ag.

The ingot solder was melted and cast into a water-cooled aluminum mold. The mass of the mold, small sample size and water-cooling ensured a fine-grained sample with a grain structure similar to that found in solder joints.

Tensile tests were carried out on an Instron Model 1011 tensile testing machine. The strain rate of 0.003 sec^{-1} was used for this work. The tensile samples had a total length of 40mm, a gage length of 15mm and a thickness of 2mm. Three samples per test configuration were used. In order to perform testing over the range from -100°C to 150°C, the tensile tester was fitted with a thermally insulated container and special, small grips. The sample was immersed in a solution of liquid nitrogen/alcohol for low temperatures of heated glycerin for high temperatures. Sample temperatures were monitored by a thermocouple.

Acoustics describes the phenomenon of mechanical vibrations and their propagation in differents materials. Determining the modulus of isotropic solids requires two experimental measurements: (1) the density of the material, ρ, and, (2) the speed of propogation of longitudinal plane wave (V_L). The test specimen must have large lateral dimensions compared with the transducer diameter to eliminate mode conversion due to reflection at a lateral surface and guided wave model effects, since the equations for calculating elastic constant is based on the assumption of linear elastic wave propagation in an unbounded half-space. The modulus of elasticity is calculated by determining the transit time taken for the sound pulse to travel through the specimen in the pulse echo technique. The equation used is:

$$E = (V_e)^2 \rho \tag{1}$$

where V_e is the extensional velocity, E is the Young's modulus and ρ is the density of the material , whose modulus is to be determined. Using this technique one can also measure shear modulus of elasticity and Poisson's ratio.

In order to perform the test at the -100°C to 150°C temperature range, liquid immersion was also used in a solution of liquid nitrogen/alcohol for the low temperature part of the test and in glycerin for the high temperature. The equipment used for this test was a Panametrics Magnetostrictive Pulser-Receiver, a KT-55 Magnetostrictive Transducer, and a magneto-strictive remendur wire wave guide. The sample was attached to the wave guide wire using a epoxy resin. The dimensions of a typical sample used for this test was 1.56 in. long, 0.17in. width, and 0.075in. thickness.

4. Test Results

4.1. Microstructure

Typical eutectic structures were observed in the 63Pb-37Sn and 50In-50Sn alloys. In the case of the Pb-Sn, the Pb-rich phase appears distributed in a Sn matrix, while for the In-Sn, it appears as a Sn-rich phase distributed in an In rich matrix. The size of the eutectic colonies were 100-200 μm.The eutectic 96Sn-4Ag is not a typical eutectic structure. The structure was observes as a distribution of small Ag-rich particles in a Sn matrix. The difference in phase volume, as predicted by the lever law, is observed in the microstructure. The 50In-50Pb alloy was a single phase solid solutions, while the 90Pb-10Sn indicated a precipitation hardened alloy structure with fine dispersed Sn –rich second phase particles, as anticipated from the phase diagram .

4.2. Tension Test

The results of the uniaxial tension tests are shown in Figs. 1 through 4. The following observations were made:

1) For 50In/50Pb, 95Pb/5Sn and 90Pb/10Sn, the strength and the elongation change linearly with the temperature.
2) For 96Sn/4Ag, the strength increases linearly with decreasing temperature above -100°C. The elongation decreases with decreasing temperature.

Figure 1 Ultimate Tensile Strength of different Solders

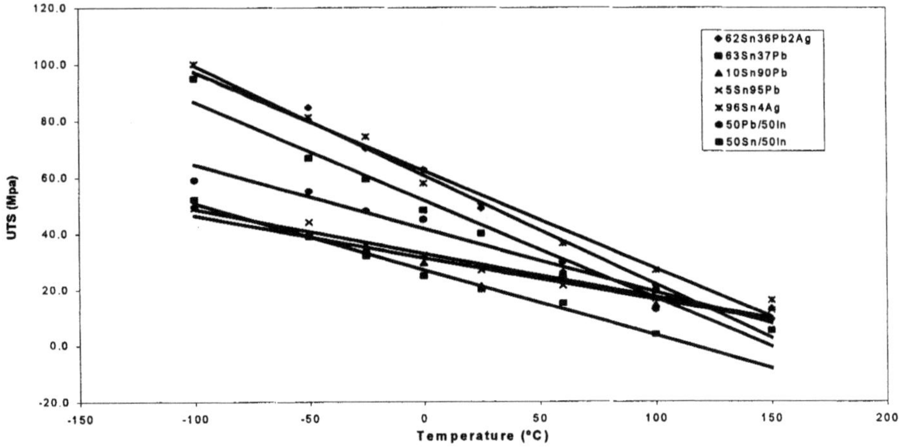

Figure 2 Yield Strength of different Solders

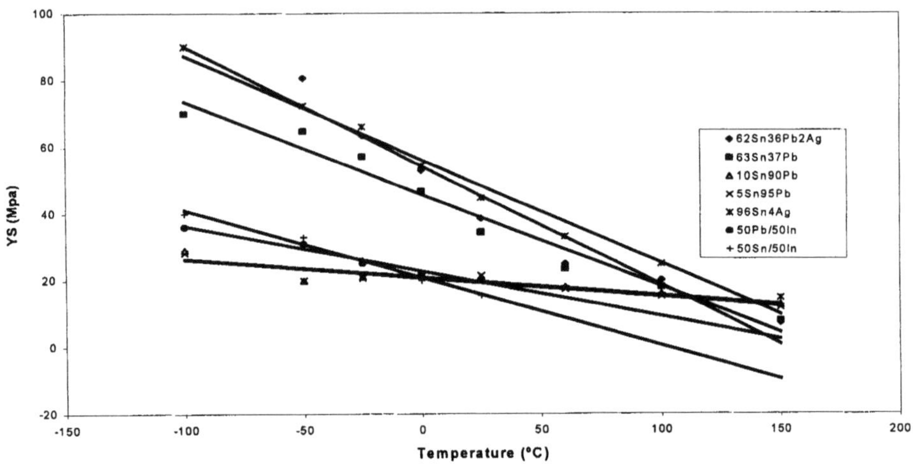

Figure 3 Total Elongation of different Solders

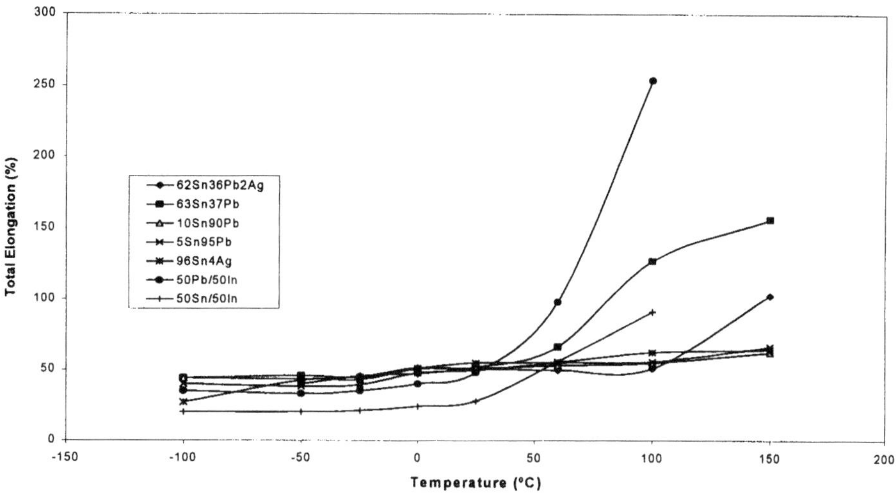

Figure 4 Uniform Elongation of different Solders

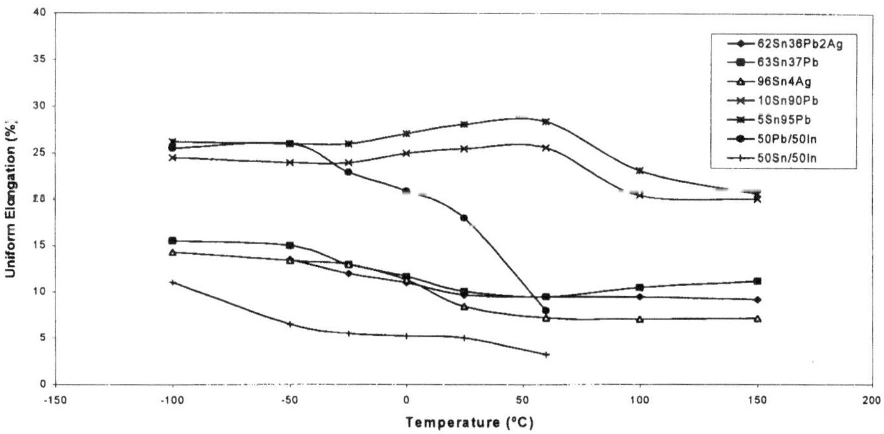

3) For 63Sn/37Pb, the strength and elongation are similar to 96Sn/4Ag at below room temperature, however, superplasticity is exhibited above 50°C.
4) In-Sn eutectic solders possess higher levels of uniform elongation at low temperatures and alloying can improve strength still further.

4.3. Acoustic Test

The elastic modulus for the five Pb-Sn solders using the acoustic method are shown in Figure 5 and Figure 6. The elastic modulus increases with decreasing temperature.

5. Discussion

1. The strength increases linearly with decreasing temperature for all alloys over this temperature range. The yield strength ($\sigma_{0.2}$) and ultimate tensile strength (UTS) in (Gpa) versus temperature T (°C) can be expressed by σ (or UTS) = $\alpha - \beta T$ as follows:

For 63Sn-37Pb

$$\sigma_{0.2} = 45 - 0.34T$$
$$UTS = 49 - 0.32T$$

For 90Pb-10Sn

$$\sigma_{0.2} = 25 - 0.061T$$
$$UTS = 33 - 0.18T$$

For 96Sn-4Ag

$$\sigma_{0.2} = 50 - 0.35T$$
$$UTS = 62 - 0.34T$$

For 50In-50Pb

$$\sigma_{0.2} = 24 - 0.13T$$
$$UTS = 45 - 0.29T$$

For 50In-50Sn

$$\sigma_{0.2} = 22 - 0.23T$$
$$UTS = 28 - 0.25T$$

The linear relationship of the strength and plasticity versus temperature indicates that the fracture mechanism does not change. This observation is compatible with the results of fracture analysis, which revealed a cup and cone dimpled structure, indicative of good ductility, on all samples tested.
In the equation $\sigma = \alpha + \beta T$, β is the temperature strengthening coefficient and indicates the degree of the strength change with temperature. Comparing the above equations, one observes higher α and β in the Sn based alloys, with minimal difference in β between yield strength and UTS in Sn based alloys and a high β in the lead solid solutions. The higher β is indicative of greater sensitivity to temperature, which depends on the crystal structure. Generally, FCC materials are not sensitive to temperature and therefore the β is low in the solid solution Pb alloys. The difference in the β between yield point and UTS relates to the capability to work harden, indicating a higher capability for work hardening for these alloys.

2. The total elongation decreases with temperature, except in the regions where superplasticity occurs. Superplasticity, with elongation greater than 100% was observed

Figure 5 Elastic Modulus of different Solders measured by Uniaxial Tension Test

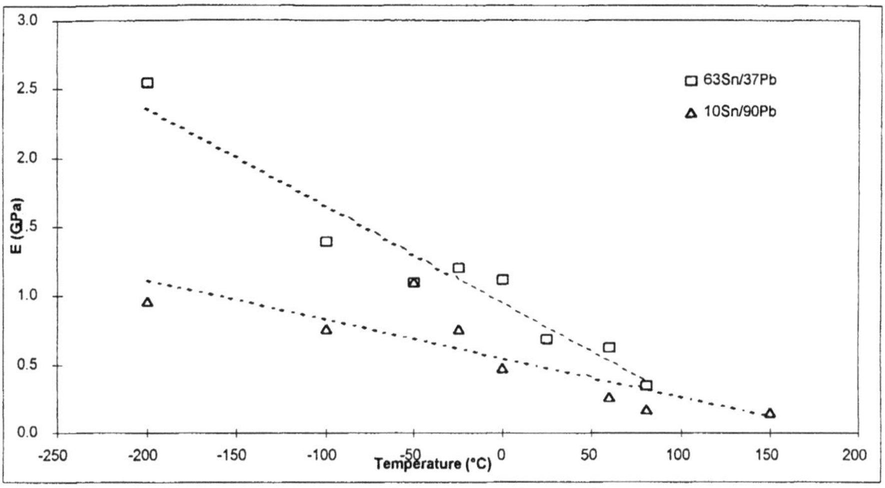

Figure 6 Elastic Modulus of different Solders measured by Acoustic Method.

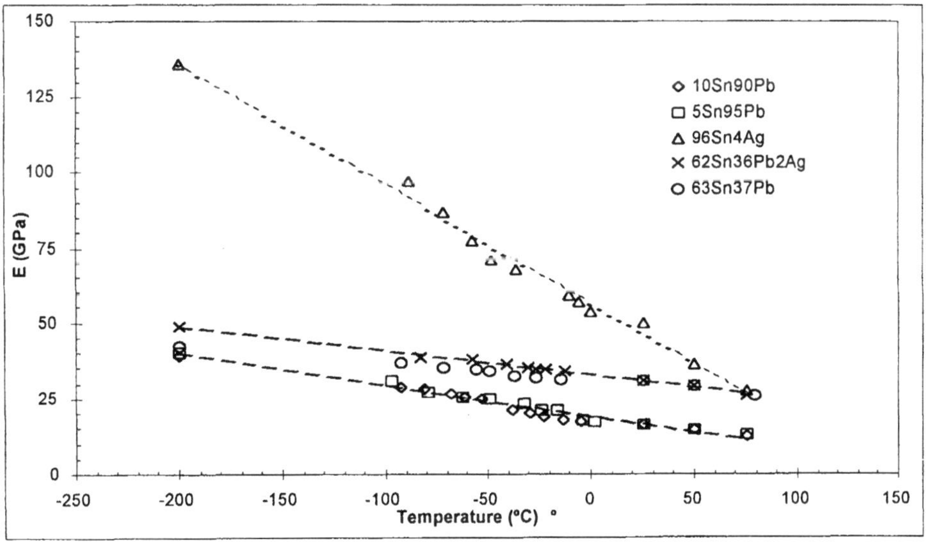

in eutectic 63Sn-37Pb and 50In-50Pb at 100°C, and 50In-50Sn and 63Sn/36Pb/2Ag above that temperature. The plastic deformation for the eutectic alloys are accommodated by the sliding of the phase boundaries between eutectic colonies, with fracture caused by sliding separation along the interface between the boundaries. For the uniform elongation, Pb solid solution solders (95Pb/5Sn, 90Pb/10Sn) possess higher δ_U, but Sn matrix eutectic solders have lower δ_U. The maximum of the δ_U appears at approximately T/Tm=0.5 for all solders. The uniform elongation means a capability of uniform plastic deformation of material and a capability of strengthening of matrix phase. The Pb matrix (FCC) alloys possess higher capability of strengthening. Meanwhile, the different fracture mechanism at low temperature for the two types of solder results in above difference of the δ_U. The peak of the δ_U results from the interaction between work-hardending and recovering.

3. The elastic modulus, shown in fig. 5and 6, exhibits a large discrepancy based on test method. The modulus observed by taking the slope of the uniaxial stress- strain curve is a order of magnitude less than that obtained from acoustic pulse. For high melting point materials, the correlation between these methods is excellent, however, for low melting point solders, large variations exist. Because of the inelastic and viscoelastic response of solders, the measurement of the elastic modulus by mechanical means includes creep and other inelastic responses. Although this measurement predicts the physical response of the solder under load, it is strain rate sensitive and is not indicative of the true elastic response. The acoustic pulse method is based on the travel of sound waves through a sample and is independent of local deformations, microstructure or dislocation motion. Therefore, it is a better method to predict the true elastic response.

6. Conclusions

1. The mechanical properties of solders are well behaved over the temperature range of -100°C to 50 °C, with linear increases in strength and decreases in ductility.
2. Over this temperature range, ductile or , at high temperature, superplastic response was observed.
3. The elastic modulus decreases linearly with increasing temperature until 70°C and then rapidly drops below 100 °C. The elastic modulus obtained by uniaxial tension testing appears to be highly erroneous or predicting elastic modulus, although it does predict the mechanical response for the tested strain rate and the strain rate sensitive inelastic responses. The acoustic pulse method accurately predicts the elastic modulus, independent of any inelastic response.

The Need for Low Cost, High Density Packaging and Interconnect Technology

Dr. Philip Garrou
Dow Chemical@MCNC
Research Triangle Park NC
919-248-9261
e-mail: pegarrou@dow.com

Abstract:

The increase of I/O on chip is driving chip, package and board interconnect schemes to alternative technologies.. Area array flip chips and BGAs are beginning to replace peripheral pad limited chips and tight pitch leads on PQFPs. The PWB industry is looking at technologies which form "microvias" to handle the interconnection of these high density chip/package formats.

Kewords: flip chip; redistribution; BGA; CSP; build up boards; LAP; microvias; build up boards; thin film

1. Introduction

Faster, smaller, lighter and more reliable are the prime drivers that are pushing the electronics industry into the 21st century. The demand for these attributes appears to be insatiable. At the same time, it is expected that these system level performance advancements will be achieved with little or no overall increase in component cost.

Semiconductor advances are directly related to a continuing decrease in feature size. Smaller features provide increased gate density, increased number of gates per chip, and increased clock rates. With reduced feature size, devices exhibit reduced parasitics, which allow faster switching. Such benefits are offset by the increased number of I/O which are needed.

A projection of IC technology into the new millennium was recently developed by the National Technology Roadmap for Semiconductors group under the auspices of the Semiconductor Industry Association as shown in Table 1. The AVERAGE size for DRAM and ASIC devices is expected to increase to 100 mm on a side during the next decade to accommodate the increases in transistors and I/O. In the recent past, transistor size has halved every six years and I/O have doubled every three years. If the industry achieves anywhere near these levels of integration in the future, the needed I/O for CPU and other logic chips will require major changes in semiconductor packaging. Many VLSI devices are already limited by the pitch of the peripheral bond pads. Figure 1 shows a report on die shipments by I/O in 1996 vs a projection for 2001. It is expected that ca. 7 billion chips will have > 208 I/O in the very near future [1]

R. Tummala et al. (eds.), Electronic Packaging for High Reliability, Low Cost Electronics, 107–131.

THE NEW SIA ROADMAP

	1995	1998	2001	2004	2007	2010
Minimum Feature Size (microns)	0.35	0.25	0.18	0.13	0.10	0.07
DRAM Bits/Chip	64M	256M	1G	4G	16G	64G
Max. No. of Input/Output Contacts						
Chip	900	1350	2000	2600	3600	4800
Package	750	1100	1700	2200	3000	4000
Package Cost (cents/pin)	1.4	1.3	1.1	1.0	0.9	0.8
µP On-Chip Clock Frequency (MHz)	150-300	200-450	300-600	400-800	500-1000	625-1100
Chip Size (mm²)						
DRAM	190	280	420	640	960	1400
µP	250	300	360	430	520	620
ASIC	450	660	750	900	1100	1400
µP On-Chip Wiring Levels	4-5	5	5-6	6	6-7	7-8
Minimum No. of Logic Mask Levels	18	20	20	22	22	24
Wafer Diameter (mm)	200	200	300	300	400	400
Desktop µP Chip Voltage	3.3	2.5	1.8	1.5	1.2	0.9
High Performance µP Power Dissipation (W)	80	100	120	140	160	180

Source: SIA

Table 1. National Technology Roadmap for Semiconductors

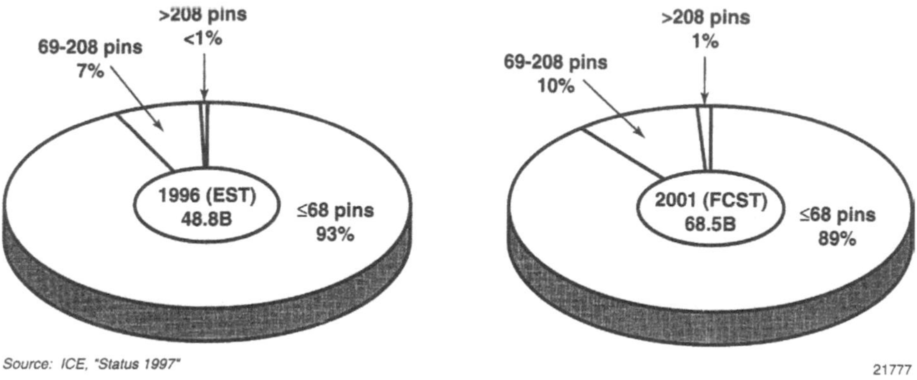

Figure 1 Die Shipments by I/O; 1996 Actual vs 2001 Projected [1].

It is felt that these high I/O die will drive technological changes for escape on chip, in packaging and on board. The requirements are depicted in Table 2. It appears inevitable that these high I/O die will be interconnected (assembled) by area array techniques. In addition to the I/O advantages, cost savings can be achieved by eliminating the peripheral pads . The resulting smaller die size have a two fold impact on cost savings, namely more die per wafer and slightly higher yields on the smaller die.

Versus peripheral bond pads, area array allows more I/O per the same area or the same number of I/O with much looser pitch. This will be described in further detail later. Dispite area array I/O allowing looser pitch for high I/O chips, I/O are increasing so fast that board routing will have to get denser (especially vias and pads) in order to handle the expected local area density generated by > 200 I/O on 0.5 mm pitch.

	ISSUE	SOLUTION
CHIP	- chips are pad limited - need for more I/O	-design for area array bumps- or - redistribute peripheral pads to area array bumps before mounting
Package	- package taking up to much real estate - package need more I/O	- area array small form factor packages such as BGA and CSP
Board	- need to route signal from 0.5 to 1 mm pitch area array bumps	- thin film on laminate (sequential build up boards) or other substrates (Large Area Processing)

Table 2. Needed Changes on Chip, in Package and on Board.

2. Area Array Chip Technologies

In this technology, solder (or for some applications gold) balls are created directly on area array pads on the top level of a chip. Since the chip is attached face down on the next level of interconnect, the name "flip chip" evolved. Early renditions of the technology, used for the last 3 decades by IBM, Delco and AT&T, involved, for the most part, ceramic substrates/packages which offered a good CTE match and a good thermal pathway.

The key to using bumping in lower cost packaging came when IBM Yasu revealed that reliable attachment to laminate could be achieved by using a reinforcing filled polymer between the chip and the laminate [2]. This "underfill" redistributes the x-y mechanical strain developed during thermal cycling from the solder into the entire package, reducing the strain in the solder balls by 5-10X.

There are a variety of approaches in use today to create bumps on wafers [3]. The major approaches include: (1) ball or stud bumping [4, 5] which uses a relatively modest modification to standard wirebonding (gold or solder wire) as shown in Fig. 2; (2) stencil printing of solder paste (limited to a minimum size of 5-6 mil diameter balls) [6,7] as shown in Fig 3 ; (3) plating of solder compositions [8] . In all cases care must be taken to insure compatibility of the pad metallizations with the bump metallurgy. For instance a sequence of Cr/Cu/Ni/Au is reportedly required as a base upon which to plate solder in the solder plating process [8] whereas a zincate treatment of aluminum pads followed by electroless nickel/vanadium and immersion gold plating has been reported as a low cost under bump metallurgy sequence for screen printed solder compositions [6].

In addition, the migration to lower cost laminate substrates also requires the use of eutectic solder on the pads of the laminate since the laminate cannot be taken to the temperatures needed to reflow the usual 95/5 Pb/Sn solders used on chip (ca. 365°C). This is depicted in Fig. 4.

In general, area array bonding technology delivers superior electrical performance, reliability and reduced footprint. Long term chips will be designed for area array bumping. In the short term, however,: (a) chip designers, especially for ASICs, build up their structures from libraries of cells which have been created for peripheral pad design; (b) the needed design tools are not widely available and (c) changes in design when the needed infrastructure is not in place would impact product time schedules.

Thus most of the chips needing foundry bumping today also need a layer of redistribution to reconfigure the peripheral pads to area array. Such bumping and redistribution costs are somewhat offset by the less complex boards that are necessary to route the broader pitch of the area array chips.

Redistribution is shown in Figure 5 in cross section and Figure 6 as a top view. Such redistribution requires thin film polymers and metallization to reroute the pads to an area array configuration. Fig. 7 compares a chip on 1 mm WB pitch and the same chip rerouted to 0.25 mm area array [11].

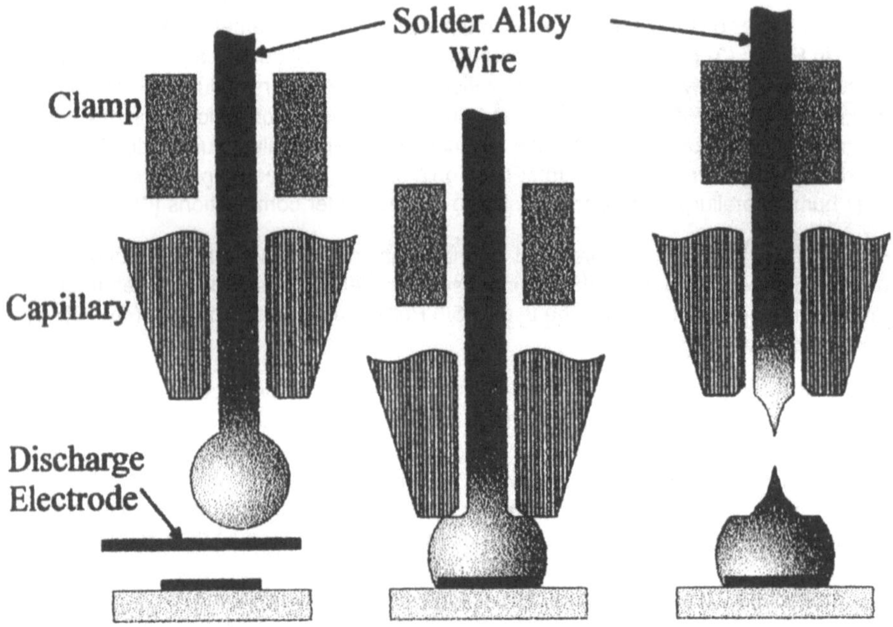

Fig 2 . Bump Formation using Modified Wirebonding Equipment [5]

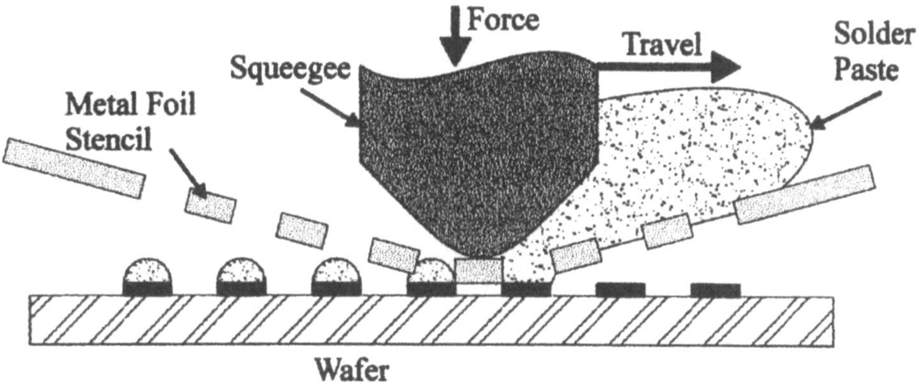

Fig 3. Stencil Printing of Solder Paste

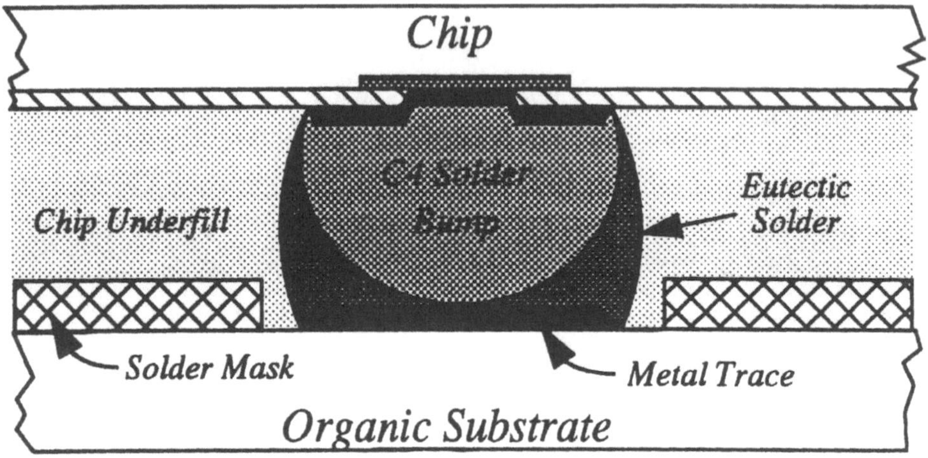

Fig. 4 Bare Chip Solder Assembly on PWB Laminate

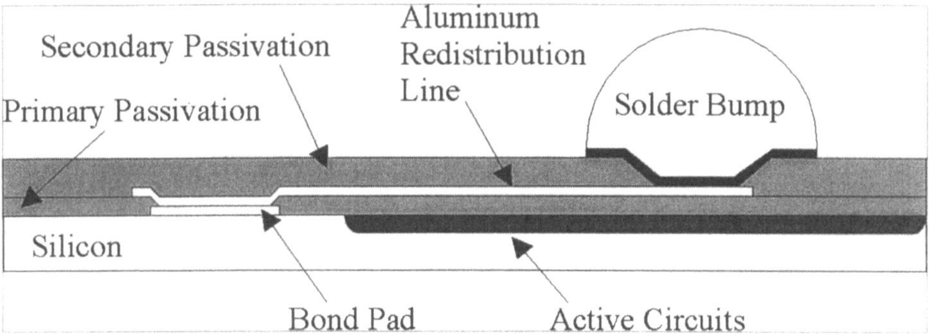

Fig 5 Redistribution of Pads for Area Array Bumping [8]

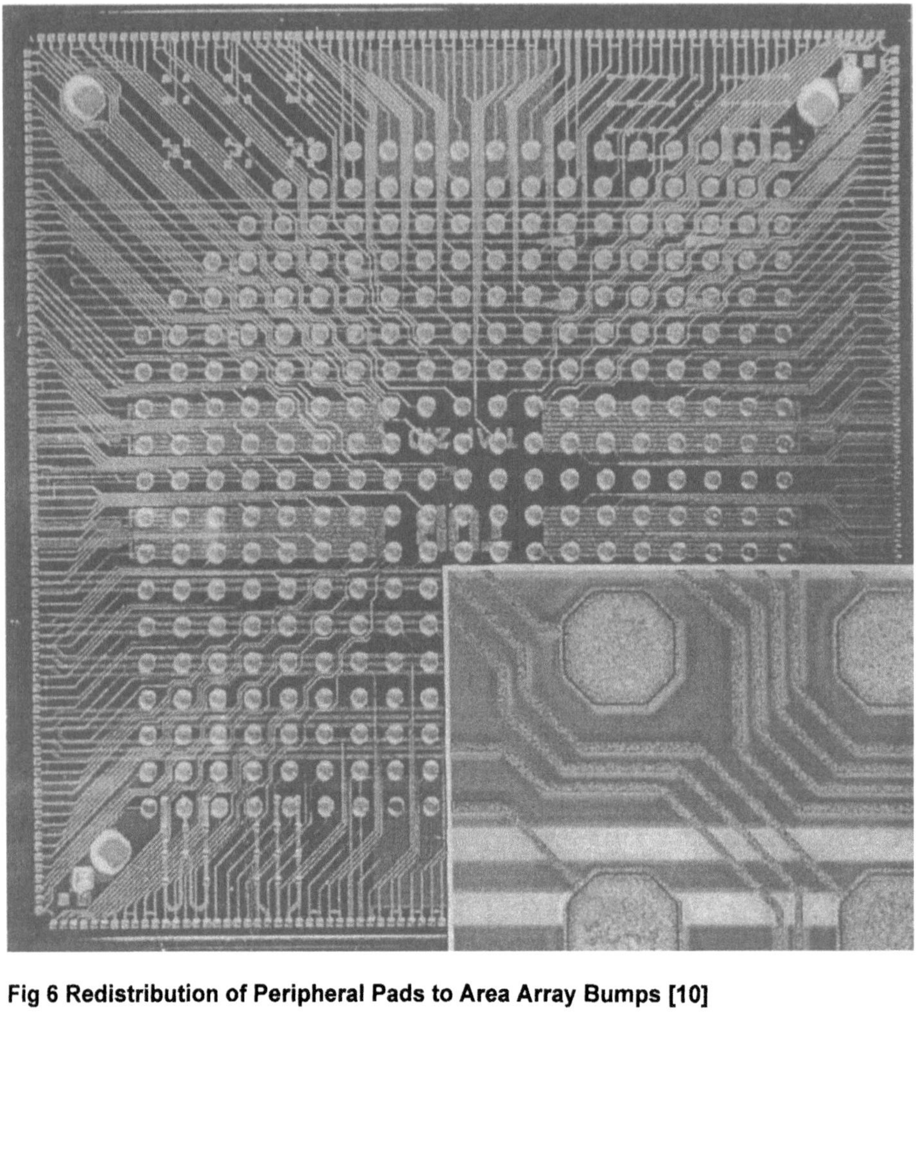

Fig 6 Redistribution of Peripheral Pads to Area Array Bumps [10]

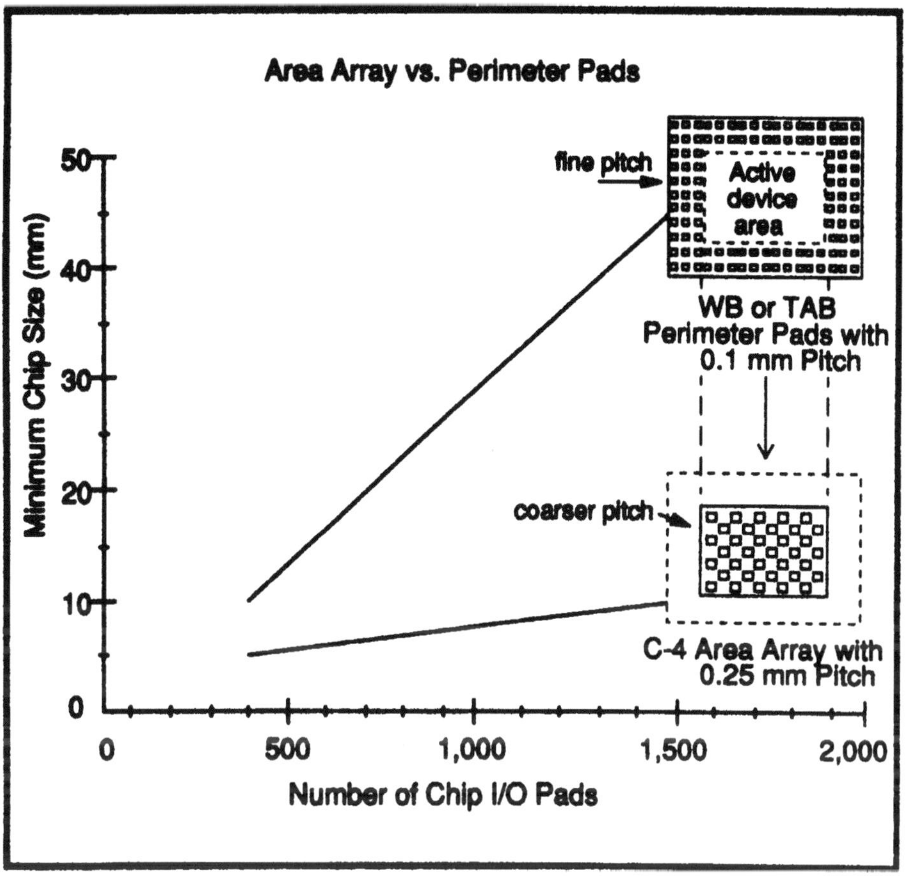

Fig 7 Advantages of Area Array: 0.1 mm WB pitch vs 0.25 mm Area Array [11]

3. Chip Packaging: BGAs and CSPs

New package types are poised to take their place alongside both the ceramic high end and low cost plastic/ leadframe packages that have served the industry since the 1960s. The two most prominent drivers are increasing I/O and chip speed. In the case of plastic leadframe packages, it is becoming clear that devices with over 250 pins will transition to BGA packages because the fine leads on the PQFPs have become to difficult to process and handle. In addition the packages become unacceptable large when peripheral leads must be used for high I/O packages (the package size becomes dependent on the number of pads, i.e. pad limited). It should be noted that fine pitch gold ball bonding is capable of as low as 70 micron pad pitch with as small as 60 micron pads [12], but such tight WB pitches will require the use of very dense laminate to escape and interconnect on PWB.

In addition microprocessors now surpassing 250 MHz in operating speed at pushing the limits of ceramic performance. Intel is leading this drive and has in fact convinced their ceramic suppliers at Kyocera and NTK to ready laminate offerings due to their cost and performance advantages .

Two levels of packaging are being developed to meet these needs: ball grid array packages (BGAs) and the smaller peripheral or area array chip scale packages (CSPs)

BGAs

BGA packaging combines the merits of area array technology and the ability to fit into surface mount manufacturing lines. BGAs offer a smaller footprint, lower parasitics [13] (more and more important at high frequencies) and more I/O s per unit area. Many kinds of BGAs have evolved. They are usually classified by their substrates; ceramic (CBGA); tape (TBGA); and laminate (PBGA). All have area I/O. JEDEC standards are evolving around 1.27, 1.0, 0.8 and 0.5 mm ball pitch.

The first CBGAs grew out of IBMs C4 flip chip process work. The TBGA is an adaptation of the tape automated bonding (TAB) technology of the 1980's.

The popular PBGAs are built on a laminated, high temperature substrate and molded after die attach and wire bonding. Motorola pioneered this type of BGA with their OMPAC (over molder pad array carrier). Cross Sections of typical double layer and multilayer PBGAs are shown in Fig 8.

Due to fine pitch limitations of the PQFP (plastic quad flat pack) and the higher cost of ceramic and tape BGAs, the sweet spot for PBGAs appears to be 250 to 500 pins.

The chip can be mounted by either area array flip chip or wire bonding into the package. For PBGAs (plastic) the CTE difference between the chip and the laminate must be addressed if flip chip assembly is used. For wirebonding, the die attach supplies the compliance needed to accommodate the TCE mismatch.

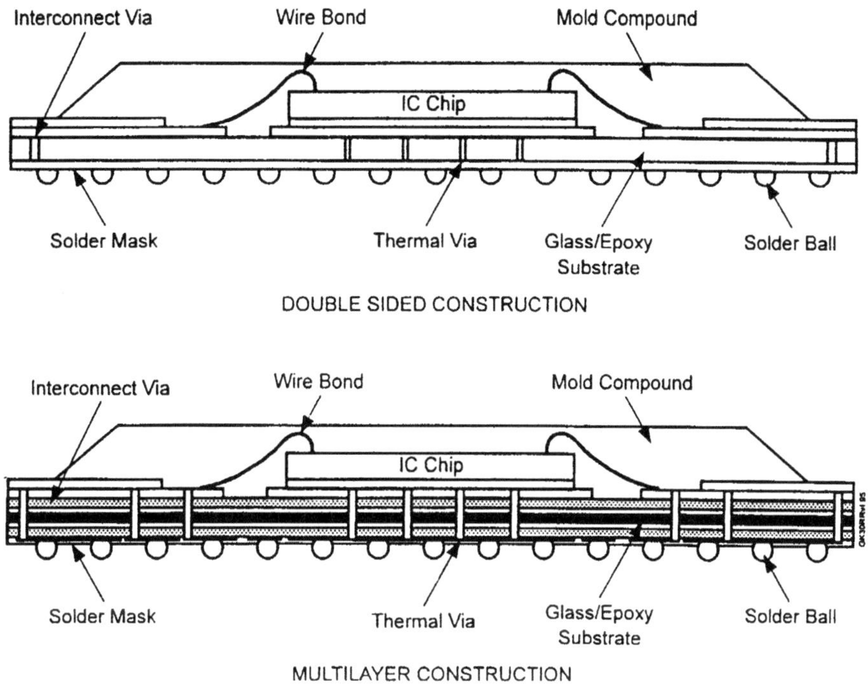

Fig 8 Double Sided and Multilayer PBGA Constructions

The concern with PBGA packages is their susceptibility to delamination and cracking during IR reflow assembly. This is due to the moisture absorption of both the BT laminate and the molding compound.

CSPs

Chip scale packages are defined as packages that are less than 1.2 times the size of the packaged chip. Their evolution can be traced to Japan in the mid 80s where MCMs were not fabricated with bare die, but rather very small "chip carriers" which were burned in and tested before module assembly. At NEC they were called FTC (flipped TAB carrier); at Hitachi they were called MCC (micro chip carrier).

There are several advantages of CSP over bare die attach (sometimes called direct chip attach, DCA). They are easier to test at speed and burn in, handle, assemble and rework. Size can be standardized so they can deal with die shrink. DCA has obvious advantages in electrical and thermal performance, size, cost and weight. While it likely that the industry will eventually move to DCA, the infrastructure (with a few large OEM exceptions) is currently more comfortable with CSP and/or BGA.

Standards for chip scale packaging have not yet been defined,. There are four basic types being developed as shown in Table 3 [14].

Figure 9 compares the available I/O and chip size for PQFP, BGA and CSP packages

It is likely that the lowest cost area array CSP technologies will be those that are done "wafer scale". The technologies described by Shellcase [15], Chipscale [16] and Technical University of Berlin [17] allow formation of the area array CSP on the wafer and subsequent dicing to create fully packaged chips. Key to the Shellcase technology shown in Fig. 10, is backgrinding of the wafer to generate a fully packaged thickness that is no thicker that the original wafer. This has obvious advantages for the PCMCIA marketplace.

Other CSPs] have been detailed in reference 14..

4. High Density Interconnect

If we look at the evolution of PWB technologies as it is depicted in Figure 11 we see that density is the driver [18]. More correctly density of vias. It has been shown that via drilling costs are rising exponentially as the demand for ever smaller holes continues [19].

Example	CSP Type	Companies
	Flex circuit interposer	General Electric NEC Nitto Denko Tessera
	Rigid substrate interposer	IBM Matsushita Motorola Toshiba
	Custom lead frame	Fujitsu Hitachi Cable LG Semicon Mitsubishi Electric Rohm
	Wafer-level assembly	ChipScale ShellCase Sandia

Table 3. Classification of Chip Size Packages [14]

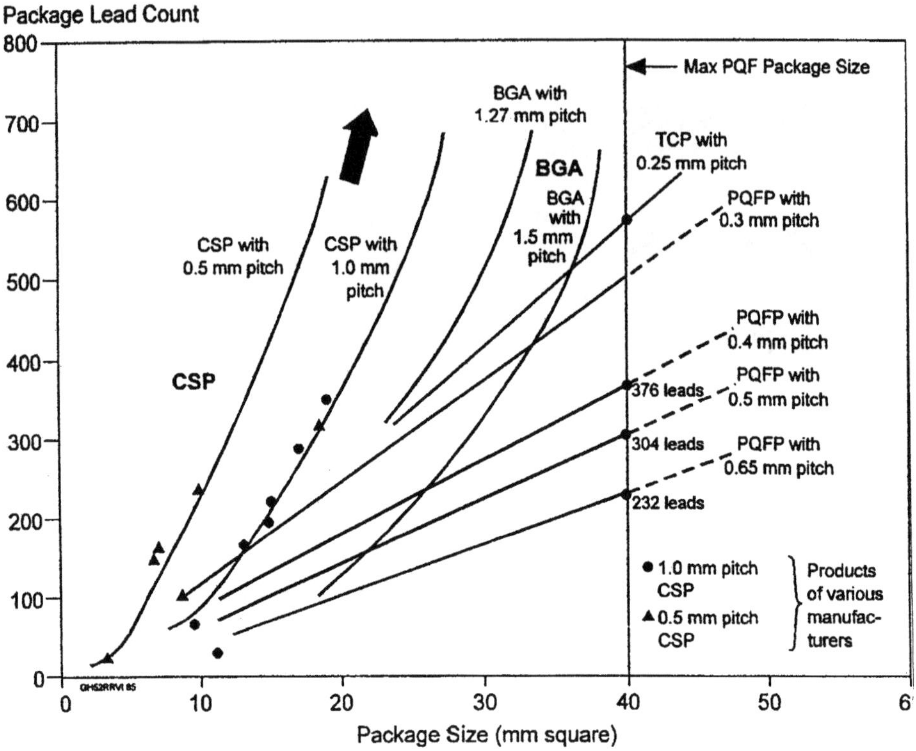

Fig 9 Comparison of PQFP, BGA and CSP: I/O vs Pkg Size

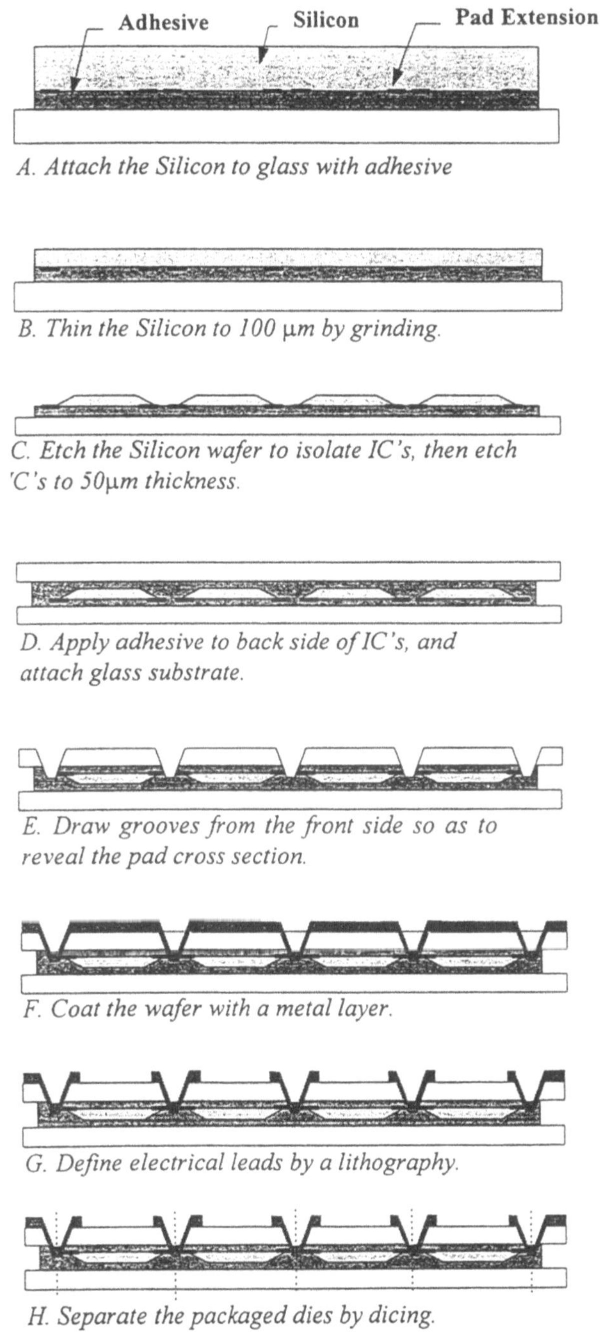

Figure 10 Shellcase CSP Process Flow

Conventional PWB constructions cannot handle grid arrays with high I/O and/or pitches below 0.5 mm. The solution is to have low cost, higher density, so called microvia technologies.

A local wirability problem arises when it is not possible to escape the pads that make connection to a PWB. This situation will exist when there is insufficient space for PTHs in the connection area and the connection demand of the device exceeds the wiring capability of the top surface of wiring. With area array, the situation is difficult since the signals from internal rows of the area array configuration must be routed between the pads closer to the edge which does not permit more than one or two conductors to pass through. In most cases the signals from the internal pads must be brought down into internal layers of the PWB.

As shown in Figure 11, many technologies have evolved to address this issue. Activity has been particularly heavy in Japan as evidenced by the list of laser and photovia technologies reported in Table 4. Recently US companies like Amp [20], Continental Circuits [21] and Hadco have initiated development programs in this technology which has been dubbed "build up board", "sequential build up board" and/or "microvia" technology by the PWB industry.

Technology	Company	Material	Via Process	Via Dia (um)	Via land (um)	line/space (um)
DV-multi	NEC	epoxy film	photo	100	230	50/50
IBSS	Ibiden	epoxy liq	photo	100	150	50/50
ALIVH	Matsushita	aramid-epoxy	laser	150	300	60/90
	Fujitsu	epoxy	photo	100	150	50/50
VB-2	Victor	epoxy	laser	100		95/75
B^2IT	Toshiba	BT laminate	paste/bump	200	300	90/90
Hitavia	Hitachi	adhesive film	drill	200	500	100/100
MSF	Shinko	PI	laser	120		20/20
SLC	IBM Yasu	epoxy liq	photo	100	175	50/50

Table 4. High Density Laminate Technology in Japan

Surface Laminar Circuitry (SLC) was the first commercialized sequential build up process [22] . It is shown in cross section in Fig 12. It initially used a modified solder mask as the dielectric, but has since changed to epoxy materials formulated specifically for this application [23]. Dielectric is usually needed on both sides of the panel to balance stresses. BGA chip carriers with 350 I/O on 50 mil grid require 6 to 8 layers just for their own escape and no through wiring is possible without exceeding 8 layers. It is reported that with SLC the escape can be accomplished on 2 to 3 layers
Higher Density versions have been developed by NEC [24] and Motorola/Dow [25] .

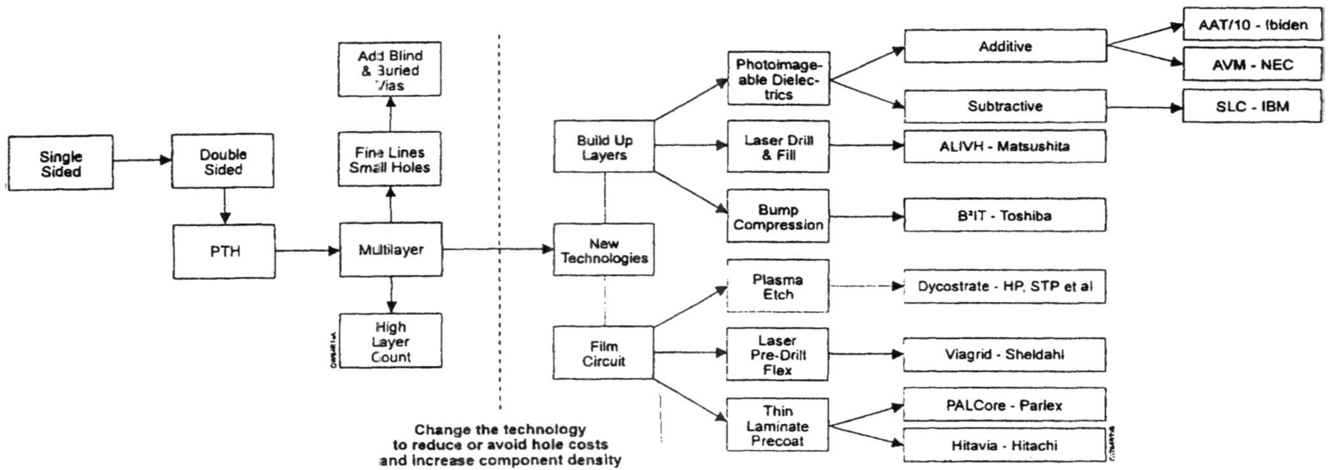

Fig 11 PWB Roadmap [18]

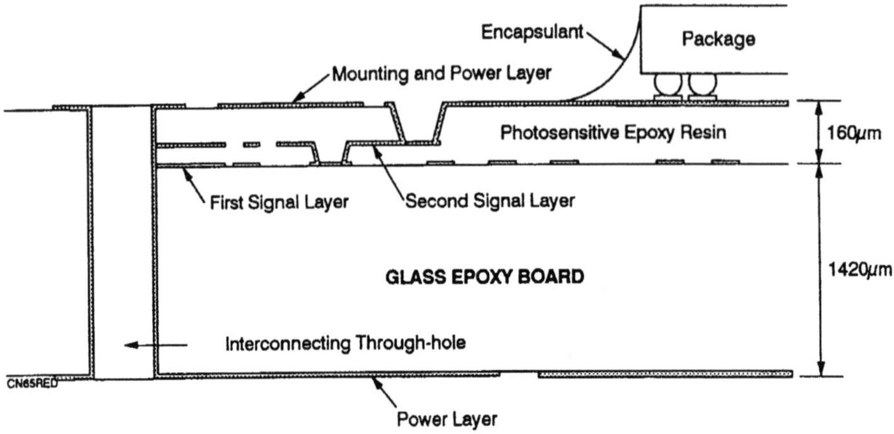

Figure 12 Cross Section of IBM Yasu SLC Technology

It is also being shown that one can achieve low cost high density packages on non laminate substrates [26].

As the IC industry learned early on, layout of multiple dies on one substrate is the path to better economics. There are two basic methods of increasing throughput : fabrication of very small structures as exemplified by the IC industry which routinely generates 100's of ICs per wafer, or, for substrates, fabrication on large area carriers which has been the cost reduction route of the PWB industry.

The economic advantages of large panel high density substrate production were first proposed by Motorola/ Dow [27] in 1992. It was deduced that the equipment advances being implemented in the flat panel display industry could also have a major impact on reducing the price of thin film MCM-D manufacturing if such large panel techniques could be adopted by the emerging thin film MCM industry. Z-Systems and IMC Microsystems published the first work on a 350 x 350 mm FPD format (Al / BCB thin film interconnect on glass) and proposed that larger panel production was possible in 1993 [28]. The impact such technology could have on the MCM infrastructure was clear to all when the IMC prototypes were shown at the Int. MCM Conference in Denver in the spring of 1993. DARPA immediately requested proposals from those interested in the development of such technology. This creation of a series of industry-government-university partnerships spurred the initial development of the large panel technology, driven by a combination of substrate foundries, equipment and material suppliers, and their customers. Programs were put in place to rapidly fund the development of LAP processes and equipment.

Cost Modeling

The very favorable economics of large panel processing are depicted in Table 5 where > 10 times more modules are fabricated on a 400 mm x 400 mm panel than is possible on a 150 mm diameter wafer. Even considering the greater cost of producing the larger panels (due to more expensive capitalization and anticipated initial lower yields) it was predicted that the overall cost of substrate fabrication would be reduced.

panel size →	150 mm round	400 mm sq.
↓ substrate size		
50 mm	4	64
25 mm	20	256

Table 5 The Advantage of Larger Panel Sizes

Several groups have addressed the economics of such processing with a more quantitative analysis. Workers at Motorola [29], a collaborative effort between groups at

Fig. 13

Dow and Motorola [30] and workers at IBM [31] have all examined the cost impact of Large Area Processing. All results support the advantage of larger format substrates in driving down typical MCM-D processing costs.

The goals of the DARPA CILAP consortium [32] and the MMS/Dow LAP line [33] has been to bring the dimensions of MCM-D thin film wiring to 400 mm glass, aluminum and laminate panels for low cost high density single and multiple chip packaging. Whereas a normal PWB can support one routing track between 0.5 mm BGA or CSP balls on 0.5 mm grid, a single BCB thin film layer an accommodate up to 17 routing tracks in the same geometry [33]. An 400 mm aluminum panel is compared to a 150 mm wafer in Fig 13.

Conclusions:

The need for increased I/O coming off chip is driving the industry to area array interconnect. Most high (> 250) I/O chips will be designed for area array in the future. In the interim peripheral pads with tight I/O will be redistributed with a layer of this film to a looser pitch area array.

The handling of >250 I/O on PQFP packages is becoming very difficult. Again area array packages with smaller footprint and looser pitch such as BGAs and CSPs appear to be gaining a market share for such applications.

These area array packages with increasing I/O have driven to PWB industry to look at several alternative technologies that can generate higher density "micro vias". Sequential build up boards such as IBM SLC have begun to gain significant share of the high density market in Japan. Even higher density capabilities are becoming commercial with programs such as the DARPA supported Large Area Processing line which combines MCM-D like feature densities and flat panel display equipment to drive down thin film costs.

REFERENCES:

[1] ICE Corp , Tempe Az.

[2] Y. Tsukada, S. Tsuchida, Y. Mashimoto, "Surface Laminar Circuits and Flip Chip Packaging", Proceed. 42nd ECTC, 1992, p. 2227.

[3] Rinne, G., "Solder Bump Methods for Flip Chip Packaging", Proceed. 47th ECTC, 1997, p. 240.

[4] Nakamura, Y. Et. al. , "Advanced LSI Packaging Using Stud Bump Bonding Technology" , Proceed. Int MCM Conference, Denver, 1995, 302.

[5] Ogashiwa, T., et. al., "Direct Solder Technique on an Al Pad and Its Reliability", Japanwese Journal of Applied Physics, part 1, vol. 31, 1992, p. 761.

[6] P. Elanius, "Flex on Cap - Solder Paste Bumping", Proceed. 47th ECTC, 1997, p. 248

[7] J. Kloeser, et. al., "Fine Pitch Stencil Printing of Pb/Sn and Pb Free Solders for Flip Chip Technology", Proceed 47th ECTC, 1997, p. 254

[8] Mis, D., Rinne, G., Deane, P., Adema, G., "Flip Chip Production Experience", Proceed ISHM, 1996, p. 291

[9] Simon, J., Topper, M., Reichl, H. , Chimel, G.,., " A Comparison of Flip Chip Technology with Chip Size Packages", Proceed. IEPS, 1995, p. 665.

[10] M. Toepper, J. Simon, H. Reichl, Chip Scale Packaging using Photo-BCB", Future Fab Int., 1996, p. 363

[11] Puttlitz, X., " "Proceed. ECTC ,1994 p. 16

[12] D'Ignazio, J., "Wirebondings Reign Continues", Semiconductor International, June 1996, p. 117

[13] Huang, C., Wong, F., Kim, D., "Electrical Performance of PBGA as Compared to MQFP in High Frequency Applications", Proceed IEPS, 1995, p. 417.

[14] Goodman, T., Crowley, R., Vardaman, J., "Chip Size Packages: Comparisons and Structure/Performance Relationships", Proceed IMC, Omiya, 1996, p. 62

[15] private communication, Shellcase inc, Jerusalem Israel

[16] private communication, Chip Scale Inc., San Jose Ca

[17] J. Simon, G. Chimiel, Lautenberger, R., H. Reichl, "Development of a Flip Chip mBGA", Proceed. 1995 Int. Flip Chip, BGA, TAB and Advanced Packaging Symposium - ITAB, San Jose, p. 54.

[18] private communication, Prismark Associates, Cold Harbor Springs, NY.

[19] Tessier, T., Garrou, P., "MCM L: Materials, Processes and Applications", in MCM Handbook, P. Garrou and I Turlik Eds., McGraw Hill, New York, 1997.

[20] Li, W., Tummala, R., Funer, R., Sullivan, C., "Materials, Processes and Reliability Evaluations of Photodielectrics for Sequential Circuit Applications", Proceed. Int MCM Conf., Denver, 1997, p. 128

[21] McDermott, B., Tryzbiak, S., "Practical Applications of Photo-Defined Micro-Via Technology", Proceed Int. Symp. Advanced Packaging Materials, Atlanta, 1997, p. 24

[22] Y. Tsukada, Y. Maeda, K. Yamanaka, "A Novel Solution for MCM-L Utilizing Surface Laminar Circuit and Flip Chip Technology", Proceed. Int. MCM Conf., Denver, 1993, p. 252

[23] Ho, I., "An Overview of Sequential Build-up/Microvia Technologies", Proceed. Int. Symp. Advanced Packaging Materials, Atlanta, 1997, p. 19

[24] Matsui, K., Shimoto, T., Shimada, Y., Utsumi, K., "MCM D/L Technology Based on Cu/BCB Thin Film Multilayers", Proceed ISHM, 1995, p. 396.

[25] Strandjord, A., Garrou, P.E., Heistand, R.H., Tessier, T.G. "MCM - LD: Large Area Processing using Photosensitive BCB", IEEE Trans. CPMT part B, Vol 18, 1995, p. 269

[26] Garrou, P.E., and Tessier, T., " Large Area Processing", chap 5 in Multichip Module Handbook, P. Garrou and I. Turlik Eds. , McGraw Hill, 1997, in press

[27] Tessier, T., Garrou, P., "Overview of MCM Technologies: MCM- D" Proceed. ISHM, 1992, p. 235.

[28]. Hodges, C.R., Hull, J., Strandberg, J., Hagel, J., Theide, H., Karlsson, A., "Large Panel Format Multichip Module Substrates", Proceed. Int. Conf. MCM's, Denver, 1993, p. 133.

[29] Tessier, T., Myszka, E., "Approaches to Cost Reducing MCM-D Substrate Fabrication", Proceed. ECTC, 1993, p. 570.

[30] Frye, D., Skinner, M., Heistand, R., Garrou, P, Tessier, T., "Cost Implications of Large Area MCM Processing", Proceed. Int. MCM Conf, Denver, 1994, p. 69.

[31] White, G., Perfecto, E., Demercurio, T., McHerron, D., Redmond, T., "Large Format Fabrication - A Practical Approach to Low Cost MCM-D", Proceed Int MCM Conf., Denver, 1994, p. 88.

[32] Wells, G.T., Chau, C.C., Garrou, P.E., Heistand, R., Cummings, S., Rehg, T., Ho, C., Chieh, E., Gibson, G., Newquist, C., Kruzek, R., Ellec, C., Vogt, L., "Consortium for Intelligent Large Area Processing (CILAP)", Proceed ISHM, Minneapolis, 1996, p. 397

[33] Chieh, E., et. al. "Development of Large Area Processing (LAP) for Thin Film Substrates", Proceed Int. MCM Conf., Denver, 1997, p. 234.

APPLICATION OF ELECTROCONDUCTING POLYMERS IN LOW COST DEVICES[1]

G. HARSANYI, R. DOBAY, M. RECZEY, ZS. ILLYEFALVI-VITEZ
Technical University of Budapest, Department of Electronics Technology
1521 Budapest, Hungary, Phone: 36-1-463-3634, Fax: 36-1-463-4118
E-mail: Harsanyi@ett.bme.hu

CS. VISY
Attila Jozsef University, Department of Physical Chemistry
6701 Szeged P.O. Box 105, Hungary, Fax: 36-62-31-1943
E-mail: Visy@chem.u-szeged.hu

Abstract

In the beginning of the 80's years an essential property of most polymers, which distinguishes them from metals, was their inability to carry electricity. These insulating properties of polymers are significant advantage for many applications of plastics. During the past 20 years, a new class of organic polymers has been devised with the ability to conduct electrical current. These conducting plastics are still under development for practical applications, such as rechargeable batteries, capacitors, enzymatic biosensors, gas sensors, etc... This paper describes some of these applications of conducting polymers especially paying attention to the research area of electrochemical sensors.

Introduction

BAND THEORY OF CONDUCTING POLYMERS

The electrical properties of any material are determined by its electronic structure. Conducting polymers are similar in that they conduct current without having partially empty or partially filled band. When an electron is removed from the top of the valence

[1] This work was carried out with financial support provided by Hungarian National Scientific Research Found (OTKA projects No. T016017 and No. T021102). Our research was supported by the Found for Hungarian Higher Education and Research (AMFK 30/96) and INCO-Copernicus (96/0156).

R. Tummala et al. (eds.), Electronic Packaging for High Reliability, Low Cost Electronics, 133–141.

band of a conjugated polymer, a hole or a radical cation is created that does not delocalize completely, as would be expected from classical band theory. Only partial delocalization occurs, extending over several segments causing them to deform structurally. The energy level associated with this radical cation (polaron) is higher than the energies in the valence band. The polaron stabilizes itself by polarizing the medium around. If another electron is removed from the already oxidized polymer, two things can happen: this electron could come from either a different segment of the polymer chain, thus creating another polaron, or from the first polaron level by removing the unpaired electron to create a special dication, called bipolaron. Both polaron and bipolaron are mobile and can move along the polymer chain by the rearrangement of double and single bonds. (Figure 1.)

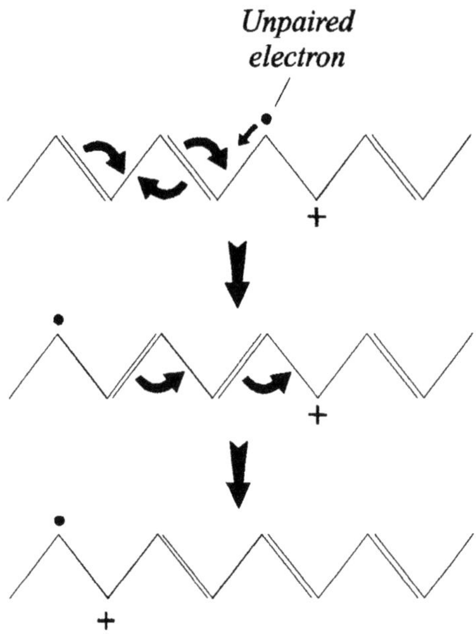

Figure 1. Propagation of a polaron through the polymer chain by rearrangement of double bonds

APPLICATIONS OF ECPS

The electroconducting plastics concerned here are not composite plastics filled with metallic particles, but organic materials that exhibit intrinsic electronic conductivity, reflecting their delocalized electronic structure along the conjugated backbone. A one-step electropolymerization forms an adherent film deposited at the surface of the metal electrode with anionic species being included by doping [1].

Rechargeable batteries

Rechargeable battery devices have an electroconducting conjugated polymer (ECP) cathode, an electrolyte, and a lithium or LiAl alloy anode. At this stage, these batteries are excellent for applications in which long lifetime, low-power, and reliable operation is required, such as backup sources for static RAM, telephones, timers for videocassette recorders, and as batteries for calculators, and wristwatches [2].

Electrochromic devices

Many conducting polymers like polyaniline (PANI), polythiophene (PTh) or Polypyrrole (PPy) and their derivatives show color changes according to the doped or undoped state of the film.

In a „smart window" device, which changes color in response to sunlight or temperature changes, both electrodes must be transparent. In the bleached state of the device, the polymer is in the undoped form. Smart windows can be used for thermal window that would work in winter, absorb the sunlight in summer, foils can be fabricated for the screen-window of automobiles, etc...

Solid-state polymer LEDs have been fabricated using thin films of blends of conjugated polymers and solid electrolytes. The cells contain three layers: the polymer blend films (as the emissive layer), and indium-tin oxide (ITO) and aluminum films as the two contact electrodes. When externally biased, the conjugated polymers are p-doped and n-doped on opposite sides of the polymer layer, and a light-emitting p-n junction is formed between [3].

Liquid crystal displays, electrochromic devices made of different ECPs can be fabricated with the large areas and unlimited visual angles. Electroconducting polymers do not change color after the applied voltage has been removed, so they exhibit memory function (bistability), thus can be used to store information [2].

Electrochemical capacitors

Conducting polymers have been investigated for use in electrochemical capacitors (EC) because of a combination of high charge density with predicted low material cost. ECs are basically the same as a battery in terms of general design, the difference being that charge storage is capacitive in nature rather than Faradaic [4].

Military application

Conductive fabrics can be prepared that could be incorporated into the plastic composites used in airframes for charge dissipation during lightning strikes, thus preventing damage to the airplane. In addition, because such fabrics distort radar signals, they could be used in military aircraft to mislead enemy radar. Latex compositions have been developed containing PPy, PTh, and PANI colloid particles. These latexes can be sprayed onto surfaces to give conducting coatings. Such coatings provide electromagnetic shielding and may be useful as a weapon against electronic eavesdropping [2].

Electrochemical sensors

ECPs appear very attractive for use in sensors either as sensing components or as a matrix for easy immobilization of specific substrates. They easily can be synthesized by a simple electrochemical method (electropolymerization), where the doping reaction modulates the conductivity reversibly by several orders of magnitude via redox interactions. Electroconducting conjugated polymers are good candidates for the elaboration of chemical or electrochemical sensors in two ways [5]: as a matrix for specific immobilization as in the case of enzymatic biosensors or as sensitive components as follows.

Suspended-Gate Field Effect Transistor (SGFET)

A promising research area of ECP-based chemical sensors is their arrangement in a FET device. The structure of SGFETs is similar to other conventional FET devices: the surface potential of the semiconductor can be influenced by changing the gate potential. The gate potential of the device can be changed by absorbing gases on the polymer matrix, thus it determines the depth of the p-channel [1].

Ion detectors

The detection of electroinactive ions by ECP-based electrochemical devices originates in the so-called 'ion-sieving' or memory effect. This means that it is possible to vary the ion-permeability of electropolymerized films via their well-controllable anodic growth [6]. A PPy film prevents anions of larger size from penetrating the matrix.

Gas sensors

A variety of devices has been constructed and tested utilizing the reversibility in redox properties of ECP films, coupled with the incorporation of electrolyte anions into the polymer film, and dissolution of the incorporated anions takes place in the redox reaction. So redox changes of conducting polymers can be used to sense gases and vapors by monitoring the change in conductance on exposure of the polymer to the sample [7-8]. The preliminary studies on these materials have shown that they exhibit fast, reversible responses at room temperature [9-11].

Experimental

URIC ACID ENZYMATIC BIOSENSOR

Amperometric biosensors transduce the diffusion-controlled flux of a substrate to an enzyme electrode into an electrical current. The application of ECPs for immobilizing bioactive compounds in biosensors seems to be one of their most promising approach. Biosensors are suitable for indicating and quantitative measuring of physiologically important, biologically active substances. The determination of these substances is realized by catalytic reactions, the redox enzymatic reaction being the most important of these [8]. In the case of uric acid, uricase catalyses its oxidation in slightly alkaline solutions, producing allantoin and hydrogen peroxide:

$$\text{Uric Acid} + O_2 + 2H_2O \longrightarrow \text{Allantoin} + H_2O_2 + CO_2$$

The change of concentration of hydrogen peroxide (H_2O_2) can be amperometrically detected. The details of the detection mechanism can be seen in Figure 2.

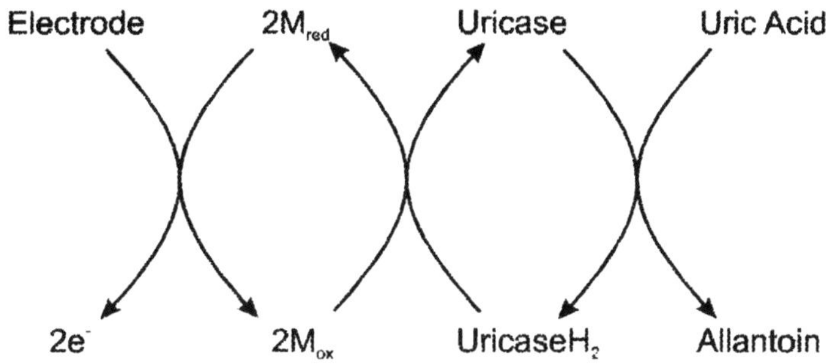

Figure 2. Electrical wiring of a redox enzyme; flow of electrons from the electrode through a monoelectric relay (M_{ox}/M_{red}) to the uricase enzyme.

N-Methylpyrrole (MPy) was polymerized to form a thin film; this electrochemical polymerization method of enzyme immobilization makes possible to regulate the film thickness electrically.

Thick film technology was used to prepare a new type of enzymatic microbiosensor. The approach of thick film technology producing biosensors offers several advantages over applying conventional microelectrodes or other integrated (thin film and monolithic) technologies, as discussed elsewhere [12].

For the substrate of the sensor, sintered alumina was used in order that Pd-Pt-Ag thick film conductor paste could be screen-printed for the working and counter electrodes, and Pt-Ag paste for the pseudo-reference electrode, respectively.

The polymerization takes place from a solution containing the monomer (0.1M MPy) and also the enzyme (0.6μM). In order to ensure the electrostatic interaction between the film and the enzyme for the immobilization, sodium lauryl-sulfate conducting salt (0.1M) was used, which raises the pH of the solution higher than the isoelectric pH of the enzyme. Moreover, during the polymerization, the anion of the surface-active material incorporates into the film, and it is too large to get out of the film. During the detection these immobilized anions do not leave the structure and prevent the urate anions to penetrate into the polymer chain. Thus it shows cation exchanging properties solely, the uric acid can only interact with the enzyme [12].

The current responses of the pure polymer film (not containing the enzyme) and the enzymatic cell were studied by changing the uric acid concentration. The results indicate that the sensor is sensitive to the concentration of the uric acid of the solution (Figure 3).

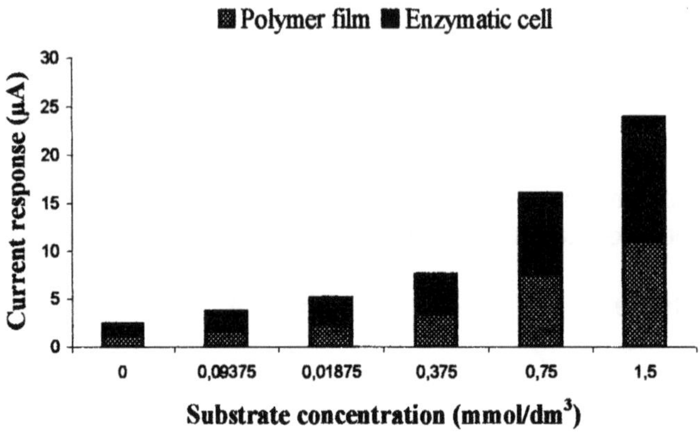

Figure 3. The response of the two films to uric acid concentrations

The prelimiary studies showed that the current of the sensor comes from two different sources: the current of the interaction between the uric acid and the polymer film (which is non-selective), and the additional current of the enzymatic reaction.

The new polymerization technique produced such a stable polymer film, which keeps its activity for a long time. The sensor was kept in Na-borate buffer solution at 4 °C. During three weeks after the preparation, the activity of the cell was measured, and the results indicated that the sensor kept its dependence on the uric acid concentration, although the usage was continuos. The lifetime-curve of the biosensor is shown in Figure 4.

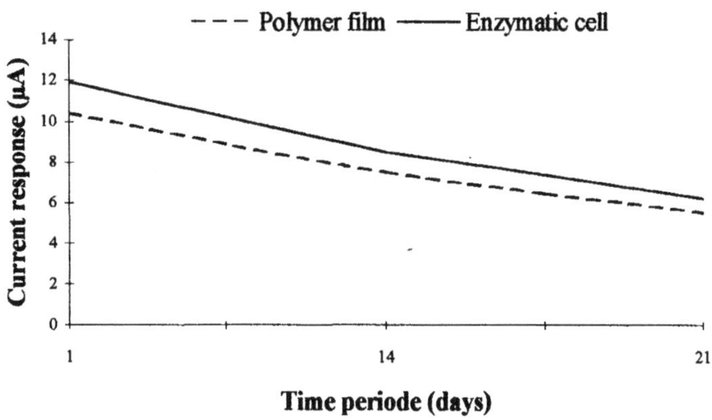

Figure 4. Current response controlled during 3 weeks after their preparation (cc= 1.5 mMol/dm^3)

The measurement of the extra current of the enzymatic reaction can be more precise and more reliable by examining the two electrodes in same conditions at the same time. To realize this method a new layout was designed containing two working electrodes.

(Figure 5.) In order to measure the current response of the sensor with two working electrodes, a special bipotentiostatic method was used, where the sensor and the polymer electrodes were uniformly polarized.

Figure 5. The layout of the sensor

GAS SENSOR

The sensor electrodes were screen-printed onto alumina ceramic carrier. Pd-Pt-Ag paste was used for the working and the counter electrode. The thickness of the Pd-Pt-Ag layer was about 10 µm. A 60 µm wide gap was cut into the screen-printed metal pad by GRAVILASER developed at the Department. Three kinds of electrode structures were studied.

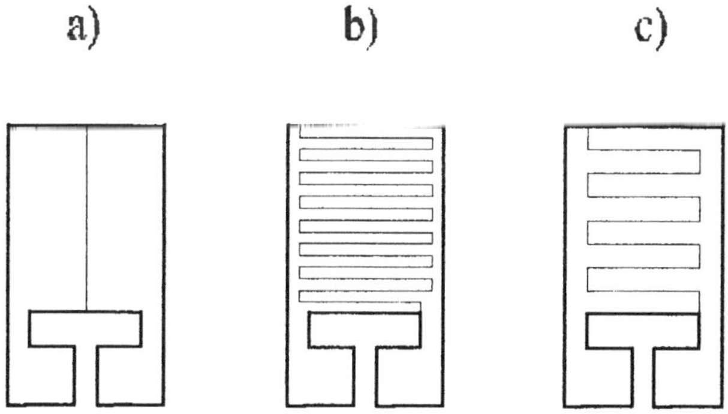

Figure 6. Shematic diagrams of the electrode structures for sensor substrates

The polymerization took place in a 0.1M pyrrole and 0.05M dodecyl-sulfate (DS) solution by applying cyclic voltammetric method using 0.1M SCE as reference electrode. The cyclic voltammograms of the growth of the polymer film can be seen in Figure 7.

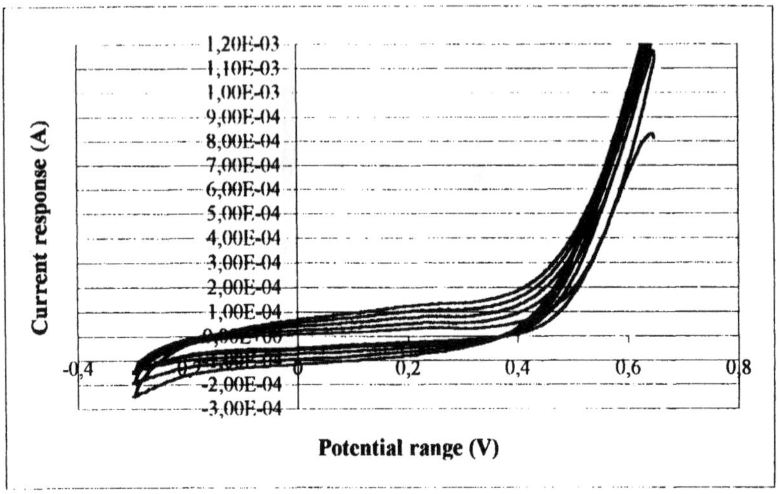

Figure 7. Cyclic voltammograms for the deposition of polypyrrole on the electrode

The preliminary results showed that the polymer film grew over the gap, and the resistance of the layer was about 2kΩ. The sensitivity of the polymer film was investigated by changing the temperature and it was found that the temperature coefficient of the examined samples $TC=1.09*10^{-4}$ /°C. The sensor produced reversible changes in the conductivity in the presence of methanol gas.

Summary

The results of the gas sensor and the uricase enzymatic biosensor showed that thick film technology is suitable to fabricate low cost electrochemical devices. Different ECP films were studied to form chemically sensitive materials.

The gas sensor produced a small dependence on the temperature and showed change in the conductance in the presence of methanol gas. The uricase electrode gave monotonous response to the uric acid concentration. The biosensor was measured repetitively during 21 days, and the results showed that it kept its dependence on the uric acid concentration, though the sensitivity decreased.

References

1. Bidan, G. (1992) Electroconducting conjugated polymers: new sensitive matrices to build up chemical or electrochemical sensors. A review., *Sensors and Actuators* **B**, pp. 45-56

2. Kanatzidis, M. G. (1990) Conductive polymers, *C&EN Special Report, December 3*, pp. 36-54

3. Pei, Q., Yang, Y., Yu, G., Zang, Z. and Heeger, A. (1996) Solid-state polymer light-emitting electrochemical cells, *ANTEC '96 Proceedings*, pp. 1332-1336

4. Rudge, A., Raistrick, I., Gottesfeld, S. and Ferraris, J.P. (1994) A study of the electrochemical properties of conducting polymers for application in electrochemical capacitors, *Electrochimica Acta*, **Vol. 39**, **No. 2**, pp. 273-287

5. Janata, J. (1990) Chemical sensors, *Anal. Chem.*, **62**, 33R

6. Wang J., Chen, S.-P. and Lin, M. (1989) Use of different electropolymerization conditions for controlling the size-exclusion selectivity at polyaniline, polypyrrole and poplyphenol films, *J. Electroanal. Chem.*, pp. 273-231

7. Hwang, L. S., Ko, J. M., Rhee, H. W., Kim, C. Y. (1993) A polymer humidity sensor, *Synth. Metals*, **Vol. 55-57**, pp. 3671-3676

8. Charlesworth, J.M., Partridge, A.C., Garrard, N., (1993) Mechanistic studies in the interactions between poly(pyrrole) and organic vapors, *J. Phys. Chem.*, Vol. 97., , pp. 5418-5423

9. Bartlett, P.N., Archer, P.B.M., Ling-Chung, S.K., (1989) Conducting polymer gas sensors Part I.: Fabrication and characterization, *Sensors and Actuators*, Vol. 19., , pp. 125-140

10. Bartlett, P.N., Ling-Chung, S.K., (1989) Conducting polymer gas sensors Part II.: Response of polypyrrole to methanol vapour, *Sensors and Actuators*, Vol. 19., , pp. 141-150

11. Bartlett, P.N., Ling-Chung, S.K., (1989) Conducting polymer gas sensors Part III.: Results for four different polymers and five different vapours, *Sensors and Actuators*, Vol. 20., , pp. 287-292

12. Harsanyi, G., Dobay, R. and Visy, Cs. (1997) Thick films combined with electroconducting polymer films: a solution for low cost disposable enzymatic microbiosensors, *Pan Pacific Microelectronics Symposium Proceedings*, pp. 383-388

PHOTOSENSITIVE POLYMER THICK FILMS

Selim Achmatowicz, Małgorzata Jakubowska, Elżbieta Zwierkowska
Institute of Electronic Materials Technology,
Wólczyńska 133, 01-919 Warsaw, Poland

Wiktor Nemesz, Siergiej Osieczkin,
Microtech KARAT, Stryjska 202, 290031 Lvov, Ukraine

A short review of advantages and disadvantages of employing different polymers to PTF's is presented. Out of commonly used in PTF's thermoplastic, thermosetting and UV cured polymers the latter group of polymers was selected for the investigation. The synthesis of broad range of the UV curable epoxy acrylates was performed in the author's laboratory.
The experiments carried were focused on the conductive and dielectric pastes. Results of the investigation of these pastes are presented. Possible application of such pastes is discussed.

1. Introduction

Polymer thick films (PTF) have got over thirty years history. PTF pastes are composed of active ingredients (metal powders for conductors, dielectrics for insulating pastes, etc.), polymers (often requiring special additives) and solvents. Organic polymers contained in the pastes, unlike in case of the ordinary cermet pastes, remain in the layer for the entire life of an electronic device. Therefore they play a very important role. They decisively affect technological processes (application of the paste, curing conditions, kind of substrates, trimming, etc.), as well as final properties of the layers (line resolution, glass transition temperature - Tg, coefficient of thermal expansion - TCE, water absorption, dielectric constant, dissipation factor, stability, etc.).

There are three main types of polymers applied in PTF's [1]:

thermoplastic (dried)- polyesters, poly(vinyl acetate), polyacrylates, etc.,
thermosetting - epoxies, phenolics, etc.,
and UV sensitive - epoxy acrylates, urethane acrylates, polyamic acid acrylates, etc.

R. Tummala et al. (eds.), Electronic Packaging for High Reliability, Low Cost Electronics, 143–150.

The latter two groups of polymers, unlike the thermoplastics, undergo a chemical reaction (polymerization, cross-linking) in the process of curing. It means that properties of thermosetting and UV sensitive resins are different after curing, due to the changes of the chemically structure, while the properties of thermoplastics remain unchanged. It causes, for example, that solubility of thermoplastics is same before and after curing, while the others become insoluble and resistant to most solvents. This is a major issue in preparing PTF compositions for multilayer structures. The comparison of the properties of thermoplastic, thermosetting and UV sensitive polymers is presented in Table 1, regarding the resins before curing, means of pastes applications, layers and commercial availability.

Table 1. Properties of thermoplastic, thermosetting and UV sensitive polymers

Properties	Thermoplastic	Thermosetting	UV sensitive
Resin and paste			
Solubility in solvents	+	+/-	+
Composition	simple	little complex	complex
Printing ability	+	+/-	+
Curing temperature, °C	80-150	120-180	<100
Curing time	0.5-1 h	0.5-2 h	seconds
Layer			
Resistance to solvents	-	+	+
Water absorption	+	+/-	+
Flexibility	+/-	+	+
Adhesion	-/+	+	+
Glass transition temperature, T_g	low	medium	high
Commercial availability			
Conductive pastes	+	+	-/+
Dielectric pastes	-	+	+
Resistive pastes	-/+	+	-/+
Sensor pastes [2]	+	+	-/+

Table 1 shows that thermoplastic and thermosetting polymers have certain drawbacks, such as high curing temperature, long time of curing and low resistance to solvents in case of thermoplastics. UV sensitive polymers do not exhibit any critical drawback. In addition most of they have relatively high T_g.

Several types of UV sensitive polymers have been widely used in many branches of industry (printing, electronics, medicine, furniture, etc.). Polyimides have very good properties (especially very high T_g) [3], however, their application is limited due to certain disadvantage. They dissolve only in a very powerful solvents (such as N-methyl-2-pyrolidone - NMP) which are harmful to most of other materials, e.g.

substrates, screen print masking. They are applied as polyimide precursor requiring final curing (imidization) at high temperature (above 240°C). That is why their application is limited to produce dielectric layers mostly on ceramics as for example in MCM-D. They can hardly find application in PTF pastes. UV sensitive polyester acrylates and polyurethane acrylates are useful in PTF. However, they are inferior to polymers obtained from epoxy acrylates.

2. Preparation of polymer thick film pastes

A limited supply of UV sensitive PTF's on the market, essentially reduced to dielectric pastes, is caused by two main factors: metal powder fillers and complexity of the organic vehicle.

Metal powders

UV curing requires good penetration of a layer by the UV radiation. Metal particles contained in the conductive pastes are a very serious obstacle for the UV beams. A great portion of this radiation is reflected.

Metal particles may act as free radical traps. It means that a free radical successfully produced by the UV beam can be destroyed before causing reaction of polymerization.

Noble metal particles are also known as promoting catalytic reactions, in this case most undesirable.

Normally, flake type silver powder, such as in Fig. 1, is most welcomed in PTF's in order to obtain the best conductivity. In case of UV sensitive PTF's one should be aware of overdosing a flaky powder. The experiments carried by the authors have proven that a carefully chosen proportion of silver spherical and flaky powders shown as in Fig. 1 and Fig. 2 is favorable.

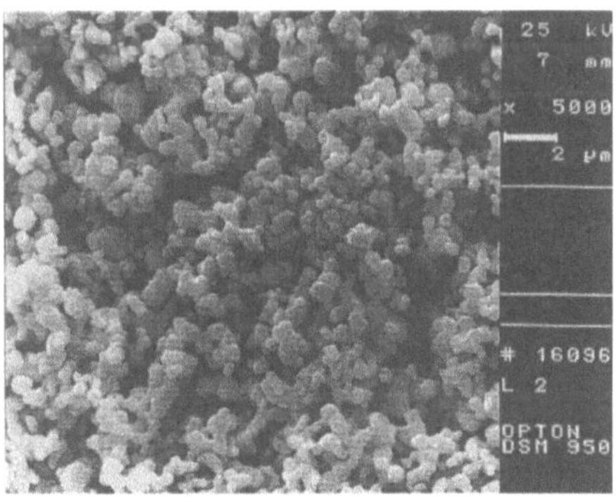

Fig. 1. Flaky silver powder

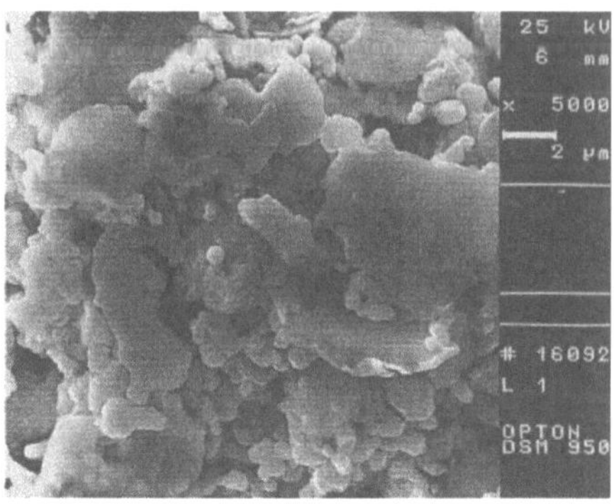

Fig. 2. Spherical silver powder

The authors were trying to overcome the other obstacles by using two different procedures. The powders, before introducing to the organic vehicle, were deactivated by immersion to a solution of stable free radical compound. Encapsulation of the powders with a very thin layer of organic substance can also reduce their catalytic and free radical trapping activity. It was found that the combination of both procedures is feasible and most advantageous.

Photosensitive organic vehicle

Complexity of the organic photosensitive vehicle is a drawback factor to a certain extend. The organic vehicle is composed of following components:

- a resin capable to undergo free radical polymerization (also called photosensitive prepolymer),
- a photoinitiator capable to produce free radical upon UV radiation,
- reactive thinners, mono- and multifunctional vinyl compounds,
- inert polymers (incapable to polymerize),
- additive commonly use in thick film paste, such as wetting agents, plasticizers, stabilizers, etc.

Selection of proper kind and proportion of all above mentioned components affects processing requirements and final properties of the PTF layers. All the components are commercially available. However, feed-back obtained results - structure of the component can be applied to a limited extend. The authors decided to synthesize the photosensitive prepolymer in order to modify its structure upon the requests coming from investigation of the layer properties. The easiness to modify epoxy acrylates, supported by the properties of the layer manufactured from these resins, make them superior over other UV sensitive polymers.
The formula in Fig. 3 shows the basic structure of epoxy acrylate molecule.

$$\mathbf{Ar_2}[\mathrm{COO{-}CH_2{-}\underset{\displaystyle OH}{\underset{|}{CH}}{-}CH_2{-}O{-}}\mathbf{Ar_1}\mathrm{{-}O{-}CH_2{-}\underset{\displaystyle OH}{\underset{|}{CH}}{-}CH_2{-}O{-}\underset{\displaystyle O}{\underset{\|}{C}}{-}C}\mathbf{R}\mathrm{{=}CH_2}]_2$$

Fig. 3. The structure of epoxy acrylate molecule

The bolded characters show the sites where the structure can be modified. Normally $\mathbf{Ar_1}$ is a back-bone of dian, which can be still modified in its aromatic and/or aliphatic part. $\mathbf{Ar_2}$ represents a variety of dicarboxylic acids. **R** is an aliphatic radical (eventually H). Nature of all the radicals affects final properties of the polymer and consequently - properties of the layer, such as solubility, rheology of pastes and flexibility, resistance to solvents, T_g, dielectric constant, etc. The proper choice of the radicals make that the polymerized epoxy acrylates inherit advantageous properties of epoxy as well as acrylic polymers.

Selection of the reactive thinners is another way where the organic vehicle can be modified. Monofunctional vinyl compounds (e.g. acrylic and methacrylic acids and their derivatives) serve to discontinue further polymerization of the prepolymer. Difunctional vinyl compounds (e.g. glycol diacrylate) can elongate the polymer or "bridge" two molecules of polymers. Trifunctional reactive thinners (e.g. pentaerythriol triacrylate) cause spherical growth of polymerizing molecules. The organic vehicle may be also modified by addition of inert polymer and broad range of other additives.

3. Experimental and results

The investigations were focused on testing photosensitive polymer pastes: conductive silver and dielectric cross-over, as well as their compatibility. All paste were base on in-house synthesized epoxy acrylate prepolymers and commercial photoinitiator, reactive thinners and other additives.

Deposition of the pastes. Screen printing was applied to deposit the pastes. Polyester screens with mesh 80 T (65 μm) and 10 μm emulsion thickness were used. The paste were deposited on two kinds of substrates: flexible polyester (Mylar type) and rigid epoxy glass. The printed layers were exposed to UV after a short period of leveling without intermediate drying.
Curing of the layers. The layers were irradiated with UV curing from Phillips lamp HOK 20/100, 2000 W, radiant power in the range 320-440 nm - 150 W. The influence of irradiation time on the layers properties was studied. The optimum of other curing conditions, such as the distance between the lamp and the layer, curing temperature (efficiency of the cooling system), were also determined.
Test patterns. Sheet resistivity of conductive layers were measured with use of simple meander 200 μm wide and 200 squares long. Tests for sheet resistivity were printed on two types of substrates: polyester foil and epoxy glass. Thickness of the layers was about 20 μm.
Migration of silver ions between two parallel silver lines (Fig. 4.) [4] was investigated:

- the two lines were "connected" with a droplet of distilled water,
- voltage (5 VDC) was applied,
- the time between applying voltage and occurrence of short-circuit was regarded as silver ion migration failure.

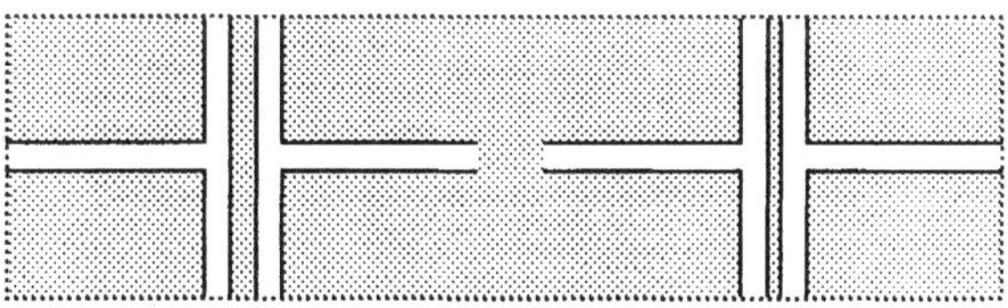

Fig. 4. The test pattern to measure migration of silver ions between two parallel silver lines; the distance between the line: 500 and 1000 μm

Vertical migration, through the screen printed dielectric layer (Fig. 5) [5], was also examined. The latter test pattern was also used to examine dielectric constant, K, dissipation factor, Tgδ, dielectric strength. The samples prepared according to test shown in Fig. 5 were stored in high humidity with elevated temperature (500 h in 40°C, 90% RH with 25 V bias). The thickness of single dielectric layer as well as conductive layer was about 20 μm. Exposure time for dielectric layer was 5 to 10 seconds.

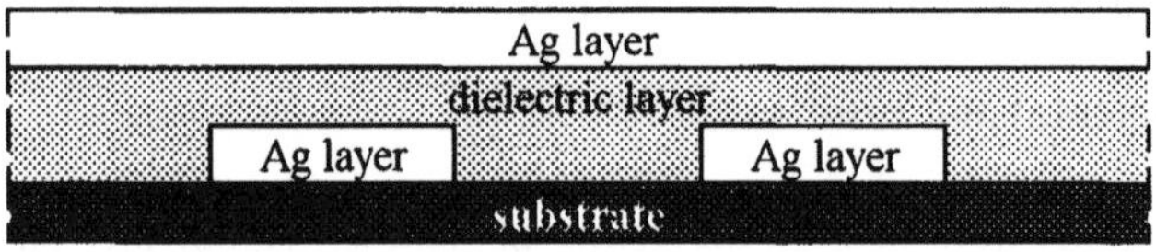

Fig. 5. The test pattern to measure vertical migration of silver ions through the dielectric layer

Results. The obtained results are shown in Fig. 6, Fig. 7 and Table 2.
The conductivity of photosensitive polymer silver paste depends on the kind of substrate and UV exposure time. Lower sheet resistivity of silver conductor was obtained on polyester substrates, after even shorter time of exposure. Further UV exposure does not affect conductivity of the silver layers, above: 60 seconds on polyester substrate and 100-120 seconds on epoxy glass. Thicker layers (>20 μm) are difficult to cure, whereas thinner layers have higher sheet resistivity.

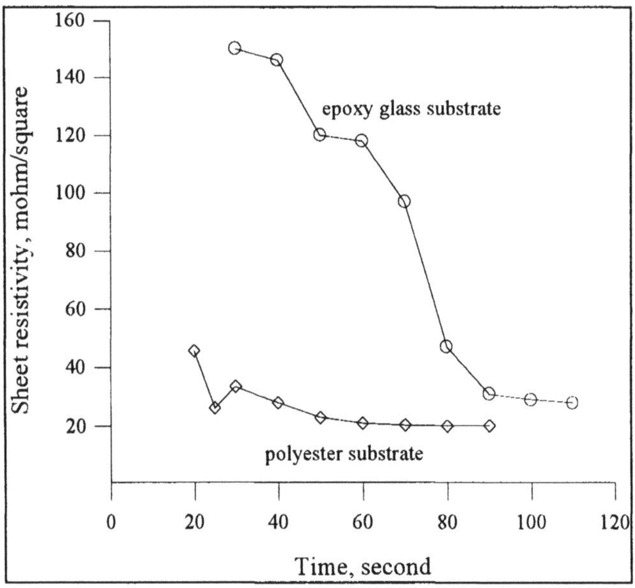

Fig. 6. Sheet resistivity of silver layer on different substrates versus UV exposure time

The applied surface migration test is a very severe, and corresponds to condensation of water vapor between the lines. The silver migration of polymer silver layer is somewhat slower then silver cermet layers. The polymer silver layers protected with the elaborated dielectric layer did not show any sensibility towards placed water droplet during measuring time (10 minutes).

Table 2. Migration of silver ions between two parallel silver lines; the distance between the line: 500 and 1000 μm

Type of tested layer	Short circuit time, seconds	
	Gap between the lines	
	500 μm	1000 μm
Fired silver paste on alumina	30	130
Fired silver-palladium paste on alumina	150	300
UV cured silver paste on polyester	50	240
UV cured silver paste protected with same UV cured dielectric paste	>10 minutes	>10 minutes

One layer of the photosensitive dielectric paste do not provide the cross-over with sufficient resistance of isolation. Double printed layer of the dielectric paste is satisfactory. There is no indication of a substantial advantage of triple-print over double-print layer.

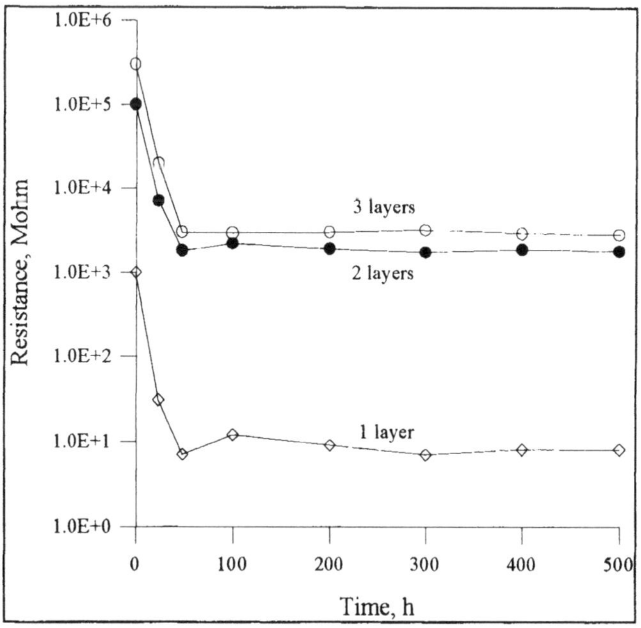

Fig. 7. The insulation resistance of dielectric layers versus test time

4. Conclusions

The proper selection of photosensitive organic vehicle, supported by in-house synthesis of epoxy acrylate resins, was the key to overcome difficulty of UV irradiation of conductive pastes. The use of suitable silver powders is beneficial as well.

A photosensitive dielectric paste compatible with the conductive one has been elaborated.

These pastes can be applied to perform electric circuitry in many devices such as membrane switches, keyboards and flexible cabling, low temperature heating elements, sensors, etc. The main advantages of their application are: low curing temperature enabling to apply even quite sensitive to temperature substrates (polyesters foils, paper, etc.), low cost and simplicity of equipment and very short production cycle.

A phtoimageable version of these pastes is under elaboration. This may lead to an uniform technology of manufacturing MCM.

5. References

1. Halbo, L. et al. (1990) Polymer Thick Film Technology (PTF), Electronik Centralen, Horsholm Denmark.
2. Harsanyi, G. et al. (1995) Polymer Films in Sensor Applications, Technical Publishing Company Inc., USA.
3. Horie, K. and Yamashita, T. (1995) Photosensitive Polyimides, Technical Publishing Company Inc., USA.
4. Szymanski D. and Szczytko B., (1982) Metody pomiaru parametrow aplikacyjnych past i warstw grubych, Prace ITME, cz. I. pp. 38-39.
5. Johnsson P., (1991) PTF Technology, Proc. of 8th European Hybrid Microelctronic Conference, Rotterdam, May 28-31, 1991, pp. 147-154.

Characteristics of Passive Ta-Resistive Planes Embedded in Al-Sheet "PCB"-compatible for integrated MCM-Substrates or Packaging Carriers

/Philipp Philippov, Milka Rassovska, Radosvet Arnaudov, Vassil Ianev/

TU-Sofia, Dept. KTPPME
Darvenitza, Bl. 1
1000-Sofia, Bulgaria
Tel (+359 2) 9876623 Fax (+359 2) 683125, E-mail fif@vmei.acad.bg

Abstract

The contemporary progress in the investigations of Al/Al_2O_3systems' material parameters leads to an increase of the possibilities, using embedded TaO_XN_{1-X} layers.

Using Al-sheets as mechanical strength carriers in combination with vacuum-deposited Al-layers and electrochemically anodized Al_2O_3 structure suitable for studies. This was found to create a periodic multilayer Al/Al_2O_3 structure suitable for the construction of this type of "PCB"-compatible for integrated MCM-Substrates or directly as Packaging. Material qualities of this system allow optimization in order to achieve high speed of data processing and signal propagation.

The existing studies on using Al and Ta combination as well and the good resistive qualities of the modified TaO_XN_{1-X} layers have shown satisfactory results, it can be concluded that the development of this new layers combination is possible in the multilayer carrier structures. Some preliminary research studies show a proper adhesion and satisfactory characteristics of the two integrated resistive planes in the multilayer combination $Al/Al_2O_3//TaO_XN_{1-X}/Ta_2O_5/Al$.

1. Introduction

Several research papers at recent times are dealing with the design possibilities of metal substrate carriers applicable, according to their mechanical strength, as packaging base, or commonly speaking, as MCM chip carriers [1, 2, 3], (figure 1).

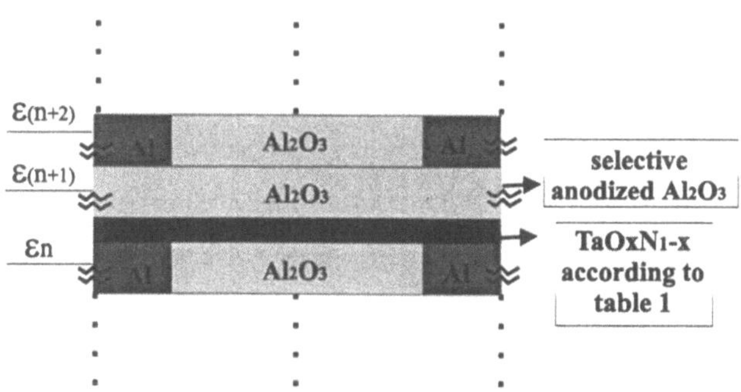

Fig. 1. Cross-section of thin-film TaO_XN_{1-X} resistors buried in the multilayer Al_2O_3 structure on a common Al metal carrier

R. Tummala et al. (eds.), Electronic Packaging for High Reliability, Low Cost Electronics, 151–156.

In the case, when we use Al-Sheet, it is shown, that the development of electrical connections is possible, in which the conducting parts consist of Al, and the isolating ones-of Al_2O_3 [2]. The peculiarity is, that they are obtained through selective anodization of vacuum deposited thin Al film. Thickness of 2-5μm is suitable, if the residual mechanical stress is monitored, during processing of the main technological steps-vacuum deposition of Al and its anodization, attended by lithography.

The comparison with multilayered printed circuit boards is possible, but numbers of the layers is limited up to 5, due to the above mentioned reasons.

Because of the anodization of thin films, there is a possibility to integrate the knowledge about material qualities of thin film resistive layers with systems, consisting of double such, The basic trends in this choice are the technological operations not only in resistive film deposition, but the chance to "embed" this film in the whole multilayer combination of carrier's structure.

2. Theory aspects

2.1. Design of the R-Level

The R-Level design contains two research activities. The one concerns the choice of the thin film. When compare NiCr and TaO_XN_{1-X} for example, which have approximately similar specific resistance ρ and sheet resistance Rs (100-150Ω), the advantage is on the Ta-layer's side, because it can be "embedded" in the complex system , by means of anodization. Besides, the Rs could be adjusted through partial anodization. Because the anodization is in the technological process basis of the investigated carrier, this circumstance determines the choice of TaO_XN_{1-X} resistive component. It is known [10], that two different resistive layers could be applied sequentially (one above the other), as the low-resistant gives the beginning of the high-resistant. Such a suitable one is the cermet ($Cr.SiO_2$) film with Rs in the range of 1000-1500Ω. The advantage of the combination $TaO_XN_{1-X}/Cr.SiO_2$ is in that case, when the resistor is realized by means of high-resistant layer and the lower-resistant is "removed" by anodization, we have "embedded" structure again (figure 2).

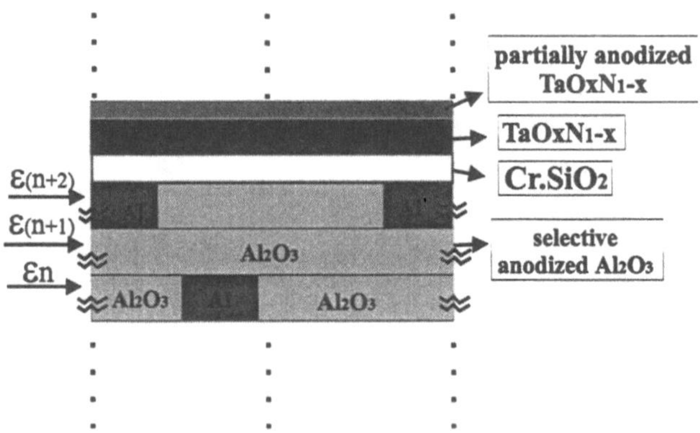

Fig. 2. Cross-section of buried thin film cermet (Cr.SiO2) and partially anodized TaO_XN_{1-X} resistors in the multilayer structure on a common Al-sheet metal substrate

The second research activity concerns the lay-out and the organization of the R-layer. Accepting, that the MCM-carrier is universal, provided with assembling level of Al/Ni-Fe, it is possible to perceive the strategy of semi-custom IC design. In that meaning, the R-Level is universal for researches of initial features and fully customized, that is, developed on customer's assignation for mass applications. An exemplary organization is depicted in figure 3. The R-layer and the R-matrix of semi-custom character are shown in table 1. For each R-complex an electro-spark trimming is applied [8, 9].

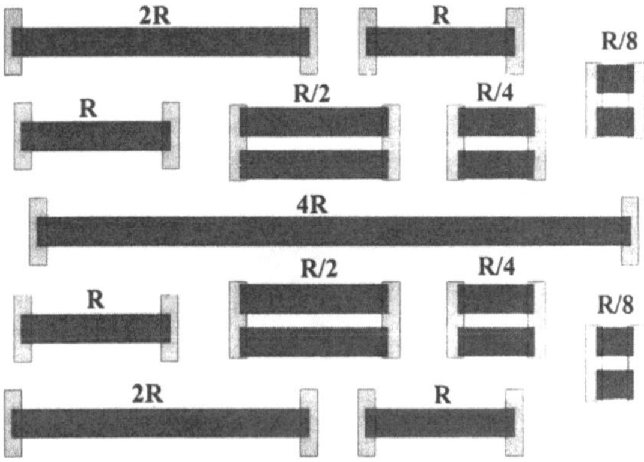

Fig. 3.Top-view of the exemplary R-matrix

Seed layer	TaO_XN_{1-X}
Layer thickness	d = 60 nm
Sheet resistance of the unetched layer	$R_F = (100\text{-}150) \pm 0{,}25\ \Omega/$
Resistance deviation	$\Delta R/R = 10\ \%$ without trimming
Time stability of the resistance after 4000h at 175°C	$\Delta R/R_0 < 0{,}2\ \%$
Transient resistance of the interface resistive film / contact pads	$R_{Ü} \leq 1\ \Omega$
Line and pitch widths of the meander	$l_{min} = 50 \pm 2\ \mu m$
Resistance density	$r = 1\ M\Omega/cm^2$
Temperature coefficient	TCR = -150 ± 5 ppm/K
Resistor to Substrate breakdown voltage	200 V
Resistor to Substrate Capacitance	-------

Table 1 : Some characteristic values of Ta thin film resistor parameters

2.2. Electrophysical parameters and deposition of the R-layer

The TaO_XN_{1-X} layers are produced by means of reactive sputtering. For this purpose, a vacuum equipment type A400VL-"Leybold AG", is used. The gas mixing system provides exact adjustment of the quantity of each gas component in the gas mixture, its homogeneous structure and precise dozing. The analysis of the ions, presented in the plasma discharge is carried out by mass-spectrometer "Q-200". For the definition of the film crystal structure is used röntgen diffraction with Cu-Kα emission.

In the conditions of depositing nitrogen-tantalum compounds, there exists the so called "plateau", where the layer resistance and its TCR does not depend on the nitrogen density contents in the sputtering gas, which circumstance is a great technology advantage. When measuring the specific layer resistance [ρ], a graphic dependence of ρ versus partial nitrogen tension is given (figure 4).

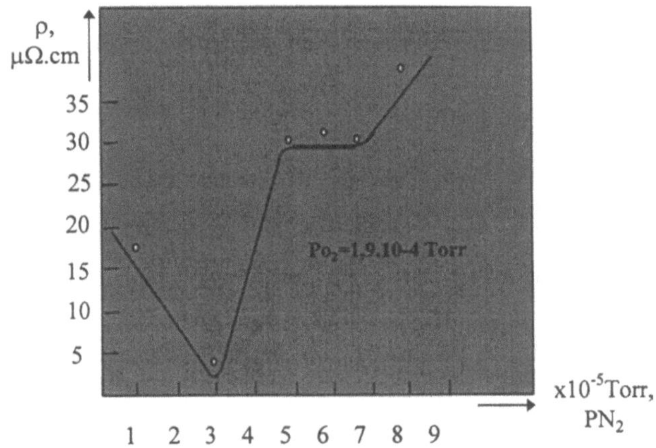

Fig. 4. Dependence of the bulk resistance ρ versus the partial nitrogen tension

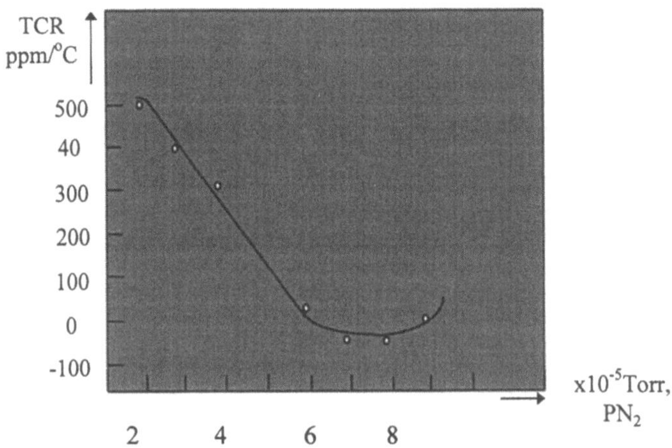

Fig. 5. Dependence of the temperature coefficient of R versus the partial nitrogen tension

Another especially important electrophysical parameter of the thin Ta films, doped with nitrogen, is the TCR. At the "plateau" region its value is minimum. The experimentally developed dependence is shown on figure 5.

It is known, that thin tantalum films, according to the conditions of deposition, possess different crystal structure and electrophysical properties. The proposed electroconducting model of thin resistive films [6], give us the ability to derive analytical dependence of chemical contents versus the structure and physical parameters, more exactly the influence of nitrogen and oxygen doping profiles upon the thin film characteristics (in their depth), (figure 6). The carried out investigations use electron Augerspectroscopy, combined with SIMS. The experimental results lead to the obtaining of doping profiles, depicting the ratio of Ta, O_2, N_2 in the complex alloy substance. Due to the strong getteric properties of Ta, a surface increase of O_2 concentration is observed.

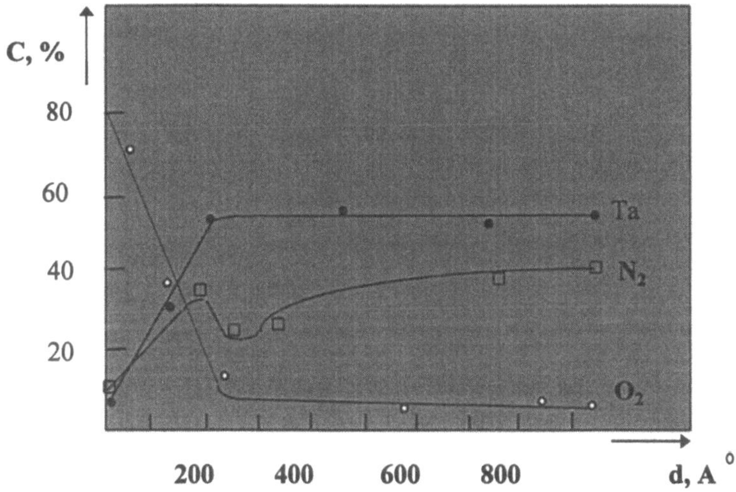

Fig. 6. Dependence of structural chemical contents vrs. Layer thickness

3. Conclusion

A novel developed multilayer MCM (D) structure, including-buried thin film resistors, selectively anodized Al_2O_3 isolated layers and Al-sheet metal substrate, is concerned. We should notice, that the versatility of the proposed structure relies on the unique properties of tantalum to produce, either resistive, or isolated layers, through reactive sputtering and selective electrochemical anodization. Moreover the resistors' contact pads are situated in the bottom layer, adjacent to the resistive one. Further investigations in this direction are conducted. Also, the sequential combination of two different resistive layers, with partial anodization of TaO_XN_{1-X} (fig. 2), obtains new features and possibilities distributed RC "embedded" structures in MCM (D) applications.

References

[1] Philippov, Ph., R. Arnaudov, S. Tzanova, "Some Investigations on Wiring in New Generation MCM Applied in High Speed LSI Telecommunication Units", 3. High Technology, Vol. 16, NATO ASI Series 0-7923-4194-5, Kluwer Academic Publisher, pp.101-124, 1996

[2] Arnaudov, R., N. Yordanov, Ph. Philippov, " Processing and Performance of High Dielectric Permeability Thin Anodized Films in Multilayer Structures for MCM-D High Speed Applications", 4. High Technology, Vol. 17, NATO ASI Series , Kluwer Academic Publisher, 1997 (in press)

[3] Philippov, Ph., R. Arnaudov, N. Yordanov, T. Genkova, "Using of Al-Sheets for Manufacturing of Electronic Packages with Consideration of High Frequency Application in MCM, Proceedings of 11th European Microelectronic Conference, Venice, Italy, May 14-16, 1997

[4] Philippov, Ph., R. Arnaudov, N. Yordanov, "The Metal Substrate - a New Challenge to Design and technology of MCM (D) for High Speed Applications", Part 1, Annual School . Lectures, Vol 17, N 1, Sozopol, Bulgaria, 1995/96 (to be found in Technische Bibl. In Hanover, Germany, ZY804)

[5] Rassovska, M., Ph. Philippov, "Untersuchung einiger Electrophysischer Parameter der mit Stickstoff Legierten duennen Tantalschichten", XI Intern. Tagung, 19-21.02.1986, Dresden

[6] Rassovska M. "Influence of Additional Gas Components of Electric-Physical Properties of Vacuum Deposited Thin Films", PhD Thesis, Sofia, 1987

[7] Philippov, Ph., "Application of Isolated Metal Substrates for Multichip Module Packaging", private correspondence

[8] Ph. I. Philippov, "Method and equipment for resistance adjustment", B.G. Patent, 87412, February 24, 1989

[9] RLA and RFA Linear Bipolar and BiFET Macrocell Array Semicustom Program, Raytheon Company Semiconductor Division catalogue

[10] Rahnev P., Philippov Ph., Parashkevov D., "Simultaneous preparation of Cr-SiO and NiCr thin film resistors on single substrates", Thin Solid Films, 1986

USE OF GRAVURE OFFSET PRINTING IN MANUFACTURING OF ELECTRONIC APPLICATIONS

S. I. LEPPÄVUORI and A. K. UUSIMÄKI
Microelectronics and Material Physics Laboratories,
EMPART research group of Infotech Oulu,
University of Oulu,
FIN-90570 Oulu, Finland

1. Abstract

In gravure offset printing (GOP) the patterns to be printed are etched in a plate. The etched patterns are filled with ink and then the inked patterns transferred to substrates by picking them up using a flexible silicone rubber pad or a roller. GOP offers flexible opportunities to produce lines narrower than those printed by conventional screen-printing. It also offers the possibility of printing on non-planar objects. These properties create a novel realisation technique for cost-effective high density circuits and multichip modules and also for the miniaturisation and integration of sensors.

There are only a few investigations reported in the literature of GOP techniques in electronics manufacturing, namely their use in applications of thermal printing heads and active matrix liquid-crystal display colour filters. Applications investigated in the Microelectronics Laboratory of the University of Oulu, Finland include applying the GOP techniques to packaging solutions such as three dimensional printing and the preparation of multilayer inductors on low temperature co-fired ceramic (LTCC) substrates. GOP has also been used to realise microstrip lines up to 50 GHz and in sensor structures such as miniature resistive temperature sensors and one-electrode SnO_2 thick-film gas sensors.

2. Introduction

There is a growing need in the electronics industry for a cost-effective, rapid technique to produce fine-line and ultra fine-line patterns (line widths from 150 μm down to 30 μm) which can be realised both on planar and non-planar surfaces. Miniaturising of electronic devices and circuits such as MCMs, integrated sensors and microwave components would derive real benefit from these novel technical advancements. Existing lithographic and etching processes of both thick and thin films, which incorporate several steps, cannot be developed to meet these demands cost-effectively. They are rather slow and expensive processes and are also environmentally unfriendly.

R. Tummala et al. (eds.), Electronic Packaging for High Reliability, Low Cost Electronics, 157–160.

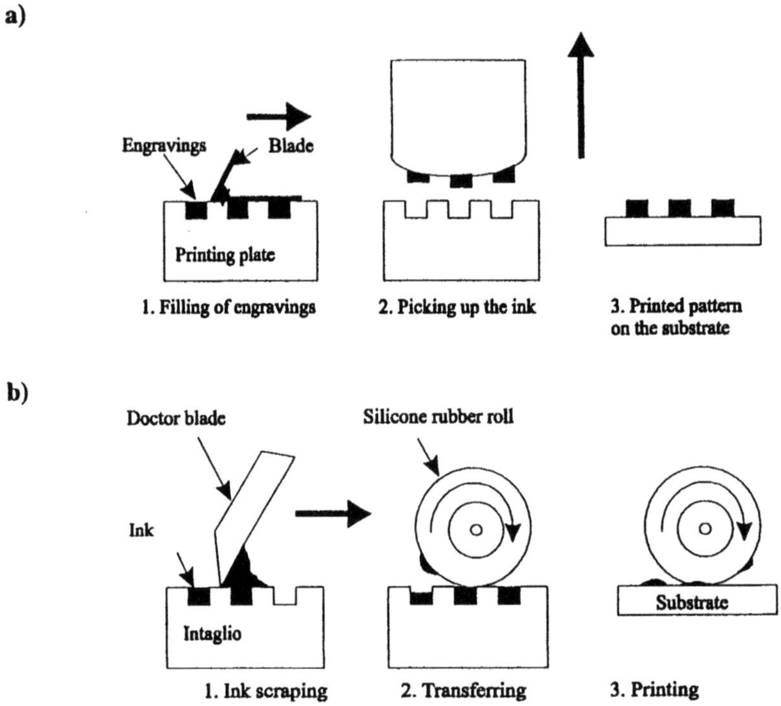

Fig. 1. a) Principle of the gravure-offset tampo printing, b) Principle of the gravure-offset roller printing

Gravure offset printing techniques (GOP), appearing as different variations, are, according to the Japanese, the most novel, flexible and cost effective printing solution for electronics manufacturing [1, 2]. The principle of the GOP technique is shown in Fig. 1. To our knowledge, one of the first to introduce this technique on a laboratory scale for electronic applications in Europe was the Microelectronics Laboratory in the University of Oulu, Finland [3, 4]. Lines with widths and spaces of 50 μm were studied. The Japanese have developed a modified GOP fine line technology, calling it a film intaglio transfer printing technology [5]. In this technology the paste is transferred directly from a plastic intaglio with heat and pressure, and, using this technique, they have developed a ceramic package BGA carrier. Special benefits deriving from this technique are fine patterns, thick films, sharp edges and fine tolerances (L/S=3D20/40 μm with an aspect ratio of 1:1 (thickness of the lines up to 20 μm)) [5]. The Japanese have also attempted to make the resist patterning of thin film transistors for liquid crystal panels using GOP with glass etched intaglio and silicone as a blanket [10]. They successfully employed a double-layer resist technique to pattern transistors down to a minimum design gate length of 10 μm. In emerging LTCC packaging technology the use of GOP fine-line patterning in multilayer structures has become an attractive research objective. Also, the surface fine-line metallization of LTCC modules (and high density PWBs) is attractive; for example, I/O pads for direct chip-to-substrate connections of LTCC modules, as outlined in ref. [6]. Gravure offset printing would have potential for these applications.

3. Applications

Planar inductors on LTCC substrate

GOP was used to prepare LTCC multilayer inductors [7]. The size of the printed coils was 3 mm x 3 mm with line widths between 80 μm and 325 μm and they consisted of one or two layers. The inductances were up to few hundred nH.

Microstrips on alumina substrate

In microwave applications narrow lines or spaces are needed. The quality of the edge of the conductor is important from the point of view of losses. For comparison, identical microstrip lines were fabricated using the different methods of three manufacturers [8]. One method was GOP using silver ink and the two others were direct copper bond (DCP) methods with subsequent etching of the patterns and thick-film screen-printing with light sensitive silver paste followed by etching and firing. Attenuation properties of the GOP silver microstrips were comparable with those realised by the other two more complicated methods. This means that the GOP technique is a competent method to use up to mm wavelengths.

Temperature sensors on LTCC substrate

In temperature sensors, a small size means savings in material and labour expenses. The response time of the sensor is also shorter. GOP printed platinum conductors were used to realise cost effective resistive temperature sensors on LTCC substrates [9]. Modified thick-film platinum pastes were used as gravure inks. The conductor line width of the temperature sensor was 75 μm and the thickness typically from 2 to 3 μm. The size of a sensor element was 3.5 mm x 3.5 mm. The temperature coefficient of resistance (TCR) was about 3560 ppm/°C with resistance values of about 100 Ω, which were reached using double printing. This miniature resistive temperature sensor has applications also, for example, in one-electrode SnO_2 thick-film gas sensor structures.

4. Future aspects on the GOP technique

In the future there is an increasing need to develop inks for a wide range of applications; today they have to be tailored from thick-film pastes. As an example, high density PWB technology including integration of high density passive components is a very attractive near-future mass production possibility. Both rigid and flexible boards with polymeric GOP inks are of interest. There are also challenges concerning the precision printing machines.

5. References

1. M. Shimada, H. Watanabe, M. Tsukamoto and T. Ishida, Manufacturing of Fineline Films by Printing Technique, Proceedings IMC 1990, Tokyo, May 1990, pp. 581-586.

2. K. Nakao, M. Hayama, K. Kuramasu, M. Shimada. and M. Tsukamoto, Giant Micro-Electronics Technology Using Gravure Printing Techniques, extended abstracts of the 1991 International Conference on Solid State Devices and Materials, Yokohama, (1991), pp.599-601

3. S. Leppävuori, J. Väänänen, M. Lahti, J. Remes and A. Uusimäki, A novel Thick Film Technique, Gravure Offset Printing, for Realisation of Fine Line Sensor Structures, Sensors and Actuators A, Nos 41-42, pp. 593-596 (1994)
4. S. Leppävuori, J. Väänänen, M. Lahti, K. Kukkola and A. Uusimäki, The Use of Gravure Offset Printing in Realisation of Fine Conductors, Proceedings Materials Research Society Symposium, No. 323, pp. 85-90 (1994)
5. H. Matsunaga, C-BGA Package using FIT printing technology, The third VLSI PACKAGING WORKSHOP of Japan, Dec. 2-4 1996 Kyoto, Technical digest pp. 161-163.
6. C. L . and D. J. Smith, Fabrication Approaches for Surface Fine Line Metallization of LTCC Modules, The international Journal of Microcircuits and Electronic Packaging, Volume 18, Number 2, Second Quarter 1995 (ISSN 1063-1674), pp. 169-178.
7. M. Lahti, S. Leppävuori, V. Lantto and K. Kukkola, Inductors on an LTCC substrate realised by the gravure offset printing technique, Proceedings of the 11th European Microelectronics Conference (1997), 436-443.
8. E. Kemppinen, P. Mikkonen and S. Leppävuori, Performance of Cu and Ag based microstrips up to millimeter wave frequencies, Proceedings of the 11th European Microelectronics Conference (1997), pp. 100-108.
9. M. Lahti, Y. and S. Leppävuori, Realisation of Temperature Sensors on Low-Temperature Co-fired Ceramic Substrates, Proceedings of the 11th European Microelectronics Conference (1997), pp. 554-557.
10. S. Okazaki, K. Asaka, K. Nakamura, T. Saito, Y. Nagae, Y. Akimoto, E. Kaneko, A TFT Fabrication by Gravure Offset Printing without Mask Aligner, SID 91 DIGEST (1991), p. 563.

INVESTIGATION OF THICK FILM RESISTORS FOR A MULTILAYER SYSTEM

M. HROVAT, D. BELAVIČ*, D. ROČAK, J. FAJFAR PLUT*

Jožef Stefan Institute, Jamova 39, 1001 Ljubljana, Slovenia
*HIPOT, Trubarjeva 7, 8310 Šentjernej, Slovenia

ABSTRACT

The electrical characteristics and stability under different ageing conditions of two series of thick film resistors designed for firing with a 30 minute cycle on a multilayer dielectric, i.e., QM 80 and QM 90, Du Pont, were evaluated. Some samples were also printed and fired on alumina substrates. The influence of repeated firing and the length of the firing cycle on sheet resistivity was tested. Temperature coefficients of resistivity and noise indices were measured.

1. Introduction

The complexity and therefore the density of thick film hybrid circuits are continually increasing. High component densities require multilayer construction of the thick film circuit and the possibility to print and fire thick film resistors on top of a dielectric layer instead of on an alumina substrate. "Ordinary" resistor materials are developed for firing on 96% Al_2O_3. If they are printed and fired on a multilayer dielectric, their characteristics, for example sheet resistivities or temperature coefficients of resistivity (TCR), are changed due to interactions between the resistor and dielectric, and in some cases can be shifted out of the required limits.

Du Pont developed the multilayer dielectric material QM-42 for use in multilayer thick film circuits with silver-rich or pure silver conductors. The material is based on a filled crystallizable glass and is densely sintered after firing ([1,2]). For the use with this dielectric other thick film materials were "designed". Some of them are the compatible resistor series QM 80 (Pd/Ag termination) and QM 90 (Ag termination), intended for firing on the dielectric layer. These materials are intended for firing with a short, 30 min. cycle to a peak temperature of 850°C for 10 min. ([3]).

In this paper some test results on both QM series are presented. Declared TCR values (see data sheets) are under 100×10^{-6}/K with the exception of 1 Mohm/sq. resistors, which have a TCR under 150×10^{-6}/K. Both series were printed on prefired QM 42 multilayer dielectric and fired with 30 and 60 min. firing cycle. To determine the influence of the substrate, resistors were also printed and fired on alumina substrates.

R. Tummala et al. (eds.), Electronic Packaging for High Reliability, Low Cost Electronics, 161–167.

Sheet resistivities, TCR (as fired and after ageing), noise indices, and resistivity drift after high temperature ageing were measured.

2. Experimental

Thick film resistor series QM 80 and QM 90 were printed and fired at 850°C on either prefired (850°C) QM 42 multilayer dielectric or on "bare" alumina substrates. QM 80 resistors were terminated by a QM 21 (Pd/Ag) conductor and QM 90 resistors were terminated with a QM 14 (Ag) conductor. A test pattern with resistors ranging in length from 1 mm to 8 mm was used ([4]). Samples were fired using 30 min. and 60 min. cycles with 10 min. at the peak temperature. Some samples were refired at 850°C (60 min. cycle) up to four times to evaluate the influence of refiring on sheet resistivities. Sheet resistivities, TRCs, and noise indices were evaluated as a function of substrate (alumina or dielectric) and firing time. Resistivity drift after high temperature ageing at 150°C for up to 5000 hours was measured.

3. Results and discussion

3.1. SHEET RESISTIVITY

Measured sheet resistivities for 1x1 mm^2 and 2.5x2.5 mm^2 resistors as a function of substrate (alumina or dielectric) and firing time are shown in Table I. Values are presented as percentages. The sheet resistivities of 1x1 mm^2 resistors fired on QM 42 dielectric with a 30 min. firing cycle are given a value of 100%. The values of sheet resistivities of resistors fired on "bare" alumina are between 50% and 70% of those fired on QM 42 dielectric. For resistors fired on dielectric, a longer firing cycle increases the resistivity of materials with sheet resistivities up to 1 kohm/sq., and decreases it for higher sheet resistivities. There is a difference of only a few percent difference between shorter and longer resistors.

Table I. Sheet resistivities vs. resistor length and firing time (%)

Resistor	R (ohm/sq.)	Dielectric			Alumina		
		30 min	60 min		30 min	60 min	
		1 mm	1 mm	2.5 mm	1 mm	1 mm	2.5 mm
QM-82	100	100	126	127	71	96	97
QM-83	1 k	100	135	137	56	88	90
QM-84	10 k	100	98	100	49	56	58
QM-85	100 k	100	89	90	69	58	62
QM-86	1 M	100	86	88	65	56	58
QM-92	100	100	105	106	55	58	61
QM-93	1 k	100	121	118	52	73	73
QM-94	10 k	100	99	90	42	45	46
QM-95	100 k	100	99	99	67	69	71

3.2. TEMPERATURE COEFFICIENTS OF RESISTIVITY

Cold (-25°C to 25°C) and hot (25°C to 125°C) temperature coefficients of resistivity were calculated from resistivity measurements at - 25°C, 25°C, and 125°C for unaged and aged (200°C, 5000 hours) resistors with dimensions of 1x1 mm^2. All TCRs of unaged resistors, fired either 30 min. or 60 min., were below $100x10^{-6}$/K. The resistivity vs. temperature curves for resistors fired on alumina and dielectric are shown in Figs. 1 and 2 for QM 83 (1 kohm/sq.) and QM 93 (1 kohm/sq.), respectively. For QM 83 the values are close together, while for QM 93 a longer firing cycle results in higher absolute tcr values.

Cold and hot TCRs of unaged and aged samples, fired on multilayer dielectric and on alumina are summarized for a 60 min. cycle in Table II. In most cases there is noticeable shift of cold TRCs to more negative values after high temperature ageing. Cold and hot TCRs of high resistivity members of the QM-80 series are more positive when fired on alumina. This is also true for QM-90, with the notable exception of QM-93.

Table II. Cold TCR (-25°C to 25°C) and hot TCR (25°C to 125°C) 1x1 mm^2, unaged and aged (200°C, 5000 hours) resistors (10^{-6}/K)

Resistor	R (ohm/sq.)	Dielectric		Alumina	
		unaged	aged	unaged	aged
QM-82	100	-20 / 20	-45 / 0	-40 / 0	-60 / -10
QM-83	1 k	-25 / 15	-35 / 15	-45 / 0	-55 / 0
QM-84	10 k	-65 / -10	-95 / -20	-60 / -20	-85 / -15
QM-85	100 k	-50 / 5	-70 / 10	10 / 65	5 / 60
QM-86	1 M	-80 / -10	-120 / -20	-35 /30	-55 / 20
QM-92	100	-30 / 10	-55 / -10	65 / 100	65 / 110
QM-93	1 k	-100 / -80	-135 / -5	-90 / -55	-125 / -75
QM-94	10 k	-80 / -30	-100 / -35	15 / 30	-25 / 35
QM-95	100 k	-40 / 20	-60 / 15	55 / 95	60 / 110

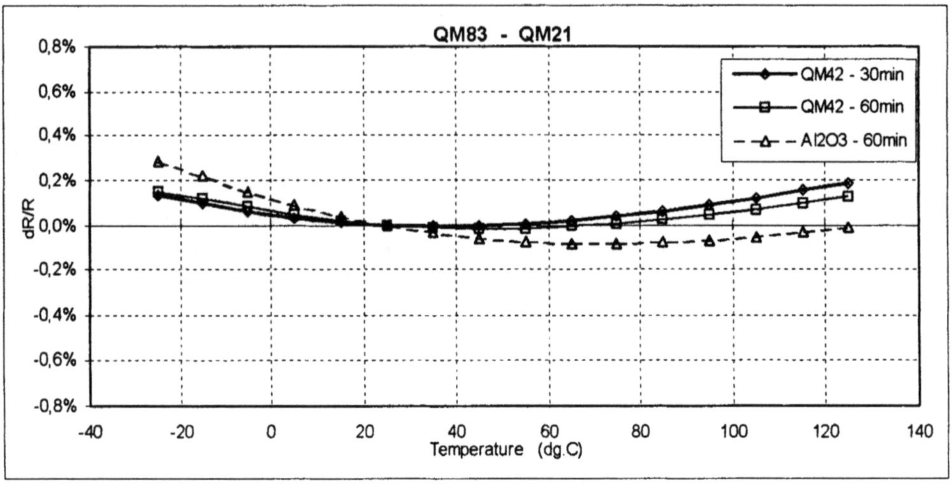

Fig. 1: The resistivity vs. temperature curves for resistors, fired on alumina and dielectric, for QM 83 (1kohm/sq., terminated by Pd/Ag, 1x1 mm^2), fired on alumina and dielectric.

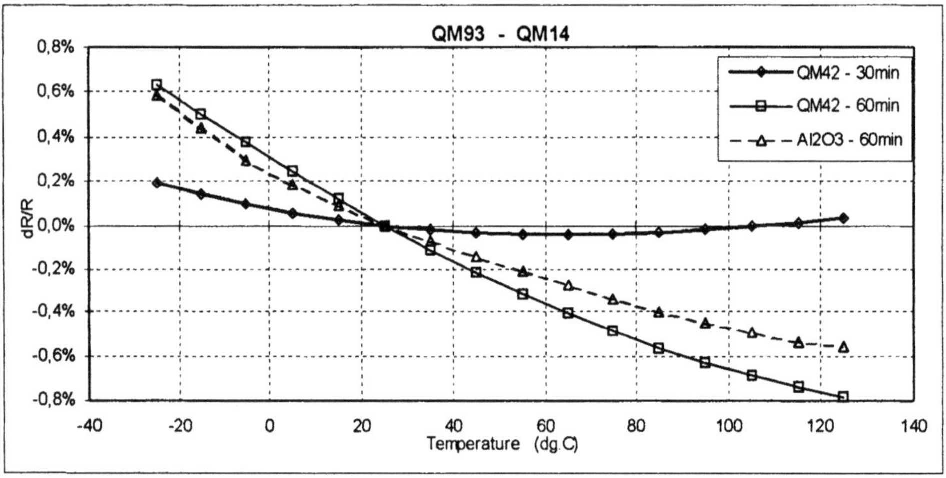

Fig. 2: The resistivity vs. temperature curves for resistors, fired on alumina and dielectric, for QM 93 (1 kohm/sq., terminated by Ag, 1x1 mm^2), fired on alumina and dielectric.

3.3. MULTIPLE FIRINGS

After repeated firing (850°C, 60 min) the sheet resistivities at first increased. Then the values on alumina tended to stabilize or decrease. It can be speculated that at least two "mechanisms" are responsible. One is the increase of sheet resistivity due to microstructural changes in the resistor, and the other is the influence of repeated high temperature reactions between the conductor and the resistor material, which presumably tends to decrease resistivity. On the other hand, sheet resistivities of resistors on dielectric increase and do not stabilize even after five refirings. A possible reason could be diffusion of dielectric material into the body of the resistor during firing.

Resistivity changes after refiring are higher for higher sheet resistivities and significantly higher on dielectric than on alumina. For example, for 100 kohm/sq. resistors the resistivity increases after 5 refirings by around 50% on alumina and 100% on dielectric. Typical examples for both 1 kohm/sq. pastes, i.e., QM 83 and QM 93, fired on alumina and on dielectric, are shown in Fig. 3.

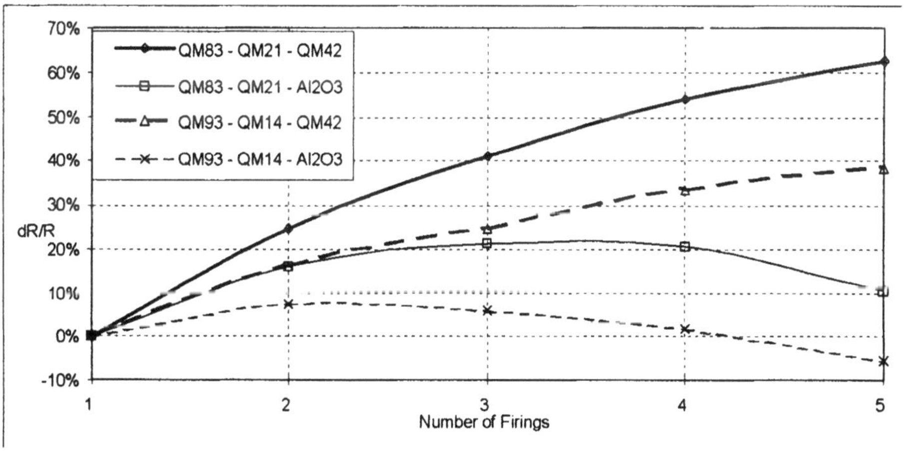

Fig. 3: Effect on resistivity of multiple firings of 1 kohm/sq. resistors at 850°C

3.4. NOISE INDICES

Noise indices are shown in Table III. The noise index of larger resistors is lower than that of smaller ones. In most cases the noise indices of resistors fired on dielectric or on alumina are comparable. A notable exception is QM-85 (100 kohm/sq.), where the noise index of resistors fired on alumina is significantly higher.

Table III: Noise index (dB)

Resistor	R (ohm/sq.)	Dielectric				Alumina	
		30 min		60 min		60 min	
		1 mm	2.5 mm	1 mm	2.5 mm	1 mm	2.5 mm
QM-82	100	-6.5	-8.5	-9.2	-10.0	-9.5	-8.4
QM-83	1 k	-8.6	-10.4	-0.5	-0.9	-10.8	-15.2
QM-84	10 k	-4.5	-7.8	-1.2	-0.7	-9	-10.6
QM-85	100 k			-2.6	-7.5	2.5	-1.4
QM-86	1 M			14.3	5	16.4	6.5
QM-92	100	-9.9	-9.4	-9.5	-9.7	-9.5	-12.5
QM-93	1 k	2.1	-5.2	-4.5	-5.9	-5.5	-7.3
QM-94	10 k	-6.7	-6.7	-5.6	-8.8	-9.6	-12.5
QM-95	100 k			-1.7	0.5	0.5	0.4

3.5. LONG TERM STABILITY

To evaluate long term stability, resistors were aged at high temperature (150°C and 200°C) for up to 5000 hours. Some results of resistivity drift after 5000 hours are shown in Table IV. The resistivity drifts after ageing at 150°C are under 1% for resistors on alumina and on dielectric, with the exception of QM 95 on dielectric. For the QM 80 series, aged at 150°C and at 200°C, resistivity changes tend to increase with increasing sheet resistivities.

Table IV. Resistivity drift after 5000 hours ageing at 150 and 200°C (%)

Resistor	R (ohm/sq.)	Dielectric		Alumina	
		150 °C	200°C	150 °C	200°C
QM-82	100	0.05	0.10	0.05	0
QM-83	1 k	0.15	0.25	0.10	0.05
QM-84	10 k	0.35	1.20	0.55	0.40
QM-85	100 k	0.35	1.20	0.75	0.75
QM-86	1 M	0.45	1.70	0.70	1.70
QM-92	100	- 0.05	0.15	0.25	0.60
QM-93	1 k	0.05	0.05	0.05	0.05
QM-94	10 k	0.95	0.60	0.05	- 0.35
QM-95	100 k	1.60	0.90	0.45	- 0.65

For 100 kohm/sq. member of the QM 90 series, which are terminated with silver, resistivity changes, after ageing at 200°C, are negative when fired on alumina. On dielectric the resistivity drift is lower after 200°C ageing than after 150°C ageing. This could be due to interactions between the silver conductor and the resistor material, presumably tending to decrease resistivity.

4. Conclusions

Electrical characteristics and stability under different ageing conditions of two series of thick film resistors, designed for firing on multilayer dielectric, i. e., QM 80 and QM 90, Du Pont, were evaluated. Some samples were also printed and fired on alumina substrates. The influence of repeated firing and the length of the firing cycle on sheet resistivity were tested.

The values of sheet resistivities of resistors fired on "bare" alumina are between 50% and 70% of those fired on QM 42 dielectric. For resistors fired on dielectric, a longer firing cycle increases the resistivities of resistors with sheet resistivities up to 1 kohm/sq. Cold and hot TCR for all resistors are below 100×10^{-6}/K. After high temperature ageing cold TRCs are shifted to more negative. values. Noise indices of resistors fired in dielectric or on alumina are similar. After repeated firing the sheet resistivities at first increased. Then the values on alumina tended to stabilize or decrease, while the resistivities of resistors on dielectric increased and did not stabilize even after five refirings. The resistivity drifts after high temperature ageing at 150°C are under 1% for resistors on alumina and on dielectric, with the exception of QM 95 on dielectric. For lower resistivity members of the QM 90 series, which are terminated with silver, resistivity changes after ageing at 200°C are negative when fired on alumina.

5. Acknowledgements

The authors would like to express their thanks to Mr. Alan Buckthorpe, Du Pont, for arranging the firing of some samples in the Du Pont facilities in Bristol, England. The financial support of the Ministry of Science and Technology of Slovenia is gratefully acknowledged.

6. References

1. W. A. Craig, C. R. Pickering, M. F. Barker, J. Cocker, P. C. Donohue, G. Vanrietvelde, High yield, high reliability materials system for silver multilayer construction, Proc. 8 th European Hybrid Microelectronics Conf. ISHM-Europe 91, Rotterdam, 1991, 188-195
2. C. R. S. Needes, Environmental durability of silver multilayer circuits, Proc. Int. Symp. Microelectronics ISHM-94, Boston, 1994, 173-178
3. Du Pont data sheets, QM System, A silver based system for low cost multilayers, Du Pont Electronics, H-34375 (2.5 M) 5.91
4. Standard DIN 41850, Part 4

RELIABILITY ASPECTS OF THE BATCH PRODUCTION BASED ON PTF TECHNOLOGY

MILOS SOMORA, JAN URBANCIK
FRANTISEK KOLESAR *),
Technical University of Kosice, KHM-FEI
Park Komenského 2, 040 01 Košice, Slovak Republic
*) INTERMIKRA s.r.o.
Vysokoškolská 4, 042 00 Košice, Slovak Republic

1. Introduction

Following of the expansion of the Polymer Thick Film (PTF) technology into electronic production a great deal of devices used the polymer materials and interconnectionsa rised everyday. The reliability of used materials and the accuracy of production technology including the treatment processes have an important influence on the electrical features, on the reliability and the quality of these devices.

This paper deals with several reliability aspects of microelectronic production based on the multilayer Polymer Thick Film technology using flexible foil as substrates. The aim of this contribution is to follow the influence of technology changes, technology, design and layout improvement upon the device characteristics and monitoring of typical device failures arised in manufacturing process. There were monitoring failures caused by technological reasons such as screen print, treatment processes, choice of used materials or implementation of the automation into manufacturing process. Also, there were evaluated failures caused by accuracy to hole shapes, printing of two conductive, dielectric and one carbon layers, electrical interconnections, crossovers, broken lines, shortcuts, high voltage shortcuts and further parameters.

This paper discuss some experiences with production process based on more than 150 thousand pieces of the manufactured PTF based multilayer keyboards and some electrical parameters and characteristics of used polymer inks. These material characteristics were obtained by measurements using accelerated life tests applicated on simply testing layout.

R. Tummala et al. (eds.), Electronic Packaging for High Reliability, Low Cost Electronics, 169–177.

2. Accelerated life tests

The polymer materials used for keyboard production are characterized by macromolecular chain structure with a lot of mutual chemical binding. Either the disintegration of the macromolecular chain or the changes of the chemical structures followed by the chemical reactions of arised fragments are phenomena going along with ageing process in all PTF structures. Consequently, the ageing process may cause negative influence to their electrical and mechanical facilities. The depolymerization process usually causes the decrease of the mechanical resistance, electrical stability, flexibility and adhesion of PTF layers to flexible substrates. Further more the unsuitable working or storing conditions may cause that ageing process is accelerated. In term of electrical properties, it represents decrease of the electric strength, increase of the dielectric loss and electrical resistance changes. Therefore it is important to have some information regarding the ageing process in used materials, too.

The speed of chemical reactions in the depolymerisation process mentioned above depends on the macromolecular structures and it is influenced mainly by the temperature and by the humidity. The accelerated life ageing is based on increasing of degradation speed which enables estimate lifetime data and determines the properties with regard to long time processes. The basic acceleration value for physical and chemical reactions is the temperature. Considering the type of the manufactured product and in compliance with another experiences with ageing process and technical standards the testing structures by figure 1 were exposed to effect of temperature 150 dgr. of Celsius at 1000 hours.

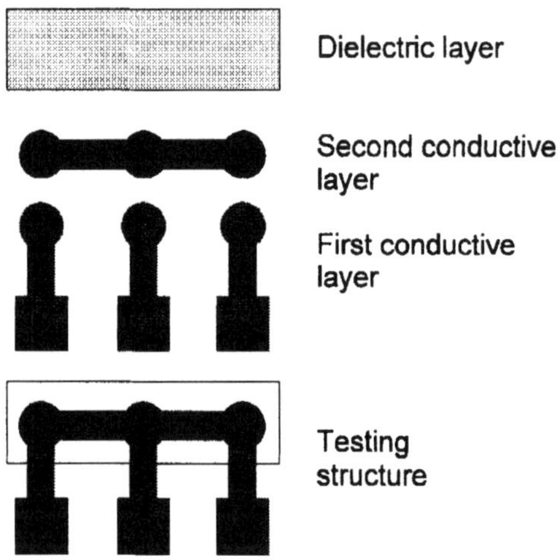

Figure 1. The layout of the tested structure.

The measurement method for testing was chosen from the group of electrical nondestructive test methods and it was measured the interconnection resistance between two conductive thick film layers using the Kelvin method. This method eliminates all of the interferences from background along with respect of the starting quality, production technology, curing parameters, nonspecified contamination and chemical influences on the active area. The electrical measurements were realized in seven ageing intervals 0, 20, 50, 100, 200, 500 and 1000 hours.

There were tested and compared two polymer inks used in the keyboard production. Both inks were manufactured by COATES, conductive ink XZ 250, 30 mOhm per square at 25 um thickness and ink which arised as mixture of conductive ink mentioned above with the well compatible Carbon cover ink in the 75% to 25% volume ratio. Acquired results in form of the graph continuance represents measured values processed by the statistics methods. Each amount on charts shapes average value from 4500 measurement. There were measured six substrates in each ageing interval. Every one of them consist of 20 basic testing structures by fig 1. and all of structures were subjected to measurement of the potential drop for 45 different values of constant current in range from 50 to 500 mA. The measurement results were represented by the AV characteristics and by the one average value of the interconnection resistance. Typical AV charts with calculated values of the resistance and its average value is drawn on figures in enclosure. Jointly during the ageing time there were obtained and evaluated more than 60 thousand measured values. The obtained results for both material combinations are displayed on the figures 2 and 3.

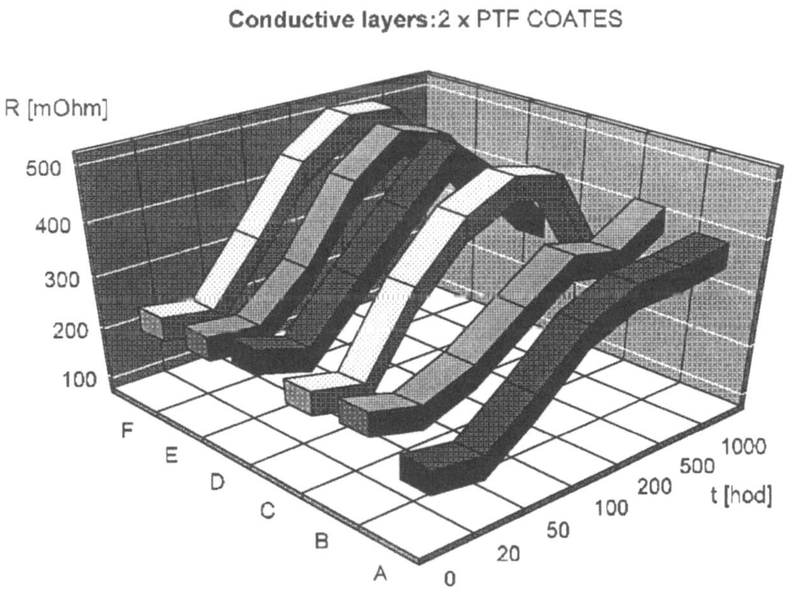

Figure 2. The continuances of the interconnection resistance for combination of two conductive PTF inks COATES.

Following these tests it can be evaluate the rate of materials compatibility and also the influence of production technology changes including the treatment process and starting quality in term of durability and reliability. Test method also brings basic information regarding choice of input material combination in term of long time processes.

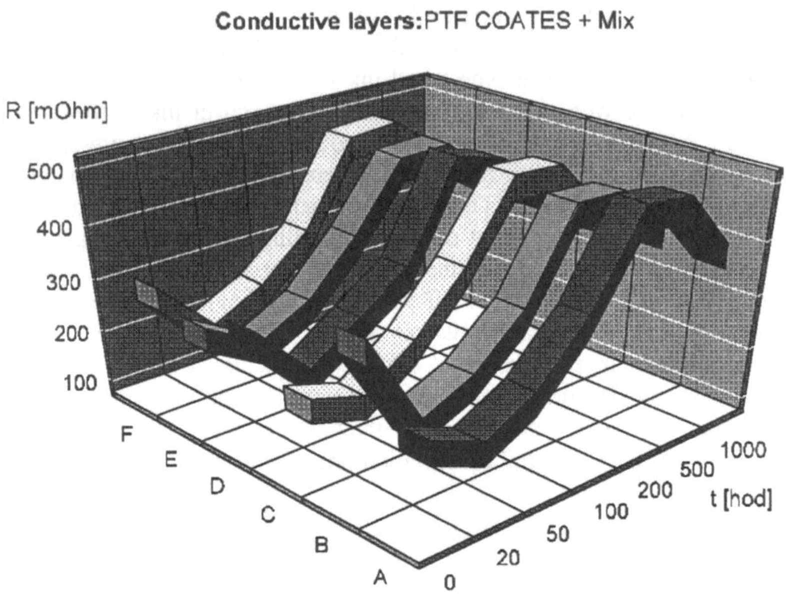

Figure 3. The continuances of the interconnection resistance for combination conductive ink PTF COATES with Mix ink.

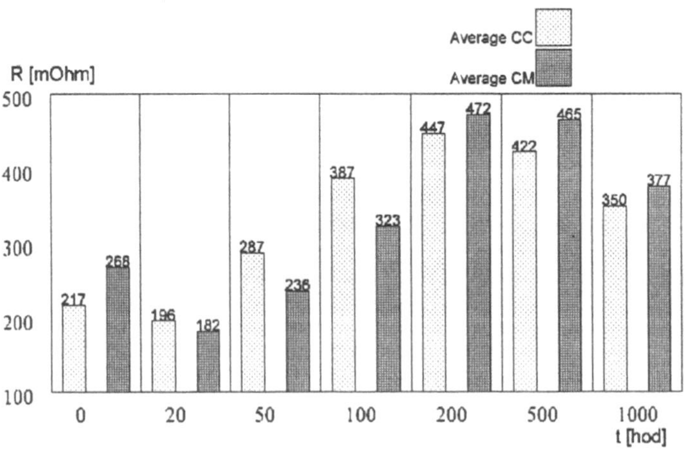

Figure 4.
The comparation of the average values for both PTF material combination.

3. Report on quality and reliability of touch switch keyboards

All described theoretical presumes and experimental results were involved and monitored in one type of the product. It was membrane touch keyboard as a part of telephone set. The type of membrane is single polyester foil covered with various PTF materials inclusive crossover and carbon protection layer. The conductive joint is ensured by embossed conductive rubber.

In generally the quality and reliability of the membranes is possible to follow from the two views:

1. During the process of manufacturing - at manufacturer where the membrane is the product.

2. During the process of assembling - at user where the membranes are components.

3.1. EXPERIENCES DURING THE PROCESS OF MANUFACTURING

As described earlier the manufacture of membranes is based on technology of screenprint. Therefore following the quality of the process, or Quality Management (QM), is to know all problems which brings that kind of technology. In generally all steps used during process are monitored and reported batch to batch, then evaluated and if statistical deviation is higher then allowed for this step, then is necessary find source of problem and to make adjustment of the process.

Batch number	**14**		
Input (pcs)	**25.200**		
	pcs	% in batch	% totally
Holes	**14**	**0,06**	**0,04**
Shape	**0**	**0,00**	**0,03**
1. conductive	**54**	**0,21**	**0,29**
Carbon	**0**	**0,00**	**0,25**
Crossover	**10**	**0,04**	**0,07**
2. conductive	**1450**	**5,74**	**1,23**
Overlayer	**0**	**0,00**	**0,11**
Shortcut in crossover	**39**	**0,15**	**0,23**
Shortcut other	**200**	**0,79**	**0,46**
Broken line	**548**	**2,17**	**2,7**
Number of rework	**1245**	**4,94**	**6,81**
Number of rejects	**1074**	**4,25**	**2,54**
Input totally	**150433**		
Rejects totally	**3815**		
Output - sold totally	**129000**		

Table 1. QM of touch switch keyboard TS 523 0168.02 during a one year.

For this purpouse was created a software supported by Microsoft Office/Windows which allows very easy to make a report of QM . In Table 1 we can see an example of quality report for batch number 14. During a year were followed 20 various batches. In table is possible to see the situation in each technology step in actual batch and next cell shows the situation in comparison to whole amount in given year. After fnishing each batch, it was evaluated and accepted certain decision. In batch number 14 we decided to reduce input amount and we found that aproximately 10.000 pcs is optimum for one batch based on given situation in company and in accordance with customer demands.

For following the process were defined several technological and check points divided into two groups:

- rejects what appear during the screenprint itself (monitored and excluded by operator). Here are excluded foils with wrong positioning, ugly print, or wrong punched... At this stage is possible not all wrong foils are excluded.

- rejects what appear during the low/high voltage tests checked by quality operator. The low voltage test which is applied on each foil (100% of the batch), exactly shows problem caused by shortcut or broken line and which is possible to rework. That means the low voltage test must be applied before printing last overlayer. If the shortcut is in crossover, such a foil is excluded without rework.

High voltage (HV) test is applied only on a part of the batch. There is made statistical choice for maximum of 25% of the batch. If rejects are less than 1% of amount then high voltage test doesn't continue, because quality of technology is confirmed, it is supposed rest of batch is in acceptable range. If rejects are higher then 1%, then HV test continue by another 25% of the batch.

3.2. DEMANDS DURING THE PROCESS OF ASSEMBLING

The tests at the user is possible divided into 3 main groups:

3.2.1. The prototype audit

3.2.2. The input controll

3.2.3. The other controll

3.2.1. *The prototype audit*

In this stage must producer of touch key switch keyboard prove his ability to provide for the customer not only quality but a lot of additional services which allow both of partners makes agreements for the future and succesfully cooperate.

Nevertheless prototype audit of product is first of all technical matter. Based on these informations were made tests in user's laboratories. Results of tests in feedback had several correction in artwork and slightly influenced technology of production.

Technical conditions and sort of tests were:

a) On - loop resistance less than 100 Ohm at switch - on by conductive rubber

Remark: That conditions was fulfilled without problem in addition it was fulfilled at blended technology too. Resistance of conductors slightly increased after dump humidity.

b) Voltage strenght 1000 V/50 Hz/ 1 min. before climate tests.

Remark: That conditions forced change an artwork (contact areas) due to problems with small distances between electrodes. Small amount of pollution or dust caused sparkling during the test. After correction we made destructive high voltage test to check insulation properties. The results showed crossover withstands at least 2.000 V/50 Hz/1 min. Higher voltage was difficult to test due to construction problems during the test.

c) Insulation resistance 20 M Ohm (at 500 V D.C.) untill 5 min after dump humidity test.

Remark: With that conditions were reported a lot of problems. In the end problem was solved by improving stencil (thinner layer of dielectric printed twice), and improving UV curing.

d) Abrasive resistance of contact areas - 1,5 mil. on-off switches without any change of surface.

Remark: Here was reported problem with insuficient abrasive resistivity caused by too thick carbon layer from the beginnig. Reduction of thicknes and improving IR curing helped to reach requested properties.

e) Rest of tests had no influence on property of membranes.

3.2.2 *The input controll*

Here is given sequence of input checking on supplied goods. For membrane was given a statistical test in accordance of Slovak national standard for double choice for P_{AQL}= 0,8 (up to 10.000 pcs) in one batch.

During input test of batch are checked functional parameters, and high voltage properties.

3.2.3 *The other controll*

That sort of tests is providing to find out average level of batch and is concentrated on appearence, resistivity of conductor and functionality.

4. Conclusions

Illustrated results reference about the fact, that described technology has a good chance to become the new standard in microelectronics production which represents cost effective and high quality technology in term of batch production and long time processes.

The main goal of this paper was to show the ability of the small company not only to help in solving certain technical and technological problems but in addition to offer its capability in production for large factory what works under strict ISO 9000 regulations.

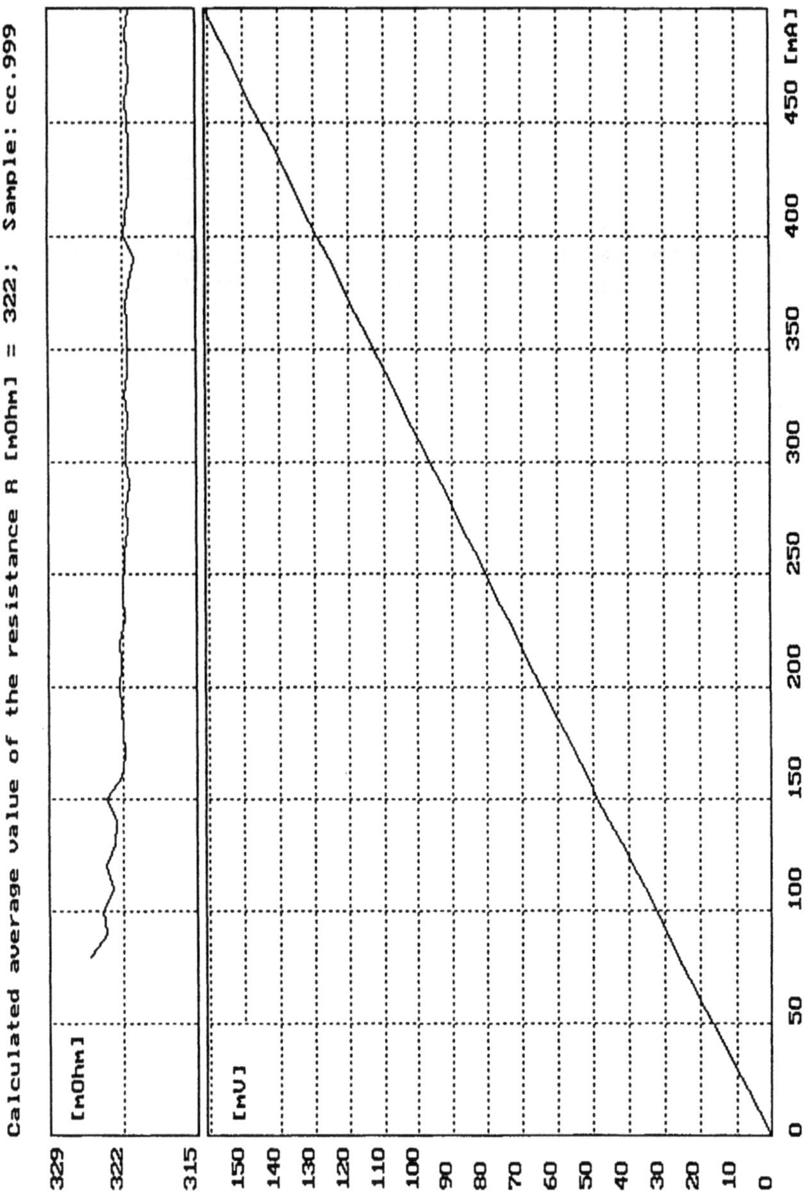

Calculated average value of the resistance R [mOhm] = 322;   Sample: cc.999
[mOhm]
329
322
315
[mV]
150
140
130
120
110
100
90
80
70
60
50
40
30
20
10
0
0
50
100
150
200
250
300
350
400
450 [mA]

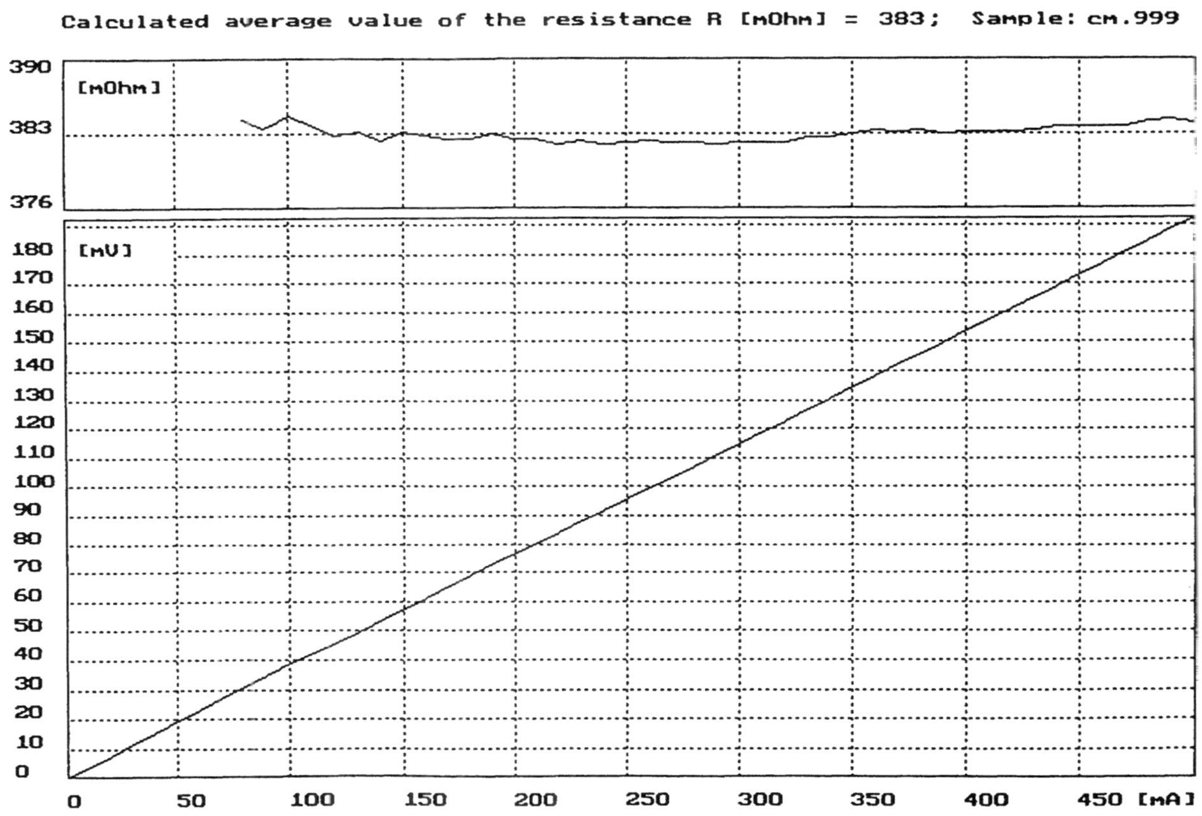
Calculated average value of the resistance R [mOhm] = 383;   Sample: cm.999
390
383
376
[mOhm]
[mV]
180
170
160
150
140
130
120
110
100
90
80
70
60
50
40
30
20
10
0
0
50
100
150
200
250
300
350
400
450 [mA]

A FIVE LAYER THIN FILM MCM-Si DESIGN USING OXYNITRIDE DIELECTRICS

JO LERNOUT, JAN VANFLETEREN, ANDRÉ VAN CALSTER
ELIS - TFCG / IMEC, University of Gent
St. Pietersnieuwstraat 41, B 9000 Gent, Belgium
GUST SCHOLS
Alcatel Mietec
Westerring 15, B 9700 Oudenaarde, Belgium

1. Introduction

Today's electronic circuits put severe demands on data transmission bandwidths, ranging from over 100 MHz for simple designs to the GHz range for opto-electronic applications.

In the past we have been presenting a two layer thin film multichip module (MCM) design on silicon, not having any impedance controlled lines nor integrated passive components. This limited the speed of data transmission to the lower clock frequency range, being typical about 100 MHz [1]. To extend this range, passives have to be placed as close as possible to the module, or even more integrated into the design, and a power and ground plane have to be added, resulting in a five layer module able to reach 500 MHz [2]. Furthermore, by careful design, this MCM can be used for opto-electronic interconnections in the GHz range. Figure 1 shows a cross-section of the proposed design.

We present a nice way to integrate decoupling capacitances by taking full advantage of the silicon substrate. Low TCR NiCr resistors are integrated into the design by using a two mask approach for the patterning of the metal layer containing the resistors.

All subsequent metallization planes are isolated from each other by a 10 μm thick PECVD high quality oxynitride (SiON) layer, optimised to a low stress content, in which via's are etched by means of a wet-dry technique. Compared to more conventional organic insulators, this dielectric has the advantage of having a higher reliability, while still being cost competitive.

The whole of the interconnect is passivated by a 5 μm SiON layer, on top of which the pad metallization is sputtered and protected by a electroless plated Ni/Au layer through a nitride mask/passivation layer, enabling chip attachment.

In the subsequent paragraphs we will present some characteristics of the five layer MCM-Si, together with a typical process flow, the design and processing of a test module (including all necessary structures for DC and HF characterisation), together with preliminary test results.

R. Tummala et al. (eds.), Electronic Packaging for High Reliability, Low Cost Electronics, 179–186.

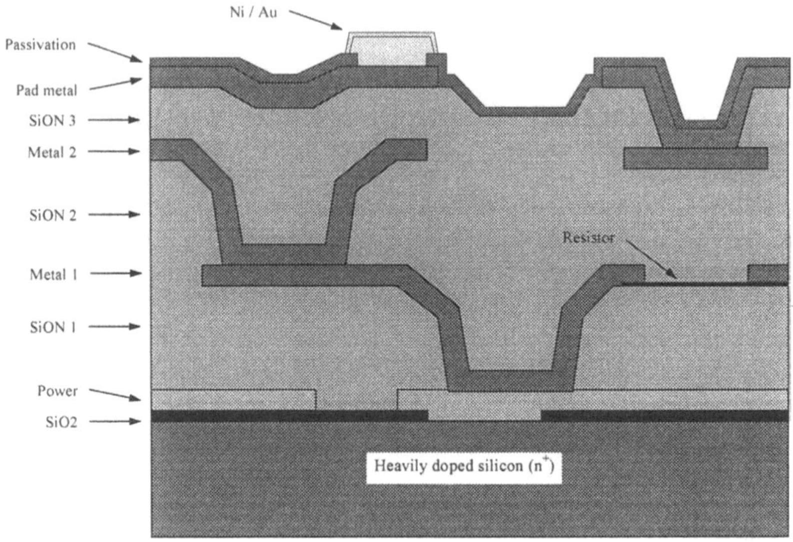

Fig. 1 : Cross-section of the MCM-Si design (not to scale)

2. Characteristics of the Multichip Module - Build Up

2.1 TiW/Cu/TiW CONDUCTOR METALLIZATION

For the sake of its high conductivity, copper (Cu) is used as conductor material. One of the disadvantages of copper is that it doesn't stick to silicon or silicon composed dielectrics. To overcome this problem titanium/tungsten (TiW) is used as adhesion metal underneath the copper. The same layer is also deposited on top of the copper to protect the copper from oxidation and to enhance the adhesion of the dielectric on top of this metal. This tri-level metallization is DC magnetron sputtered in one deposition cycle with thicknesses of 50 nm for both TiW layers and 1.5 μm for the copper, resulting in a conductor layer having a sheet resistivity of about 13 mΩ per square. Patterning is done by wet etching in hydrogen peroxide at 50°C for the titanium/tungsten and in a ferro-chloride based solution at room temperature for the copper [1]. The underetching of the copper (6 to 8 μm at both sides) has to be taken into account when designing the lines, leading to a minimal achievable line width of about 10 μm in a 30 μm pitch.

2.2 SiON DIELECTRIC

As dielectric material, a PECVD silicon oxynitride dielectric (SiON) is used. We have opted for an inorganic insulator to overcome reliability problems associated to organic insulators, knowing to be : moisture absorption, out-diffusion of copper, etc.
One of the problems of PECVD layers is their inherent high stress content. There exist several ways to overcome these problems, but most of them are quite costly (e.g. using a mixed frequency RF generator). A nice solution to form layers with a virtual zero stress

content and without any costly modification of the present equipment is by forming oxynitride layers instead of the more traditional oxyde or nitride layers. This layer is known to have an intrinsic low stress content, which can be adjusted by changing the process parameters [4]. Using Taguchi methodology, we have been able to tune the deposition process to form layers with a low compressive stress content ($-0.18 x 10^8 N/m^2$) [1], which we have already deposited in one cycle to up to 20 µm! Patterning of via's in this layer is done by using a wet-dry etch technique as illustrated in figure 2. After depositing the 10 µm SiON dielectric layer, a 10 µm thick photoresist (AZ4562 from Hoechst) is applied and patterned (fig. 2a). To remove stresses from the edges of the patterned photoresist and to enhance the step coverage, the via is first slightly wet etched in BHF resulting in a 5µm underetch of the photoresist at both sides (fig. 2b). The rest of the via is then RIE etched in a CHF_3/O_2 discharge (fig. 2c) and the photoresist removed (fig. 2d). The RIE process has also been optimised using Taguchi Methodology, resulting in a 80 degrees slope of the via wall referred to the bottom and a SiON/photoresist etch selectivity of 2.2 to 1.

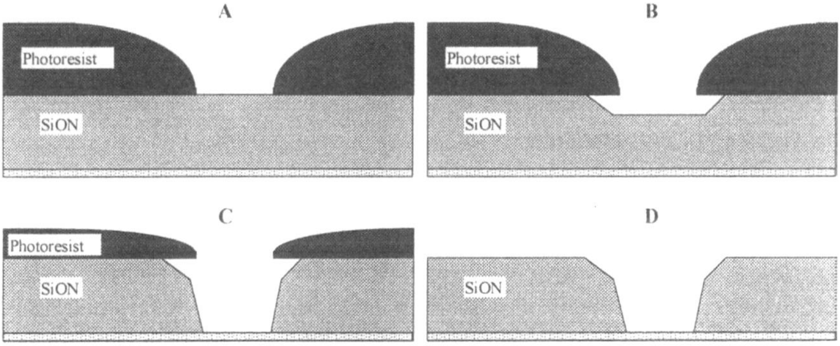

Fig. 2 : Wet-dry etching technique for SiON.

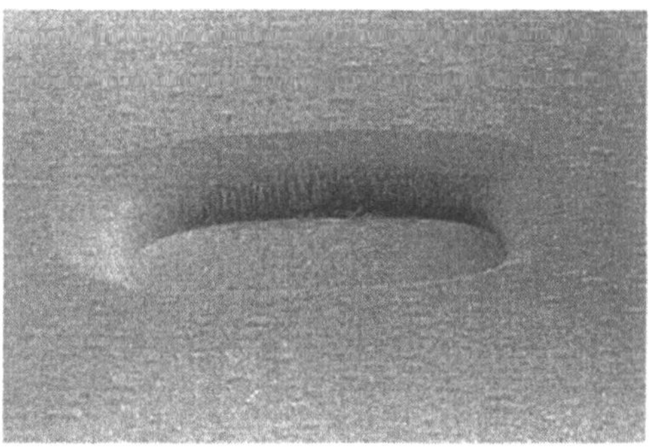

Fig. 3 : SEM photograph of a 50 µm wide wet-dry etched via.

A SEM photograph of a 50 µm wide wet/dry etched via is shown in figure 3. Using this process we are able to etch 10 µm via's in a 30 µm grid.

2.3 TOP METALLIZATION

To protect the top metallization layer and to enable chip attachment, an electroless plated nickel/gold (Ni/Au) layer (2 µm thick) is used. To achieve this, the top metallization (being the normal TiW/Cu/TiW system) is covered by a 300 nm nitride layer, which is used as passivation layer for the interconnect as well as mask for the electroless plating. The advantage of electroless deposition is that no extra mask nor seed-layer (and subsequent removing of it) is required.

2.4 INTEGRATED CAPACITORS

One of the advantages of silicon as substrate material for thin film multichip modules, is the ability to integrate active, silicon based, components. Yield problems have so far restricted the possibilities of this option. Nevertheless, two of the CMOS processing steps can be used to realise integrated capacitors into our silicon hybrid interconnect. In a first stage the substrate is heavily doped by using ALCATEL-MIETEC's deep n-dope + drive process. Subsequent thermal oxidation creates a high quality SiO_2 dielectric which serves as the decoupling between the n^+-Si ground plane and the first metallization [3]. The prime advantage of thermal oxidation on a pure silicon surface is the ability to form thin layers with a pinhole count approaching zero. Taking the size of typical multichip modules into account (in the order of one to several cm^2), a value of 50 nF/cm^2 would be sufficient to provide enough decoupling between the power and ground. Higher values would yield too thin layers and thus more changes for pinholes in the oxide : to create 50 nF/cm^2 decoupling using a thermal oxide dielectric (dielectric constant 3.9), a 700 Å thick layer is needed. The metallization used to form the top plate is either the tri-level metal system (TiW enhances the contact resistance to silicon) or aluminium (when pinholes really form a problem, the self-healing effect of aluminium can be used to solve pinhole problems).

2.5 INTEGRATED RESISTORS

A material often used to form resistors is nickel chromium (NiCr) which is known for its low TCR. We integrate these resistors by adding a thin NiCr layer underneath the tri-level metal system and use a two-mask approach for the patterning (either by lift-off or by wet etching). Optimisation of the sputter process resulted in resistance layers with a TCR of 75 ppm/°C having sheet resistances ranging from several to 50 Ohms per square with uniformities around 95% [3].

3. Processing steps

Underneath is a typical processing sequence described (refer also to fig. 1) :

1. Heavily dope (deep n dope + drive) the silicon substrate
2. Thermally oxidise the silicon to form 70 nm thick SiO_2
   note : these two steps are performed at *ALCATEL-MIETEC*
3. Etch via's in the thermal oxide to form the contacts to the ground
4. Deposit and pattern the power metallization
5. Deposit 10 µm of SiON and pattern via's in it using the wet-dry technique
6. Deposit and pattern the NiCr resistor layer
7. Deposit and pattern the first interconnection layer (TiW/Cu/TiW)
8. Deposit 10 µm of SiON and pattern via's in it using the wet-dry technique
9. Deposit and pattern the second interconnection layer (TiW/Cu/TiW)
10. Deposit 5 µm of SiON and pattern via's in it using the wet-dry technique
11. Deposit and pattern the pad interconnection layer (TiW/Cu/TiW)

12. Deposit 300 nm of Si_3N_4 and pattern via's in it ; etch the TiW in the via's
13. Electroless plate 2 µm of Ni/Au
14. Saw the wafer into the different modules
15. The modules are ready for chip-attachment and testing

As can be seen from this sequence, to process a complete run, in total 1 doping cycle, 1 thermal oxidation cycle, 4 sputter cycles, 4 plasma deposition cycles, one electroless plating cycle and 10 patterning steps (10 photo-masks) are needed.

4. Test layout

A complete five layer MCM-Si test layout has been designed for DC and AC (high frequency) testing of the module. The DC part included structures for measuring the via resistance and yield, cross-over yield, meander structures in all metallization layers and resistor structures. The AC part is composed of capacitors (both decoupling and capacitors having a dielectric of 10 µm oxynitride) and, for high frequency measurements, of microstrip- and striplines, both single and coupled, designed for a characteristic impedance of 50 Ω. Every structure is, for the ease of measuring, connected to probe-pads of constant dimensions and pitch, arranged in a repetitive array. The layout was made for processing on 4" silicon wafer, divided into individual modules of 1"x1". Four complete modules can thus be processed on a single wafer. To complete the testing, flip chip and chip&wire mounting of testchips is incorporated. Figure 4a shows a photograph of a completed 4" wafer, while figure 4b shows a detailed view of some teststructures.

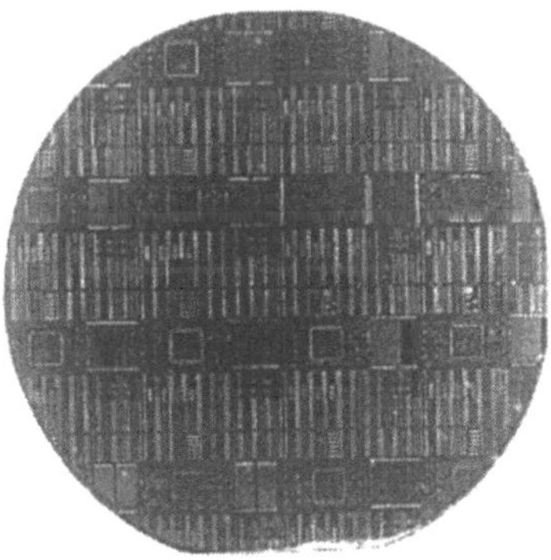

Fig. 4a : View of a completed wafer

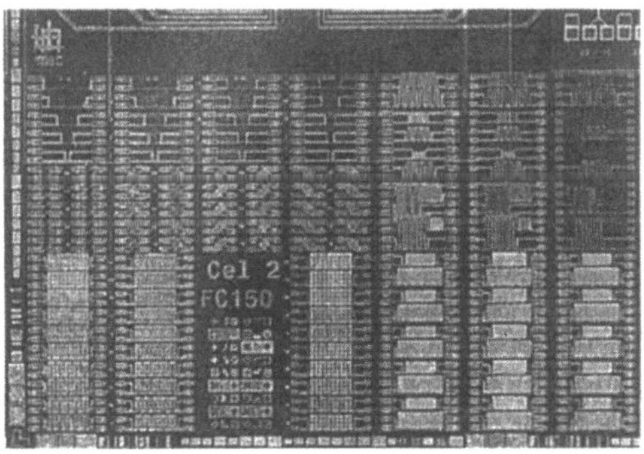

Fig. 4b : detailed view of some DC teststructures

5. Measurement results

Two runs of the test layout described in the last paragraph are finished and have been completely DC measured. At the time of this writing, only some preliminary high frequency measurements are finished.

5.1 DC MEASUREMENTS

Via measurements were performed on structures for four point resistance measurement of individual via's and on via-chains for the determination of via yield. Via sizes ranged from 50 μm down to 10 μm in all metallization levels. Four point measurements revealed a resistance as low as 4 mΩ for a 50 μm via, going up to about 35 mΩ for a 10 μm via. In total, for each via size, 15.000 via's were tested in chains of 40 via's. This resulted in a chain yield of 99.7% for all via sizes down to 20 μm and 98% for 10 μm via's. The statistical total yield can be calculated to be 99.993% for the first and 99.95% for the latter. The contact resistance to the doped silicon, using the tri-level metallization, is about 850 mΩ for a 50 μm via, going up to 3.5 Ω for a 10 μm via. Here no chains can be tested, but from the four point measurements a 100% yield was extracted for all via sizes. Please note that in a normal design, the contacts are made as big as possible to reduce the effect of the high sheet resistivity of the doped silicon to the performance of the decoupling capacitances [3].

To test the DC performance of the SiON dielectric, cross-over structures between all SiON dielectrics were included into the design. In total, 340.000 cross-overs were tested, in groups of 102. These measurements revealed a 99% group yield for all dielectric levels, resulting in a 99.99% statistical total yield. The SiON capacitances (324 in total, having areas of 0.75, 1.5 and 3 mm^2) were also DC measured for shorts, revealing a 99.4% yield. The insulation resistance was measured to be over 10^{13} Ω.

Meander structures were measured to extract the sheet resistivity of the metal system (determined to be about 13 mΩ) and the difference between the designed and actual dimensions of a track, which was determined to be about 9 μm at each side. The underetching of a tri-level TiW/Cu/TiW is larger than for a bi-level TiW/Cu system, since in the tri-level system, the top TiW will act as mask for the copper etching, such that the total underetching is the sum of those of copper (measured to be 6.5 μm) and titanium/tungsten.

Finally measurements on the NiCr resistors revealed an average sheet resistivity of 49.2 Ω with a uniformity of about 94%. The patterning of the resistors can be done either by lift-off or by etching in a cerium sulphate based solution at elevated temperature. The disadvantage of the wet etching technique is that the PECVD layer is slightly attacked, leaving a rougher surface. Nevertheless, no difference in performance was observed between the two techniques.

5.2 AC MEASUREMENTS

To extend the bandwidth of the connections in our multichip module, we have to concept the interconnection lines as transmission lines with a well defined characteristic impedance Z_0. The easiest solution to accomplish this is by adding separate power and ground planes and design the lines as (micro)strip lines. The test layout contains both strip and microstrip lines, single and coupled. At the time of this writing these measurements are still going on, such that only preliminary results on single microstrip lines in a test layout of IMEC vzw. [5] can be presented. Figure 5 shows the measurement of the S12 (transmission, top) and S11 (reflection, bottom) parameters in the 0.45 to 18 GHz range for a 7mm long straight line and a 17mm long curved line terminated at 50 Ω.

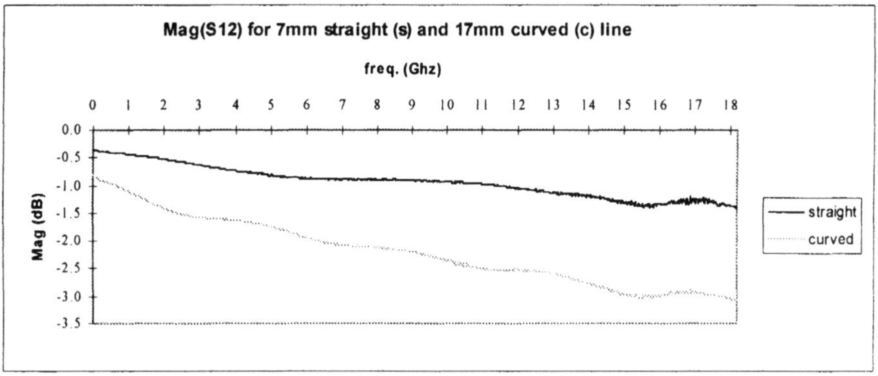

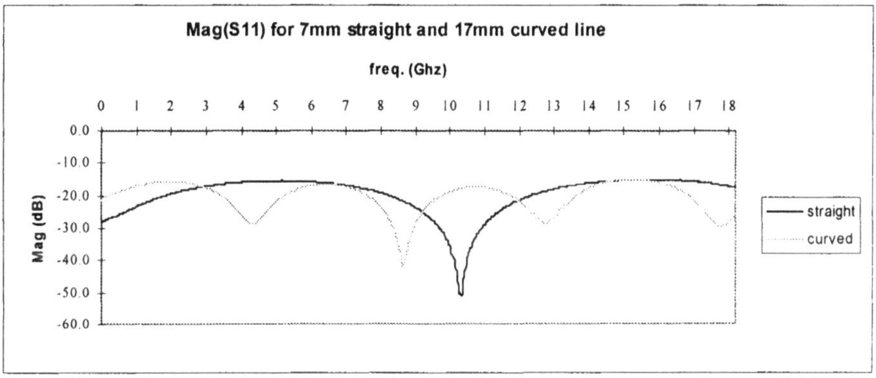

Fig. 5 : Measured magnitude of the S12 (top) and the S11 (bottom) parameters for two 44.2Ω lines of different lengths (7mm straight and 17mm curved).

It can be seen that for the longest line, the attenuation at 18 GHz is about 3 dB. Forthcoming AC measurements on lines especially designed for 50 Ω will result more information on the high frequency characteristics of the SiON dielectric.

5.3 CHIP ATTACHMENT

Currently, we are investigating the ability to flip chip mount chips using conductive adhesives, which are known to be environmentally friendlier than the more conventional solder bump technique. Both anisotropic (type paste and foil) and isotropic conductive adhesives are investigated. This work has only recently started and results will be published later on.

5.4 RELIABILITY TESTING

The aim of this test design is also testing the reliability of the interconnect by submitting the processed wafers to 85/85 tests, thermal cycling (-40°C to 85°C) and thermal storage

(85°C) and remeasuring the electrical characteristics. First reliability tests after 250 hours/cycles showed no change at all in electrical performance of the dielectric.

6. Conclusions

In this paper we have presented a five layer multichip module on silicon using low stress PECVD oxynitride dielectrics for the isolation of the different TiW/Cu/TiW metal layers. The design includes separate power and ground planes, such that impedance controlled transmission lines can be realised. Furthermore, capacitors for decoupling purposes and resistors are integrated into the design by using the silicon as ground plane and grow a thermal oxide on top of it for the first and by including a thin NiCr layer for the latter. In this way we have realised a multichip module capable of handling clock frequencies up to 500 MHz.
A test layout has been presented, together with DC and preliminary AC results. These revealed a high yield for the interconnections (via and cross-over yield approaching 100%) and a good high frequency performance. First reliability tests after 250 hours showed no change in electrical performance of the dielectric. Further tests will have to prove that SiON has not only a comparable or even better electrical performance than organic dielectrics, but that reliability problems associated with the latter can be solved by changing to the proposed inorganic oxynitride insulator.

7. Acknowledgements

The authors would like to thank Suixin Zhang for the electroless plating of the wafers, Herbert De Smet for the sawing of the modules and Eric Beyne and Philip Pieters of IMEC vzw. for the high frequency measurements.

8. References

[1] Lernout, J., Van Calster, A., Vereeken, M., and Schols, G. (1995) Two-layer silicon hybrids interconnect technology, *Proc. of the 4th Interntioanl Conference on Multichip Modules (Denver. Colorado)*, 375-380.

[2] Messner, G., Turlik, I., Balde, J. W., Garrou, P. E. (1992) *Thin Film Multichip Modules*. ISHM. Reston.

[3] Lernout, J., Van Calster, A. and Schols, G. (1996) On the effective integration of resistors and capacitors into a MCM-Si design, *Proc. of the 1996 International Conference on Multichip Modules. (Denver, Colorado)*, 391-396.

[4] De Baets, J., Van Calster, A., Capon, J., De Rycke, I., De Smet, H., Doutreloigne, J. and Vanfleteren, J. (1992) Silicon oxynitride layers for device passivation, *Thin Film Transistor Technologies. Symposium, Book of Abstracts (Toronto)* **92-2**, 454-455.

[5] Peeters, J. and Beyne, E. (1994) Broadband Modeling and Transient Analysis of MCM Interconnections, *IEEE Transactions on Components, Packaging and Manufacturing Technology. Part B: Advanced Packaging* **17-2**.

AN ASIC VIEW OF LOW-COST IC PACKAGING

by
Paul R. Van Loan
Integrated Circuits Business Division
Hewlett Packard
Palo Alto, California, U.S.A.

INTRODUCTION: The field of Integrated Circuit Packaging has grown rapidly more complex over the last ten years, and nowhere has this had a greater expression than in the Application Specific IC (ASIC) market. The reason for this is that the typical ASIC manufacturer must address a wide range of customers, whose application spaces are driven by diverse combinations of performance, cost, reliability and life cycle requirements. HP's Integrated Circuits Business Division (ICBD), for example, carries more than 700 IC part numbers, and serves HP Manufacturing Divisions in markets as diverse as High End Servers and Work Stations, Computer Peripherals, Test and Measurement Systems, Medical Products, and more.

Fifteen or so years ago (admittedly, a "geological epoch" in the microelectronics business) this was not an issue. With relatively few exceptions, IC's were packaged in DIP's, PGA's or COB hybrids, period. But with the inception of SMT and the growing functional power of IC's came an evolutionary explosion in IC packaging technology, driven chiefly by two very different forces: the desire for low-cost, industry standard package formats; the need for application-specific driven solutions, whether for performance, form factor, reliability or other purposes. The result has been a remarkable, if not bewildering variety of package types and price points, from SOIC's and TSOP's at less than a penny per pin, to high-pincount ML ceramic grid arrays at 10X that cost. The result, for a typical ASIC house, is the generalized package offering shown in Fig.1. This plot is not specific to a particular firm; it

R. Tummala et al. (eds.), Electronic Packaging for High Reliability, Low Cost Electronics, 187–195.

represents a hypothetical but not exaggerated situation in today's market.

The argument of this paper is that the packaging industry has reached a turning point which will result in both a significant simplification of package types and a progressive reduction and levelling in cost across a wide range of product applications. This "sea change" is underway, and will be the dominant feature of IC packaging by the year 2002. The goal of this paper is to identify the drivers behind the changes, and predict what the ASIC packaging landscape may look like five years from now. To this end, I have drawn liberally from the roadmap studies of SIA and Sematech(1), industry studies (2) and of course the available literature.

THE DRIVERS

There are a number of influences driving today's ASIC environment, but it is the thesis of this paper that four are of paramount importance.

1. The Shrinking Cost and Space of IC Functionality

The well-known Gordon Moore plot has implications for the powerful influence of Si technology on packaging,which is now intensifying from at least three directions.

Increasing IC functionality enables products in lower performance markets to invade higher ones. A current example is the migration of PC's into the Work Station market. The relevance of this to IC packaging is that performance can no longer be decoupled from cost, almost regardless of application.

A second factor is the steady increase in clock speeds and signal rise times as Si feature sizes shrink from 0.35 to 0.25 and 0.18 microns. The challenges raised by this march of performance far exceed the scope of this paper, but we shall discuss one or two of the ways in which this will radically alter the nature of IC packaging.

A third influence bears on the package's function as a space transformer. The evolution of finer Si feature sizes noted above gives rise to increasingly pad-limited IC designs. This in turn drives first level interconnect technologies to their limits and beyond if wasteful "white space" silicon is to be avoided. The far-reaching effects of this are discussed below.

2. The Shrinking Form Factor of Electronic Products

The Japanese have long understood the relationship between product volume, cost, performance and mobility,and even coined a phrase for it: "smaller, cheaper, faster, lighter" (kei haku tan sho). Form factor reduction is an end in itself for many products such as disc drives, portable and handheld devices. However, its advantages in promoting lower cost (reduced materials content) and improved performance (by reducing time of flight between components) are strongly influencing the evolution of package and interconnect geometries throughout the industry.

3. Shrinking Time to Market (TTM)

Ten years ago, the development cycle time of a new, complex electronic product was about three years, sometimes more, and the market window was broad. Today, TTM can be as short as 9-12 months, and a product generation life is proportionately shorter. The effect on IC packaging has been to reduce major component product rolls, and to place a heavy premium on "picking the winner" in emerging technologies. This will not change. Instead, we now see an increasing frequency in networking within the industry, using leverage among consortia, contract houses, industry, academic and government organizations to converge wherever possible on the best packaging and interconnect alternatives.

4. Shrinking Vertical Integration

Increasingly, only the largest ASIC and System houses maintain core competencies in package and board assembly. This move to outsourcing will continue, with increasing expectation on the part of the OEM that contract assemblers will behave as partners, not vendors, even to the point of carrying out process customization and technology development that once was the province of the OEM.
The motive in this migration is of course cost reduction, and the effect will be to focus the end product developer on his true core competency: integrated system design, from chip to product box. The consequences of this for packaging are noted below.

THE RESPONSE

The drivers noted in the preceding section are of course already evincing responses from the packaging world. In the past two years, development of IC package and interconnect technology, while apparently continuing to throw up a broad diversity of solutions, has actually embarked on a course of convergence which will, broadly speaking, result in three unifying principles:

1. dominance of area array interconnect at the first and second level;
2. emergence of what we shall call Isoplanar packaging, based on low-cost substrates for high-performance packaging;
3. Integrated System design, from Si to product box.

To be sure these are, as noted, broad generalizations and there will continue to be significant application- and product-specific exceptions to all three. Nonetheless, they do define the future for mainstream ASIC packaging.

Bumped Area Array First Level Interconnect

Today, the percentage of chips mounted by solder or other bumped flip chip interconnect means is minuscule in comparison to wirebonding. For many low pincount applications, that situation will continue unchanged, perhaps indefinitely. Nonetheless, the dynamics of bumped flip chip almost unanimously favor its rapid growth. The chief reasons are:

1. Wirebonding pitch is approaching a hard limit. Despite the ability to bond on staggered pads, and omitting the special case of LOC packaging, wirebonding is a perimeter technology, and even with reductions in wire diameter, capillary diameter, and ball size, a limit of about 50 microns probably exists for high-speed wirebonding.

2. Wirebonding is a sequential operation; flip chip is a gang bonding operation. This means not only improved costs at the chip attach operation, but -- as wafer diameters increase -- significant reductions in bumping cost per chip.

3. Despite the existence of C4 technology in IBM for nearly 30 years, only in the past 3-4 years have meaningful efforts been undertaken throughout the industry to reduce costs--

through new bumping processes, material cost reductions, increased throughput, and other initiatives.

4. The ability to place bumps throughout the surface area of the chip. The current limitations here are susceptible to mitigation, if not removal: Pb-free bumps avoid the alpha particle problem, and new tools for standard cell design will ultimately permit far greater freedom in placing pads.

Today, the boundary at which flip chip is favored over wirebonding is somewhere in the 500-600 pin region (making allowance for a wide variety of special conditions). Figure 2 is a schematic representation of how and why that boundary may decline over the next 5 years.

Second-Level Area Array Interconnect

In the past year, the generic BGA has at last moved from a promising newcomer to a production reality that is striving for high-volume. In this paper, I wish to make only the following points:

1. PBGA proliferation will be favored by growth of the fabrication and assembly vendor base, and by board material cost reductions and therefore by growth of that supplier base as well;

2. CSP's-- a fine-pitch organic BGA, in most cases -- will continue to make inroads where form factor is paramount, but will not dominate the low pincount market until they reach price parity with existing molded plastic packages. By 2002, they will be a significant portion of the BGA product mix.

3. BGA's which are CTE-matched to the PCA will displace high-performance ceramic packages, once development and qualification experience eliminate the reliability and cost issues. The future for the fine-line, ML substrate systems on which these are based is very promising (see Isoplanar Packaging, below).

In sum, CTE-matched BGA's will ultimately be the package of choice across the full spectrum of ASIC packaging.

Isoplanar Packaging

It is already notable that BGA's and other area array packages are, by their nature, thinner than many of the older generation packages they are replacing.This is not accidental; it is based upon the use of thin laminate or flex circuits, usually incorporating co-laminated heatspreaders for thermal management and maintaining package stiffness and with low-profile solder balls (~20 mils high after reflow). Thus,a large area 500+ pin package may be only 2-3 mm thick, including the metal heatspreader, and thinner versions may approach 1.0 mm. An excellent example of this is the Flex BGA package developed at Hewlett Packard (3,4),and now available from several package vendors. Its structure for a wirebond interconnect is shown in cross-section in Fig.3. There are several advantages to this construction besides low materials and fabrication cost: the fine line geometry permits dense routing of signal and power traces, often facilitating high electrical performance on one or two conductive planes; the 1-2 mil square trace cross-sections enable matched impedance conditions; the structure has an inherently low thermal impedance. This package represents an incremental, but very significant, advance over existing PBGA's.

The next step lies in extending this planar, thermally efficient structure to truly high pincount, high performance BGA packages at significantly lower cost than today's ceramic grid array packages. Virtually all of the conditions exist to realize the goal. Detailed discussion of the topic is beyond the scope of this paper, but the elements are as follows:

1. A fine-line, multilayer substrate system capable of flip chip attachment of large, moderately high power dissipation chips, and with a CTE matched to the PCA motherboard so that standard BGA attachment is reliable. As this audience knows, advanced development and qualification of several contenders are well underway.

2. Fabrication of the substrate at low cost. Several promising approaches are in early production stages, involving large area patterning (5) and/or continuous roll processing (6), for example.

3.The final step in making such a package a truly powerful isoplanar system element lies in embedding components within the laminate or flex structure. Development of embedded passive components in organic substrates is now underway (7), though not as well-characterized as in ceramic or silicon-based systems. This capability obviates the solder attachment of discrete passives. In ASIC systems, its first utilization will probably be in wireless and portable

products, but any system whose performance is compromised by parasitics or by RFI is a potential candidate.
The longer term significance of isoplanar packaging may be that, as an active/passive subsystem, it lends itself well to integrated system design. Why is this important?

Integrated System Design

Both Si and package technologies are approaching performance walls. In future CMOS generations, on-chip metal interconnect and passivation must be chosen carefully so as not to compromise the performance gains conferred by feature size reduction and transistor design. By the same token, the chip layout and routing must make optimal use of the Si architecture, while interfacing seamlessly with the package if maximum performance is to be extracted. Increasingly, an integrated set of design tools will be required to ensure this goal. A properly designed and fabricated isoplanar package can provide the chip designer with an important degree of freedom and flexibility not currently envisaged.

An ASIC Package Offering in 2002

We began this discussion with a plot of a "typical" ASIC package offering in 1997. Marshalling the arguments adduced above, a similar plot for the mainstream ASIC packaging industry in 2002 is shown in simplified form in Fig.4. The most obvious differences are: the significant reduction in package types; the dominance of area array interconnect, both first and second level; a much gentler slope of package costs with increasing pincount and performance. The estimates shown represent "die-free" packaging costs-- package plus chip assembly materials and labor.

MCM's or SCM's?

This paper does not attempt to answer the question: "What portion of the ASIC mainstream market will be represented by MCM's five years from now?" This is a difficult question to answer in any case. More pertinently within the context of our forecast, each of the package types shown in Fig.4 lends itself to both SCM and MCM utilization.

CONCLUSION

While it is always a questionable indulgence to engage in forecasts and speculations of the type contained in this paper, I believe the conclusions are responsibly based on observable trends in our industry, and on data in the published literature, as referenced herein. In addition, I wish to cordially thank several of my colleagues for helpful discussion and criticism, particularly Amit Agrawal, Bahram Afshari, James-Yu Chang, Kim Chen, Ed Middlesworth and Raj Pendse. Of course, any errors in the text or in the opinions here expressed are entirely the author's.

REFERENCES

1. Chi Shih Chang and Oscilowski, A., "U.S. Packaging Trends", Semicon Japan, Tokyo, Dec.3, 1996

2. Mark V. Christensen, Prismark Associates, "Benchmark Market Pricing for Single and Multichip Substrates", 6th International Conference on Multichip Modules, April, 1997

3. R.D. Pendse, Hewlett Packard, "Method for Making High Pincount Package for Semiconductor Device", U.S.Patent #5,376,588, Dec.27, 1994

4. R.D. Pendse, "High Pincount Package for Semiconductor Device", U.S. Patent # 5,468,994, Nov.21,1995

5. Eric Chieh et al, Micromodule Systems; and P. Garrou et al, Dow Chemical, "Development of Large Area Processing for Thin Film Substrates" Proceedings of the 6th International Conference on Multichip Modules,Denver, April 24,1997, p.234

6. G.Matijasevic et al, Toranaga Technologies,Inc., and Dick Pommer et al, Litronic Industries, "MCM-L Substrates Fabricated Using Patterned TLPS Conductive Composites" 6th International Conference on Multichip Modules, Denver, April 1997, p.64

7. J.Parkerson et al, "Design Considerations for Using Integrated Passive Components", 6th International Conference on Multichip Modules, Denver, April 1997, p.345

PACKAGING TECHNOLOGIES FOR THICK FILM SENSORS

D. Belavič, S. Šoba, M. Pavlin, S. Gramc, D. Ročak*
HIPOT, Trubarjeva 7, 8310 Šentjernej, Slovenia
*"Jožef Stefan" Institute, Jamova 39, 1001 Ljubljana, Slovenia

1. Introduction

Assembling and packaging aspects have an important role in the production of sensors and transducers. Due to the need to detect different physical or chemical quantities an appropriate approach to assembling and packaging must be chosen. For example the packaging mode, and the mass of the housing or the substrate have an important influence on the response time, the thermal time constant, and the dissipation constant of temperature sensors. On the other hand, the mechanical and thermo-mechanical properties of materials for assembly (bonding and protecting) and housing have a crucial influence on sensors for mechanical quantities (pressure, load, ...). For each specific sensor application the construction design and properties of the materials used for assembly and housing are important ([1,2,3]).

Through many years of experience in thick film technology and sensor production we have developed many sensors for various physical or chemical quantities. In this paper an account is given of some specific assembly and packaging aspects for sensors where thick film technology is used to produce the sensor elements themselves and/or the electronic circuits for signal processing. Some typical examples of different sensors are presented.

2. Temperature Dependent Resistors

We have produced several hybrid circuits with integrated thick film TDR (temperature dependent resistor). Most of them are used for temperature compensation of one or more electronic parameters; and some of them are used as temperature sensors. The influence of packaging and assembling on the characteristics of TDRs are numerous and complicated in the case of integrated thick film temperature dependent resistors and must be carefully observed and studied for each particular case. The influence of packaging on the characteristics is easy to determine for individual temperature dependent resistors.

R. Tummala et al. (eds.), Electronic Packaging for High Reliability, Low Cost Electronics, 197–204.

Packaging mode, and the mass of the housing or the substrate have an important influence on the following parameters:

- The response time, i.e., the time the sensor needs to reach 63.2% (1-1/e) of the total temperature difference when subjected to a temperature change from 25°C in still air to 85°C in silicon oil.
- The thermal time constant i.e., the time in seconds required for a TDR dissipating zero electrical power to change 63.2% (1-1/e) of the difference between its initial and final body temperature.
- The dissipation constant i.e., the amount of power (expressed in mW/K) required to self-heat the TDR body in still air 1K above the environment temperature.

In Figure 1 two products with the same temperature sensor element (a thick film NTC thermistor) with and with out housing are shown. The parameters are presented in Table 1. Room temperature resistivities are the same, while the time constant and the dissipation constant differ due to the packaging.

TABLE 1. Characteristics of TDR with different packaging

	TDR without housing	TDR final package
Resisitivity at 25°C	10 kOhm	10 kOhm
Beta constant	1330 K^{-1}	1330 K^{-1}
Time constant	10 sec	22 sec
Dissipation constant	4.5 mW/K	20 mW/K

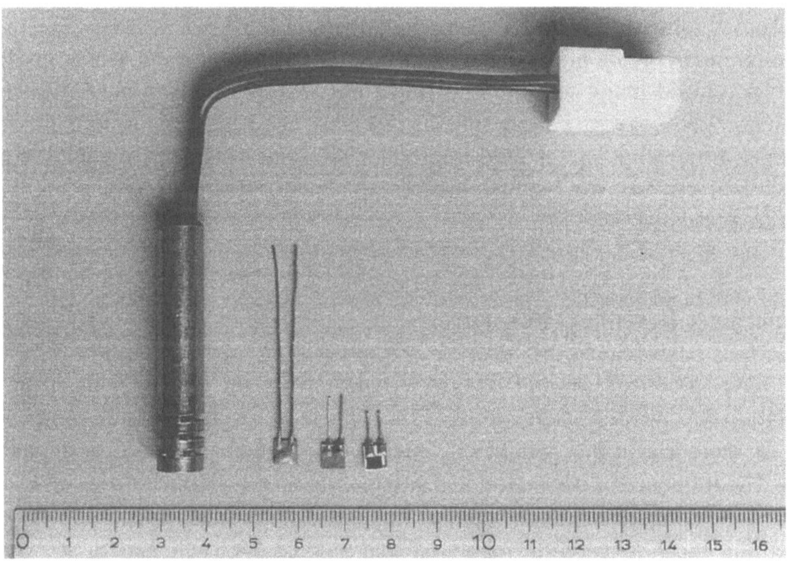

Figure 1. Thick film thermistors with different packaging

3. Load cells

We have developed a low cost force sensor (Figure 2) which is intended for the use in a kitchen electronic scale with a weighing range from 10g to 3000g, and an accuracy of ±0.2% of the full range. The sensor element was realized by taking advantage of the piezoresistive effect of a thick film resistor printed and fired on a ceramic substrate. This substrate is bonded on a metal double bending beam ("binocular" form). With the chosen materials, construction and procedure, the required characteristics were obtained. The sensitivity of the transducer is 0.4÷0.5μV/V/g, the stability of sensitivity ±0.5%, the temperature coefficient of stability lower than 100×10^{-6}/K, the offset maximum 4mV, the stability of offset ±0.5mV, and the uncompensated creep of the output signal less than 0.5%.

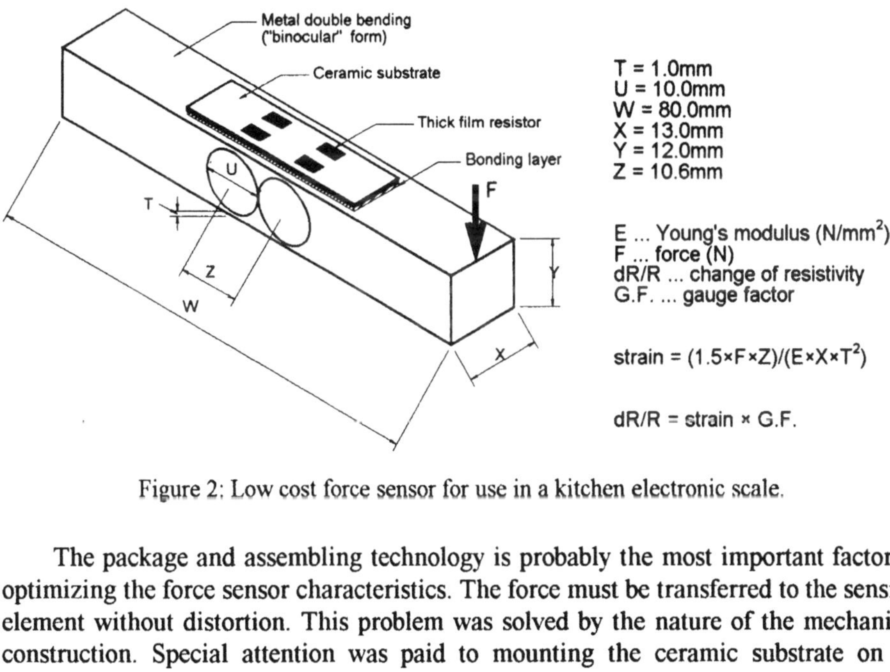

Figure 2: Low cost force sensor for use in a kitchen electronic scale.

The package and assembling technology is probably the most important factor in optimizing the force sensor characteristics. The force must be transferred to the sensing element without distortion. This problem was solved by the nature of the mechanical construction. Special attention was paid to mounting the ceramic substrate on an aluminum beam and one the most critical parameters, called "creep", was investigated.

If the force transducer is subject to a static strain, its resistance changes with time despite the constant strain in the sensor element. These changes in the measured signal take place very slowly and is said to be subject to "creep". The cause lies in the characteristics of the bonding layer between the ceramic substrate and the aluminum beam, and also in the beam itself, which transfers the strain. The creep of the bonding layer depends on the bonding material and the thickness of the layer. The creep of beam depends on its material and construction. With the right combination of dimensions and materials in the bonding layer and beam, it is possible to compensate one creep with another.

For a low cost force sensor we investigated four different bonding materials. The results of the investigation on creep are shown in Figure 3. The force sensor was loaded for 10 minutes with 30N and then was unloaded for 10 minutes. During this test the output signal was automatically measured. All tested materials were polymeric adhesives, but the result are quite different. We chose the material denoted P01 with very low creep and good long time stability.

The second step of the investigation was to establish control over creep. We found that the thickness of the bonding layer is a convenient way to control creep. The creep of the same bonding material (denoted CB) of different thicknesses is shown in Figure 4. The thickness was controlled by varying the volume of the polymer adhesive. The thicker layer has a larger creep than the thinner layer. This dependence is not linear, but it is significant enough to compensate the creep of the beam with the creep of the bonding layer. This compensated creep is less than 0.2% after 1 hour constant loading with a maximum load of 30N.

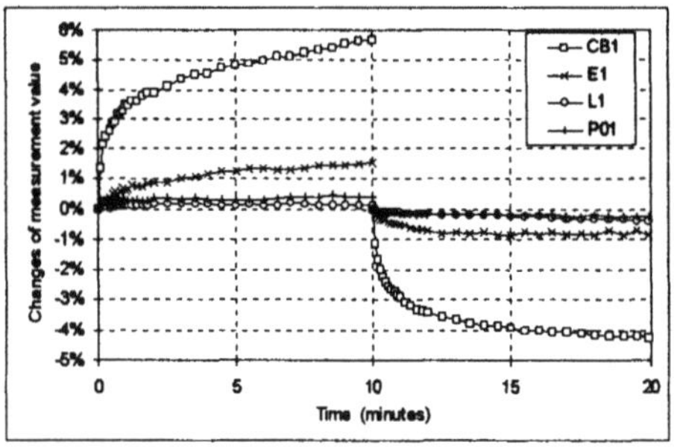

Figure 3:

Creep dependence on different bonding materials. (left-loaded, right-unloaded)

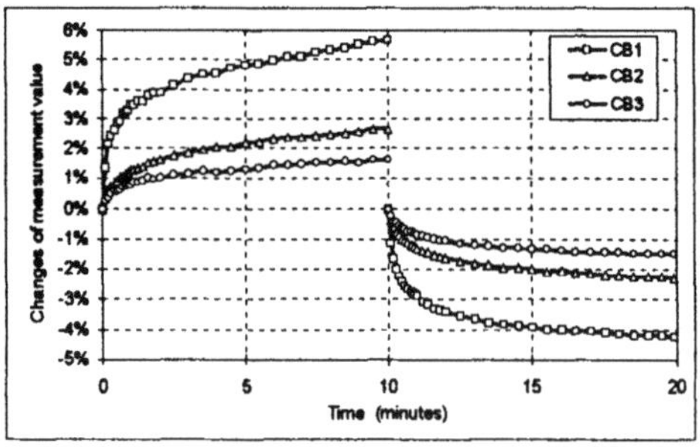

Figure 4

Creep dependence on thickness of the bonding layer. (left - loaded, right - unloaded)

4. Proximity switches

Several signal processing circuits for proximity switches were produced in thick film hybrid technology. The packaging principle is a combination of the moulded ([4]) and the metal package. The moulded encapsulating process has three steps (Figure 5). First step is to coat the bare chip and to cure the material. The second step is to coat and to dry just one side of the hybrid circuit. In the third step the second side of the hybrid circuit is coated and the second and third coatings are cured. This special moulded encapsulating process was used to bypass the problem of the different temperature expansion coefficients of alumina and epoxy. Different temperature expansion coefficients can generate external stresses, which can strain the structure. These external stresses, combined with any residual stress resulting from previous thick film processing, can induce failures, for example, cracking and bending of the ceramic substrates or loss of continuity in the metallization between the film components and the substrate. In addition, electrical properties are degraded as a result of residual stresses even when mechanical failure is not clearly observed. Therefore, it is important to detect the presence of residual stresses and to understand at what point during the production process they were introduced.

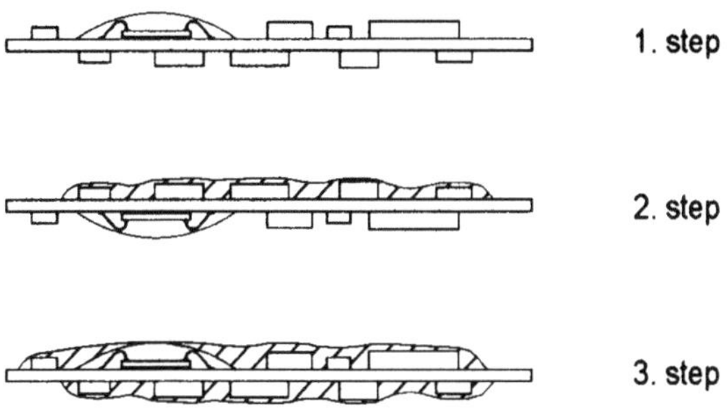

Figure 5: Moulded encapsulating process for proximity switch circuits

Another very important package mode for proximity switches is the metal housing. The functions of metal package are its mounting function, mechanical protection and protection from electromagnetic interference (EMI) in industrial environments. Figure 6 shows inductive and optical proximity switches.

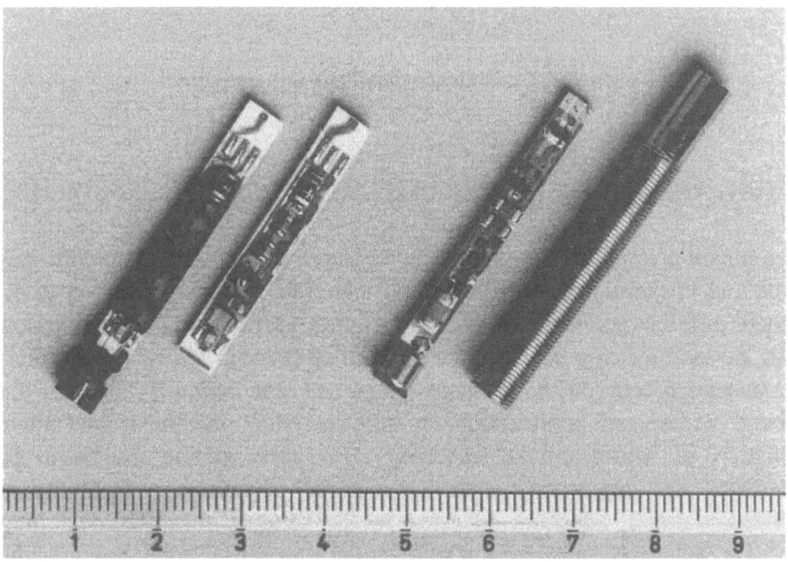

Figure 6: Inductive and optical proximity switches

5. Pressure sensors

Pressure sensors for medical and industrial applications usually use gauge silicon piezoresistive pressure sensors as sensing elements ([5]). The electronic circuits for adjustment of sensor parameters, for temperature compensation and signal processing were realized in thick film technology. Die attachment, local encapsulation and housing are the most critical steps in pressure sensor production since mechanical stresses can be induced in the silicon die. Because of the above requirements the silicon pressure sensor die is mounted on the ceramic substrate where the signal conditioning electronics are implemented. A typical construction solution is shown schematically in Figure 7.

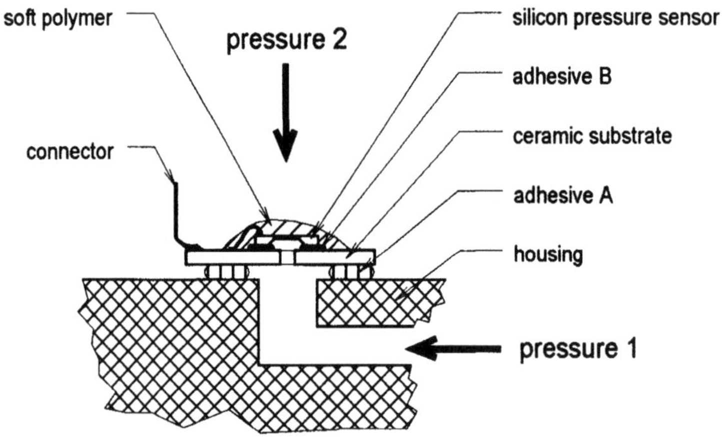

Figure 7: Construction detail of a pressure sensor

MOUNTING AND ENCAPSULATING A SILICON PRESSURE SENSOR DIE

The silicon pressure sensor die is bonded onto a ceramic substrate with polymeric adhesives and encapsulated with polymeric materials ([6,7]). According to the sensor application, different polymers are chosen to satisfy user demands. Flexible polymeric materials are used for chip bonding and encapsulating to minimize stresses due to any thermal mismatch between the pressure sensor and the ceramic substrate. Polymers have found widespread applications for silicone chip attachment and their local encapsulation in hybrid circuit technology. For this specific use some polymer characteristics are important, such as material cleanliness, flexibility of the polymerized material, adequate adhesion to the ceramic substrate, and long term stability. The adhesives must have a sufficiently high initial bond strength and must retain adequate bond strength at maximum use temperature, after exposure to commonly used solvents and high humidity, and after extended ageing.

INDUSTRIAL PRESSURE TRANSDUCERS

In industrial pressure transducers the signal condition circuits are more complex. For this reason the transducers were built in two parts. The first part was a silicon pressure sensor mounted on a ceramic substrate (dimensions 12.7 × 12.7 × 0.64 mm) where the thick film circuit with thick film resistors and thick film thermistors for temperature compensation are realized. The second part was a thick film hybrid circuit which integrates two amplifiers, a reference for offset nulling, and a stable power supply. The block diagram of the first and second part are shown in Figure 9. The thick film hybrid circuit and thick film substrate with silicon pressure sensor were packaged in a plastic housing with dimensions 35.0 × 25.0 × 13.0 mm and two connecting tubes for pressure supply.

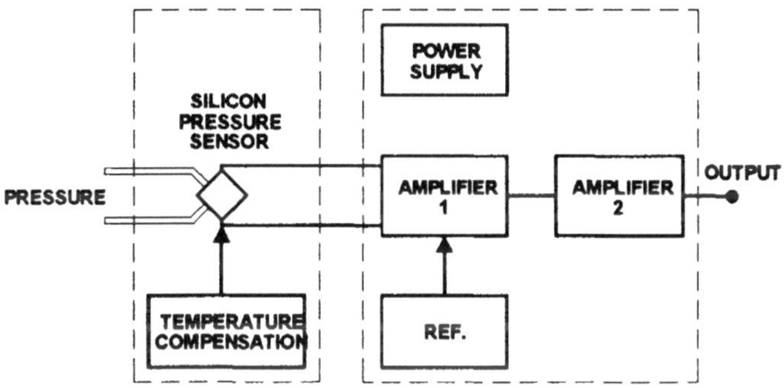

Figure 8 Block diagram of industrial pressure sensor

For successful production of pressure transducers compatibility in the following built-in materials must be achieved: the thermoplastic housing, the ceramic substrate, the silicon sensor element, the encapsulant for the sensing element, the adhesive between the housing and ceramic substrate, and the adhesive for die attachment on the ceramic. The compatibility of these materials (temperature expansion coefficient, residual stress, ...) must be effective for the use in relatively aggressive industrial environments.

MEDICAL SENSORS

In the field of medical sensors ([8,9]) the housing of the sensor must protect the sensor from the patient, and as an additional function, to protect the patient from the sensor. This new demand is called bio-compatibility. On the other hand, the operating temperature range can be reduced to that of the body or room temperature. The definition of bio-compatibility should be used to denote the collection of phenomena that are associated with the iteractions between bio-materials and physiological environments in which they have to coexist. As such, the term encompass both directions of the interfacial reaction, that is the effects of the physiological environments on bio-material and of the materials on the environments. The traditional bio-materials are titanium, alumina and some polymers.

6. Conclusion

To take a full advantage of sensor elements and transducers, they must be constructed with the packages which can handle different types of physical or chemical quantities, applications, environmental conditions, mechanical and thermal stresses, etc. The packaging and assembling technology have an important rule in the world of sensors.

7. References

1. R. Dell'Acqua, Sensors: A great chance for microelectronic technologies, Proc. 22nd Int. Conf. Microelectronics MIEL-95 / 30th Symp. on Devices and Materials SD-94, Rogla, 1994, 25-43
2. D. Belavič, S. Šoba, M. Pavlin, S. Maček, M. Hrovat, D. Ročak, Design of thick film hybrid circuits for sensor applications, Proc. 24rd Int. Conf. Microelectronics MIEL-96 / 32st Symp. on Devices and Materials SD-96, Nova Gorica, 1996, 237-242
3. N. Sinnadurai, M. Mullins, A. Medler, d. Court, R. Bayford, A. van Putten, C. Patel, Research into novel sensors on silicon substrates, Proc. Int. Symp. on Microelectronics ISHM'94, Boston, 1994, 444-449
4. A. Kahnert, New Casting Compounds and Conformal Coatings for Sensor Technology, Proc. 7th Int. Exhibition with Congress for Sensors, Transducers & Systems SENSOR 95, Nürenberg, 1995, 691-696
5. M. Pavlin, S. Amon, U. Aljančič, Temperature properties of silicon piezoresistive pressure sensors, Proc. 23rd Int. Conf. Microelectronics MIEL-95 / 31st Symp. on Devices and Materials SD-95, Terme Čatež, 1995, 373-378
6. J. S. Bergstrom, W. H. Teat, Package evaluation for silicon pressure sensors, Proc. Int. Symp. on Microelectronics ISHM'87, Minneapolis, 1987, 89-94
7. M. Pavlin, D. Ročak, S. Šoba, S. Amon, Evaluation of polymer adhesives for use in silicone pressure sensor bonding on ceramic substrates, Proc. 24rd Int. Conf. Microelectronics MIEL-96 / 32st Symp. on Devices and Materials SD-96, Nova Gorica, 1996, 243-248
8. D. Gee, J. Bryzek, New generation of disposable blood pressure transducers, Proc. 7th Int. Exhibition with Congress for Sensors, Transducers & Systems SENSOR 95, Nürenberg, 1995, 39-43
9. J. C. Lesho, Biomedical packaging from the package designer's perspective, Proc. Int. Symp. on Microelectronics ISHM'89, Baltimore, 1989, 477-484

RADIATION HARDENED ASIC DESIGN FOR SPACE APPLICATIONS

JANEZ TRONTELJ, LOJZE TRONTELJ
University of Ljubljana
Tržaška 25, 1000 Ljubljana, Slovenia

1. Introduction

The effects of radiation in CMOS integrated circuits are numerous, however the dominant are the effects of the ionizing generation of both electrons and holes.
In the semiconductor this effect generates photocurrent, so the circuit must be designed to be capable to handle such currents. Usually no permanent damage is associated with these currents.
In the silicon dioxide on the contrary the situation is different. The less mobile holes which do not recombine can be trapped in the insulator. With the applied electric field these holes move to the surface. The resulting charge build-up causes permanent shift of the surface potential. For PMOS devices this means an increase of the threshold voltage and for NMOS devices it means a decrease of the threshold voltage.
The objective of the presented work was to design a mixed signal standard CMOS ASIC for absolute optical angular encoder for accurate positioning needed in space applications. The function of the ASIC was analog and digital processing of the photo-diode current determining the absolute angular position being digitized to the maximal resolution of ten angular seconds. The task of the ASIC was to provide eight bit resolution from sine/cosine photo-diode current. The required radiation tolerance was 100kRad. The radiation hardened design approach was divided into analog design considerations, digital design and device isolation and latch-up optimization.

Fig.1: Thin oxide extension of the N-channel MOS transistor to prevent drain-source leakage

R. Tummala et al. (eds.), Electronic Packaging for High Reliability, Low Cost Electronics, 205–210.

2. Radiation tolerant analog design

The selected technology was N-well 2μm standard CMOS process with minimal radiation hardening features as minimization of the gate oxide thickness below 30nm, and the use of thin oxide extension of the channel for NMOS transistor, as shown in fig.1.
Experiments on test transistors proved that the threshold shifts for 100kRad dose were below -0.4V for N-type transistors while for P-type transistors the threshold shift was below +0.3V.
P-type transistor showed also less degradation of 1/f noise.

In the design we have addressed the following issues:

1. Degradation of matching
2. Increase of the noise
3. Degradation of high impedance nodes
4. Operating point shifts due to increasing difference of the absolute values of the PMOS and NMOS threshold voltages.

The performance of the circuit was determined mostly by the following subcircuits:

- Operational amplifiers
- Comparators
- Bias subcircuits

Fig.2 shows the schematic diagram of the operational amplifier. Due to the noise and offset voltage behavior the P-type transistors were selected for the input differential amplifier stage.

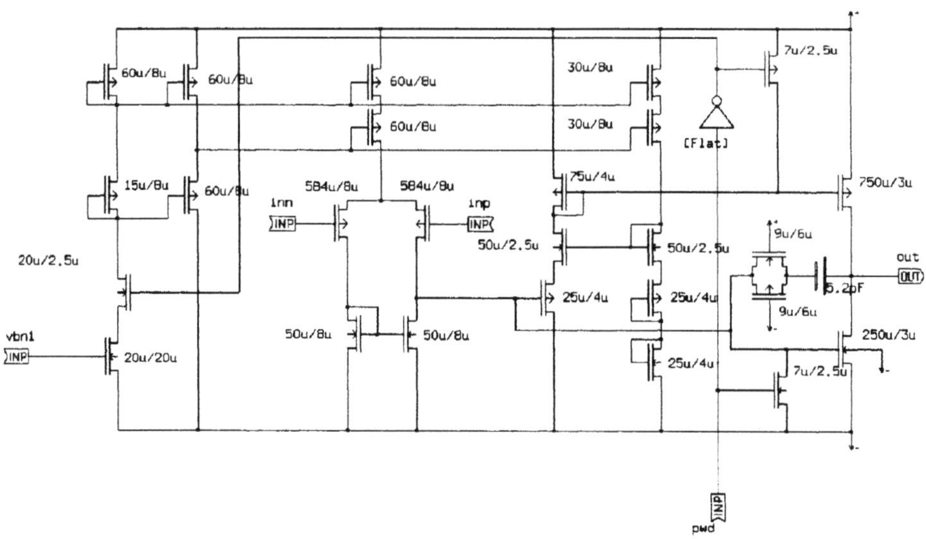

Fig.2: Schematic diagram of the operational amplifier

Degradation of matching was compensated with the minimization of the sensitivity for random asymmetry of the devices in the differential stage. The operating point and the device sizing was optimized and characterized by the use of extended corner analysis (additional threshold voltage shifts and mobility shifts were added to the normal corner extremes). The layout introduced asymmetries were also minimized by careful layout.

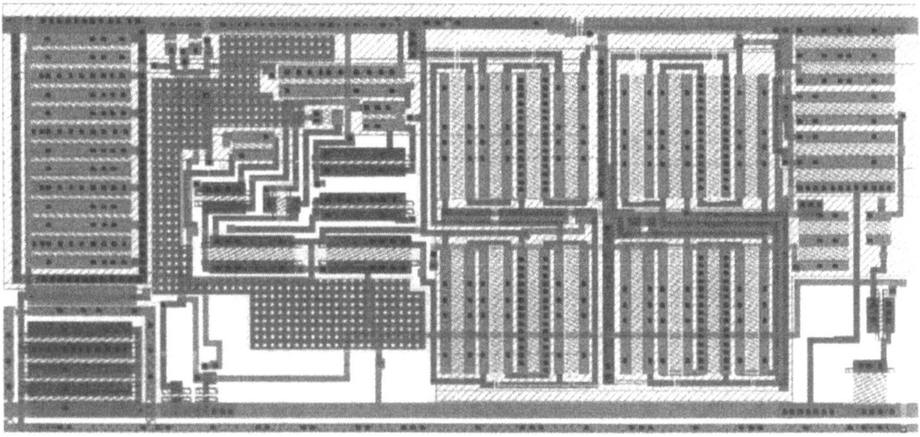

Fig.3: Layout of the operational amplifier

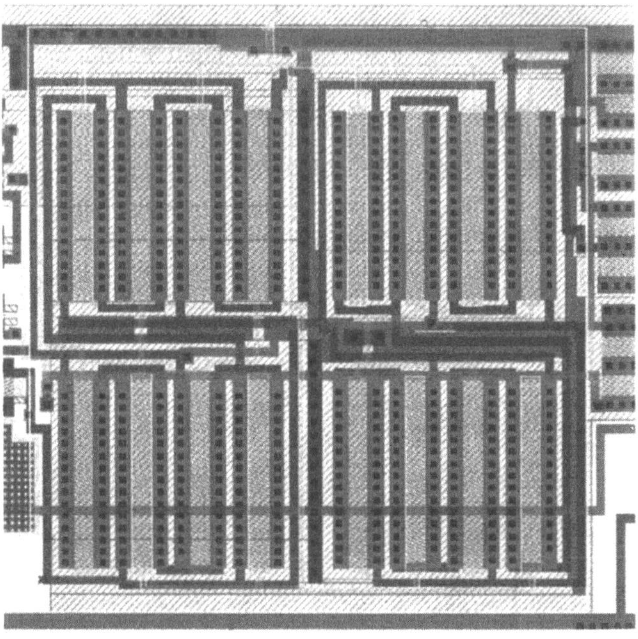

Fig.4: Layout detail - input differential stage of the operational amplifier

Figs. 3 and 4 show the layout of the operational amplifier used. As a result we received very good preradiation offset voltages, not exceeding 0.5mV. After radiation only up to 1mV offset voltage was measured, which caused no performance degradation.
1/f noise was optimized by increasing the gate area of the input differential transistors and sizing the load devices taking into account post radiation noise figures.
The preradiation noise was below one tenth of the required noise limit. Due to this fact no performance degradation was detected after radiation. In the design phase we have selected such amplifier topology which is insensitive to high impedance degradation due to the leakage current, photocurrents or parasitic bipolar transistor currents, so no problem was detected regarding these phenomena.

Operating points were strictly controlled by constant bias current generator which was in principle temperature compensated PTAT current source. This approach proved to be very effective; the maximal current deviation of preradiated ASIC total current consumption to radiated one was +10%.

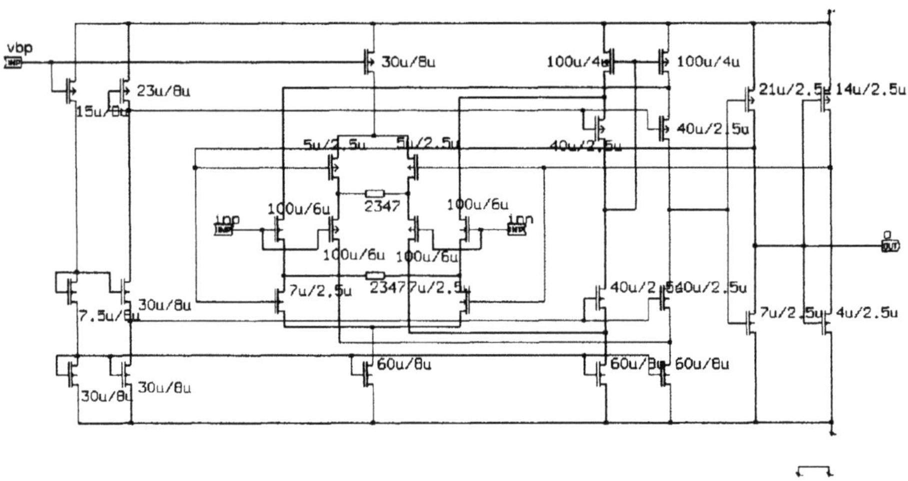

Fig. 5: Schematic diagram of the comparator

The specification for the offset voltages of the comparators were less strict compared to the amplifiers when we decided to use rail to rail common mode input voltage range. For this reason P and N type input differential stage was utilized. Similar approach as for operational amplifier design and layout were utilized also for comparators. The comparator layout is shown in fig.6.

3. Radiation tolerant digital design

Digital design is less problematic compared to the analog due to the nature of the signals. Nevertheless the following problems were taken into consideration:

1. Increase of quiescent current
2. Digital noise immunity degradation
3. Degradation of the charge in the circuitry using dynamic logic
4. Shift of the "off" state to subthreshold operation.

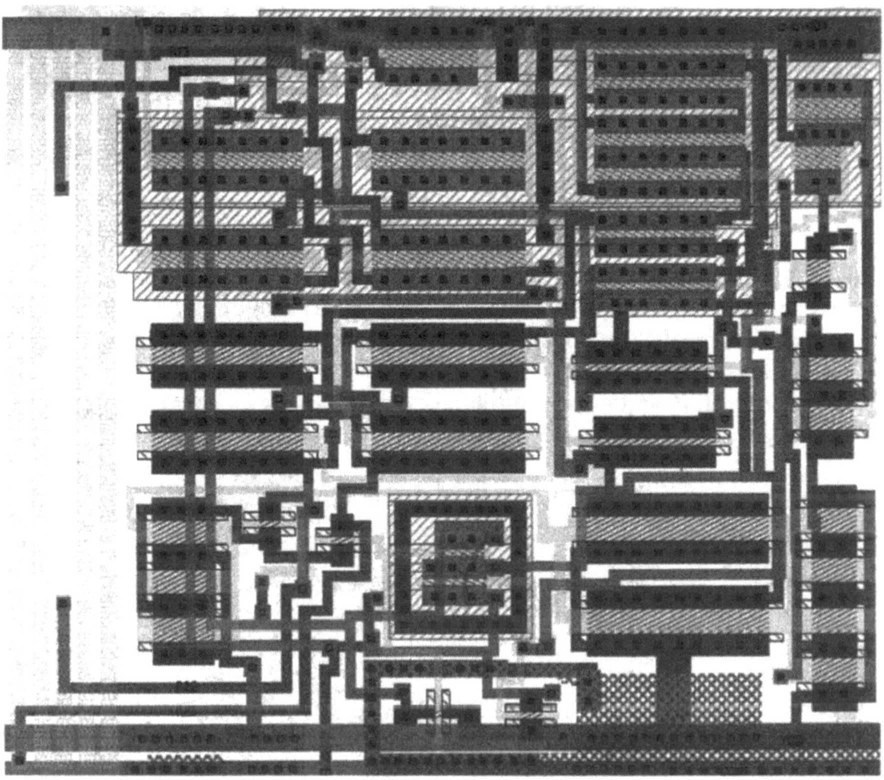

Fig.6: Layout of the comparator

The increase of quiescent current was reduced by the use of longer channel devices. The most problematic issue in the presented design was digital noise immunity degradation which was corrected in the redesign by the use of signal conditioning such as Schmitt-triggers and hysteresis in comparators where possible and to prevent crosstalk by layout and by the use of controlled switching currents. Unfortunately all the logic had to be asynchronous since no system clock could be used. To avoid charge degradation we did not use dynamic logic.
N-channel devices have a tendency to operate in subthreshold region when in the "off" state. Since the number of digital transistors in the ASIC was relatively low we did not see more than 10% current consumption increase after radiation.

4. Other considerations

Devices isolation and latch-up are often reported problems for the circuit exposed to more than 10kRads.

Design and layout can help in these areas by the use of extensive guard bars. In our case we have followed this rule by sacrificing some silicon area. Epi wafers were used to help minimizing latch-up problems.

5. Results

Fig. 7 shows the photomicrograph of the ASIC. As it can be seen the majority of the ASIC is analog circuitry. The main problem which required single redesign step was degraded noise immunity in the digital processing of the signal. When we improved this by more robust processing the ASIC fulfilled all specification before and after radiation.

Ten parts were exposed to radiations up to 100kRad in steps of 10kRad, 30kRad, 50kRad, 100kRad and after each radiation the parts were characterized.

For every part and every radiation step the functional test showed full compliance with the specifications, while some of the parameters, like the supply current and angular position shifted slightly showing some tendency to partly recover after few days. The ASIC is in production and is working in space applications for more that five years. The authors have no report of malfunctions.

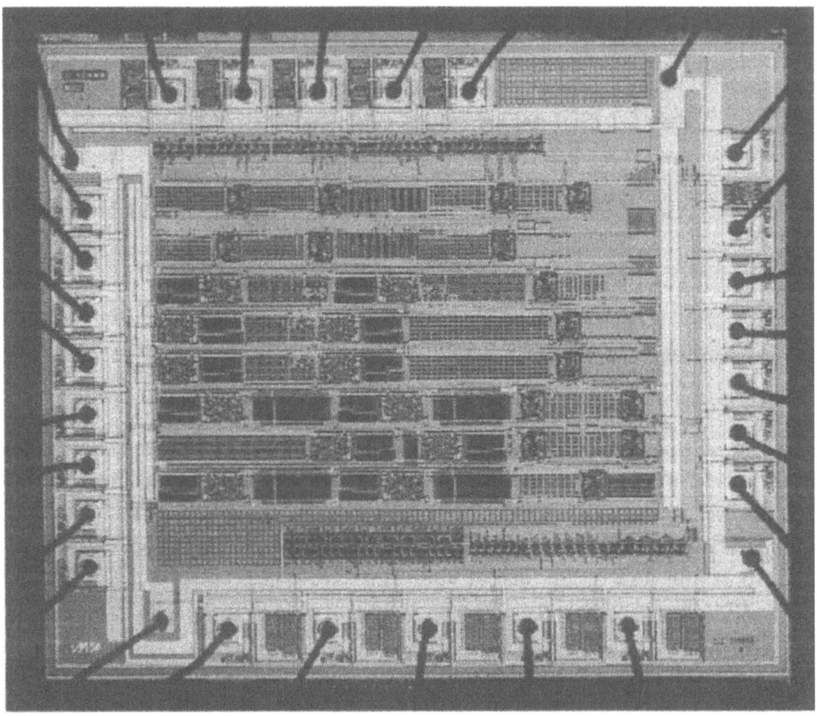

Fig. 7: Photomicrograph of the ASIC

LOW COST PACKAGING ENCODER CIRCUITS

S. Amon, D. Resnik, D. Vrtacnik, T. Sladic*
Laboratory for Electron Devices,
email: slavko.amon@fe.uni-lj.si
Faculty of Electrical Engineering, University of Ljubljana
Trzaska 25, 1000 Ljubljana, Slovenia
* ISKRA - EMECO / R&D, 4000 Kranj, Slovenia

1. Introduction

Low cost high reliability electronic packaging circuits are gaining growing attention from many R&D people as well as producers. This is a consequence of new possibilities, based on innovative concepts related to improved materials and technologies, having impact from chips design and realization to PCB fabrication.

The main goal of the presented work was to realize a low cost and high reliability packaging for optical encoder and similar circuits. Basic constraints for these optoelectric circuits were low optical loss access to the photochips, good mechanical compatibility of the involved materials, high hermetic properties of the package and application of low cost materials. To meet all this demands, a "complete fabrication controll" approach from photodiode chips design and fabrication for improved environment resistance to PCB fabrication was introduced. Presented paper gives a review of works necessary to fulfill the main goal for low cost yet reliable circuits.

Concepts developed and reported in this work are applicable also to other (opto)electronic circuits with similar constraints.

2. Photodiode Chip Fabrication

Photodiode chips with improved environment resistance were developed and fabricated on silicon planar pilot production line in authors Lab. Photodiode chip design was based on existing standard planar p^+nn^+ silicon photodiodes with oxide antireflective layer in the active area of the device[1]. Special care during chip design was devoted to the demand for improved environment and humidity resistance of the chip while not to degrade electrooptical properties of the photodiode. This was achieved by a proper

R. Tummala et al. (eds.), Electronic Packaging for High Reliability, Low Cost Electronics, 211–221.

design and fabrication of oxide and nitride antireflective/protective layers as described in the following.

In order to achieve good photodiode optical properties, it is necessary to maintain low minority carrier recombinations as one of the most important limiting factors. Material quality and critical process steps maintaining high bulk lifetime are usually adequate, nowadays the dominant photogenerated carriers loss is surface recombination. Most common approach of emitter surface termination in today's high efficiency photodiodes fabrication is to grow thin high quality thermal passivating silicon dioxide. Such an oxide exhibits very low density of the fixed oxide charge and Si-SiO2 interface states, and thus the low surface recombination velocity. To improve environment resistance of the photodiode, a second layer of low pressure chemical vapor deposited (LPCVD) silicon nitride Si_3N_4 was introduced[2].

Therefore the first layer of antireflective/protective coating on p^+nn^+ photodiode chip was thermally grown thin SiO_2 in dry oxygen atmosphere with or without addition of TCA, which has primarily the function of passivating the doped emitter surface by reducing the surface states at Si-SiO_2 interface. As a consequence, lower surface recombination velocity is achieved and finally higher photocurrent via considerable rise of spectral response in short wavelength region. On the other hand it serves also as an antireflective layer, but unfortunately with an unmatched refractive index to the silicon. Therefore another layer is needed on the top to reduce the reflectivity losses. The ideal top layer would have to exhibit the refractive index of 2.3 without any significant absorption. The refractive index of low pressure chemical vapor deposition silicon nitride Si_3N_4 is quite close to this value and in addition it satisfies some other criteria which our device has to fulfill (such as superior barrier against moisture and alkali ions from the ambient) [2].

Short circuit current density J_{SC} was observed as the most adequate parameter to evaluate the efficacy of different double layer antireflective/protective coatings. As shown in Fig.1, six different oxide and eight different LPCVD silicon nitride thickness (48 combinations) were fabricated and measured.

Short current density J_{SC} has a peak for every curve of selected thickness of oxide layer at one specific thickness of nitride (Fig.1a). Curve (and peak) is shifted toward thicker nitride top layers as the oxide thickness decreases. This imposes the interference of trapped light within the layers. Peak values remain quite constant (+/-1%). The highest peak in J_{SC} was achieved with 28.5 nm SiO_2/60 nm Si_3N_4. If the thickness of both layers are not matched, the J_{SC} variations can be as high as 26% within the chosen range.

As expected, antireflective coating has negligible influence on the open circuit voltage Voc as presented in Fig.1b (<3%). The maximum V_{OC} for every chosen thickness of oxide is given at 60 nm of nitride. At this point V_{OC} is equal for all oxide thickness. Since all the oxides were grown in one run, the surface passivation is the same for all samples. The only factor for changed V_{OC} is then the changed spectrum of absorbed light due to different antireflective layers.

To evaluate different antireflection coatings in order to distinguish between the effects responsible for different photocurrents obtained, the spectral response was measured on different samples as shown in Fig.2. The results are presented in Fig.3a.

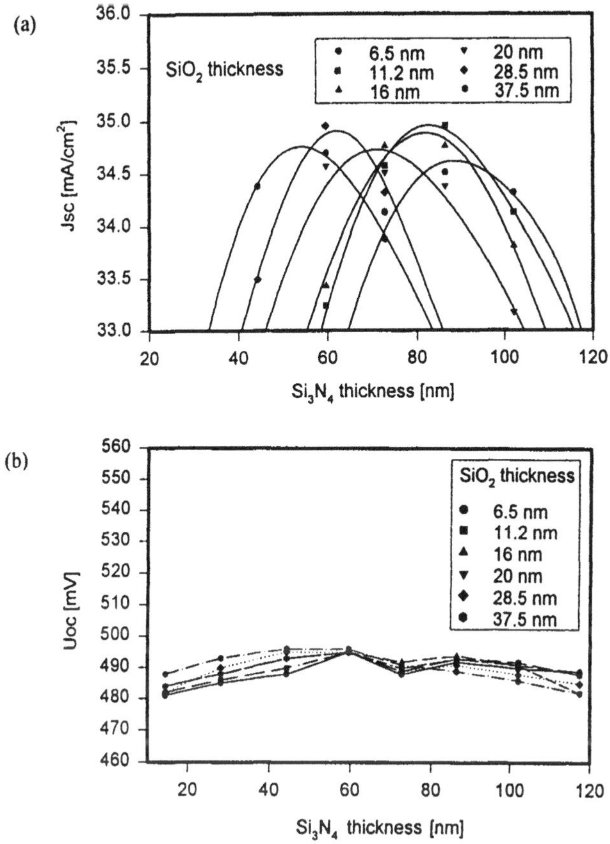

Figure 1. Short current density J_{SC} (a) and open circuit voltage V_{OC} (b) as a function of different SiO_2/Si_3N_4 thickness combinations

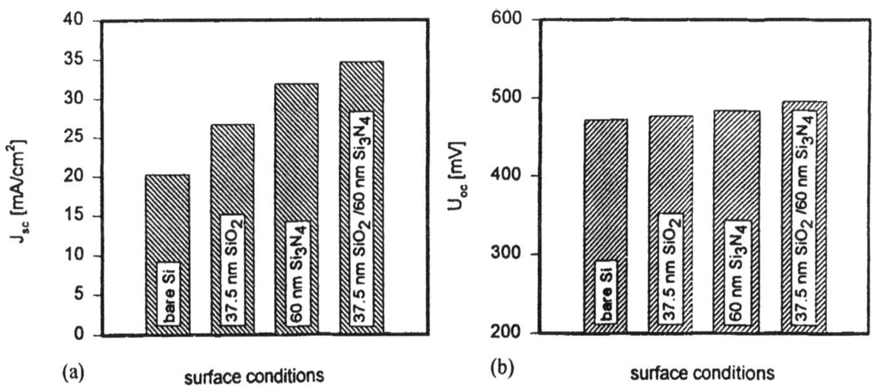

Figure 2. Influence of different surface conditions on short current density J_{SC} (a) and on open circuit voltage V_{OC} (b)

Bare silicon surface has poor response across entire wavelength range, particularly in shortwavelength portion due to high surface recombination velocity. This is improved by passivating the surface with thermally grown SiO_2. This single layer as well decreases the silicon surface reflectance over the entire range. Also Si_3N_4 layer passivates the Si surface effectively (but definitely not as good as the oxide). Unfortunately nitride has relatively high absorption in shortwavelength region and the passivation effect is not seen directly. In visible and infrared portion, Si_3N_4 shows dominant action as antireflective layer and therefore higher spectral response.
With double-layer antireflective coating we gain better response in infrared portion due to the lower reflectance i.e. better light trapping. In ultraviolet portion we loose some response due to the dominant absorption of the top layer despite of the good passivation effect of oxide. High red response also implies a good lifetime of the base as well as an effective back surface nn^+ field and reflectance from metal back surface. As all the samples have the same treatment, also the response above 1000 nm is almost the same.

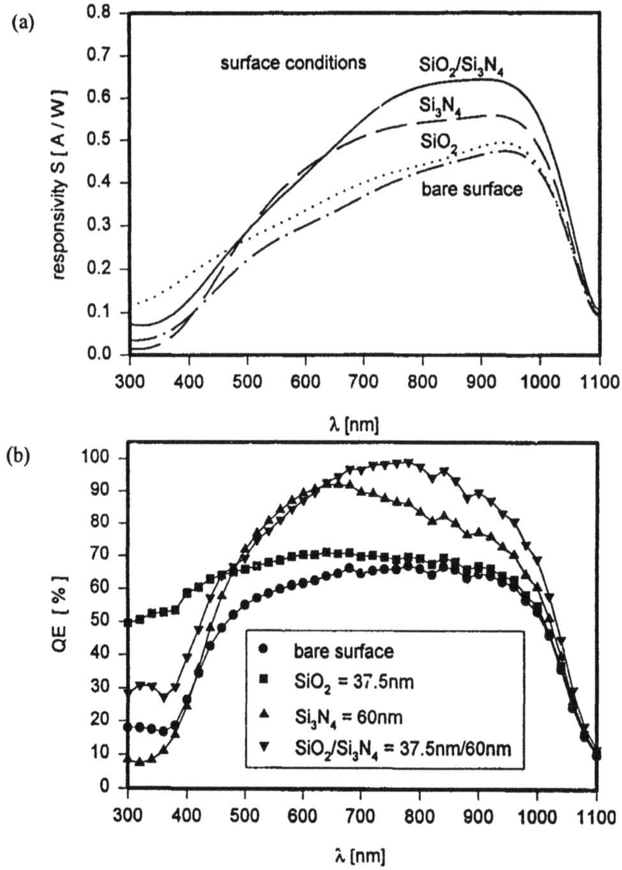

Figure 3. Measured spectral response (a) of the samples with different surface conditions and adherent external quantum efficiency (b)

Extensive characterization and analysis were carried out to provide guidelines for an optimized double layer antireflective/protective coating, taking into account the results from illuminated IV characteristics, spectral response and model data. The most significant conclusion is that thermally grown SiO_2 in O_2-TCA atmosphere is inevitable as passivating layer in the first place and in the second place as an antireflective layer. Silicon nitride Si_3N_4 layer alone does not function so well as a passivating layer, but acts better as an antireflective layer, particularly in infrared portion. The combination of both layers revealed superior optical results over single layers and optimum was found at values 28.5 nm SiO_2/60 nm Si_3N_4.

3. Assembly and Encapsulation Technique of Module

The aim of this R&D part of the presented work was to develop a module for optical position sensing. The PCB layout which was adopted is presented in Fig.4.

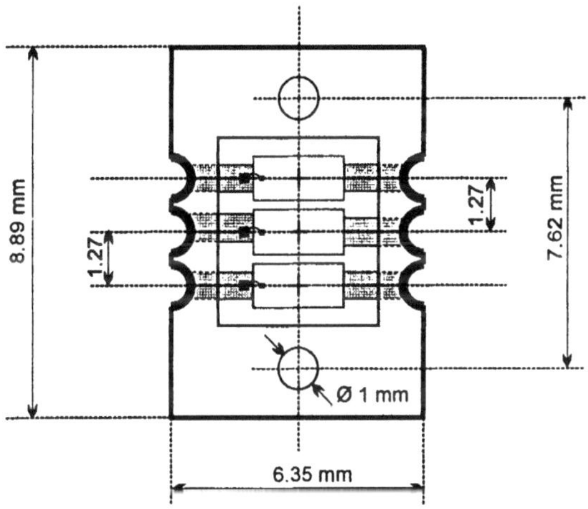

Figure 4. Encoder module design with three photodiode chips

Three photodiode chips are mounted on exact position on the PCB and contacts to the module are satisfying the demand for SMD mounting and for soldering from the side. In principle when the technology of mounting and assembling is solved we can extend it to various more sophisticated layouts. Following our layout, masking films were elaborated for the realization of PCB and CNC milling machine.

3.1. PCB REALIZATION

The basic idea was to fabricate double side PCBs with photolithographical process for higher throughput and lower price. This approach enables realization of more encoder modules at a time, as can be seen in Fig 5. After the copper etching, the plating of 2μm of nickel layer, followed by electroplating of 0.2μm of gold was made. The through holes, which are connecting both sides, were carefully examined after metal plating to avoid any inconsistencies of the process. This metal layers have to ensure a reliable bonding of photodiode chips to the PCB and also good backside attachment of the chip to the PCB by silver epoxy glue.

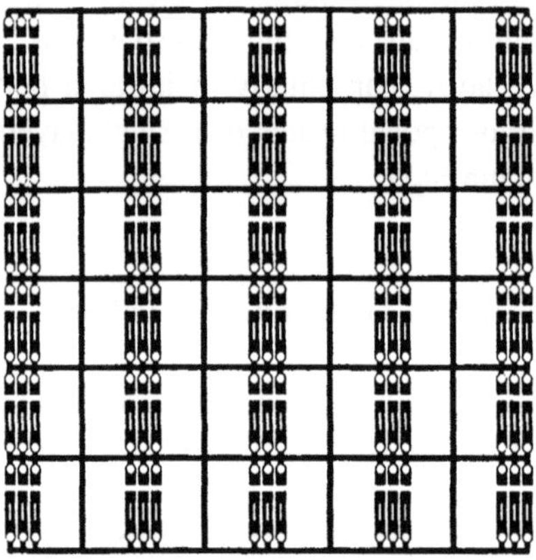

Figure 5. Photomask for front side of PCB for attachment of backside of photodiode chips and frontside bonding connections

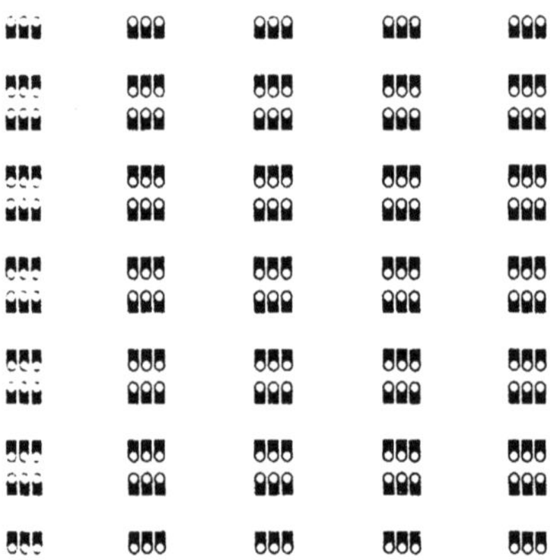

Figure 6. Photomask for back side of PCB, enabling the SMD assembly

3.2. BACKSIDE ATTACHMENT OF Si PHOTODIODE CHIPS ON PCB

Conductive silver epoxy glue was deposited on precleaned golden pads by means of semiautomatic dispensing unit (or manually with syringe in the early stage).
The conductive epoxy silver glue TRA-DUCT 2958 of Tracon company was chosen among several suppliers. A close attention has to be paid during glue deposition on the

proper quantity of the glue applied on gold plated PCB pad to avoid cross-connection between the chips.

The positioning of the photodiode chips is one of the most severe and demanding operations in the whole process of encapsulation, due to very narrow specifications of the customer. Tolerances concerning the location of chips on the PCB and distances between them were 20μm or less. In case of wider tolerances an experienced operator would be able of placing chips within 50μm with vacuum tweezers and microscope. For more precise and accurate positioning in a large scale production, pick-and-place machines are commonly used to satisfy criterion.

In our case, a special mask was developed (Fig.7) that enabled us to do the positioning of our chips.The mask was made of thin stainless steel foil, into which the openings were etched following photolithographical steps, that exactly matched the size of the chips and define as well their mutual distances. The tolerances are in the range of dicing saw track fluctuations. Chips have been manually placed backside up into this mask opening and afterward the PCB from previous section was laid on. The alignment of both is defined by three aligning holes. In this way we have achieved the satisfactorily positioning although the loading of the chips was done manually. For small scale production this is probably the most viable solution as it demands very low investment. The tray with positioned and attached photodiode chips is subsequently placed in an oven at temperature of 125^{0}C for 15-20 minutes to ensure the proper curing of epoxy glue. After this operation the positioning mask is removed and is ready for the next loading.

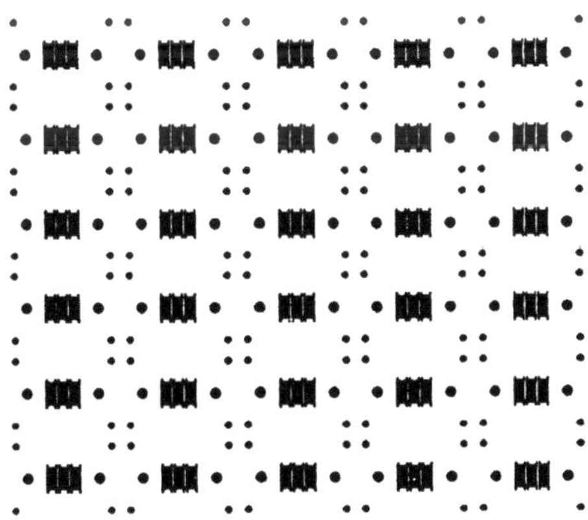

Figure 7. Photomask developed for precise loading and positioning of the photodiode chips on the PCB

3.3. BONDING

After backside attachment, the wedge bonding with aluminum 30μm diameter wire as a standard operation was applied to connect the anode of the photodiode chip to the PCB. To ensure the bonding reliability, 5gram pull test was performed at the end.

3.4. ATTACHMENT OF THE TOP FRAME

Top frame is also made of FR4 laminate by means of CNC milling machine, which makes the openings as defined by mask on Fig.8, which actually defines the opening size which will constrain the transparent fill representing the optical window of the module. The main purpose of this top frame is to avoid spreading of the fill all over the top surface.

Thin layer of thyxotropic glue is applied on the rear side of the top frame. The alignment of the top frame tray and the bottom tray of positioned and attached photodiodes is provided via aligning holes. The glue is afterward dried and hardened in air at room temperature for 60-90 minutes.

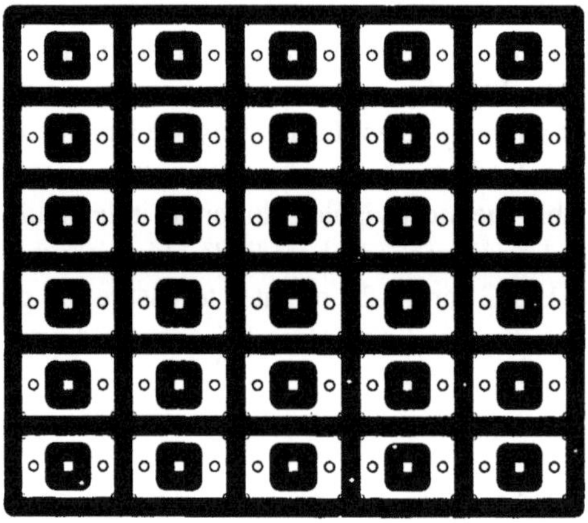

Figure 8. Top frame mask, defining the optical window of the encoder module

3.5. OPTICAL WINDOW FORMATION

The purpose of this operation is to mechanically protect the photodiode chips and the bonding wires as well as to ensure the final environmental hardness against humidity and dirt shadowing effects. In addition the transparent layer should not degrade the basic optical response of the photodiode chip itself. After performing several tests, the OS4000 "Dexter Hysol" transparent epoxy was chosen. Semiautomatic dispensing unit was utilized for dosing. Due to surface tension, a slightly concave shape develops if the dose is lower than as-enough-to-fill the opening.This case can be observed on Fig 9, showing the cross section of the encoder module. We should remain within the height limits determined by the thickness of both laminates and not to exceed this by overdose of the transparent epoxy content. Furthermore, the meniscus would become convex in that case.

The other important aspect in filling the openings is that the formation of air bubbles should be avoided, due to possible optical diffraction or even malfunction of the assembled module. The curing of the transparent epoxy was done therefore at temperature of 125^{0}C for 30 minutes in air.

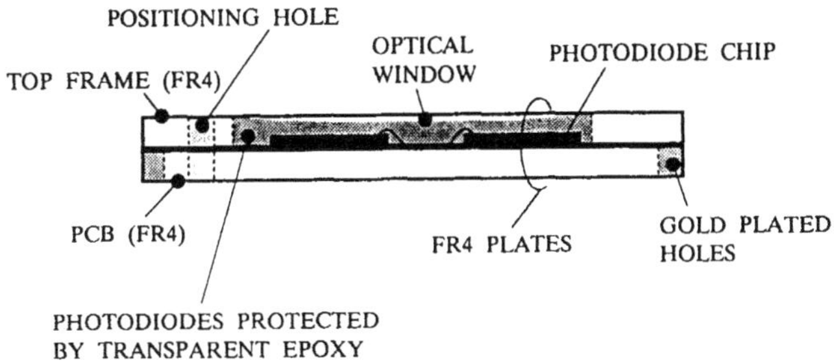

Figure 9. Cross section of the encoder module

3.6. SEPARATION OF ENCODER MODULES

Separation of the whole array of modules into single dies is done either by milling machine or with dicing saw. The position of cutting lines are enabling separation in a manner that half of the metalized through hole is left free for eventual side soldering. Although the milling machine offers more acceptable solution of separation compared to dicing method, it requires more cutting space and is more time consumable.
More sophisticated way of separation is to attach the array ready for separation onto the adhesive dicing tape and then perform the cutting completely through the die. In this case dicing showed to be the preferable method.

4. Measurements and Testing

4.1. THERMAL CYCLING

The purpose of this operation is to preliminary locate and eliminate the potential malfunctional modules which will fail at fast thermal cycling. This would include cases such as bonding connections reliability, backside chip to module effects, effects of transparent epoxy fill and its thermal expansion compatibility to photodiode chips or/and laminate FR4.
A set of modules under test was immersed into vessels filled with FC43. First vessel with cold media at -30^{0}C and then hot one at 110^{0}C. This was done alternatively in time intervals of 10 minutes. After the procedure was completed, rinsing in IPA removed the traces of FC43. In addition, this was also a rigid test for hermetical properties of the encapsulation, which showed that the compatibility of chosen materials and technological approaches were adequate.

4.2. AGING OF THE MODULES

Aging was performed as the last step of the encapsulation process with the aim to provide long term stable electrical and mechanical parameters of the finished module.

Procedure was performed at 120°C for 24 hours. Afterwards the electrical test was made and almost no variations were observed.

4.3. MEASUREMENT OF THE SPECTRAL RESPONSE

Spectral response of the encapsulated photodiode chip was measured in order to determine the influence of the transparent optical window on the module performances. Fig. 10 is showing the two curves of responsivity , one for the photodiode chip without epoxy window filling and one for fully encapsulated chip (final performances). There is a small decrease in red part of the spectrum, less than 10%, while in visible region there can be observed even a slight increase of the response, probably due to the enhanced light trapping. In UV region there is a natural decay of Si chip response which is actually not of interest because optical sources applied are commonly emitting in red region, around the response peak.

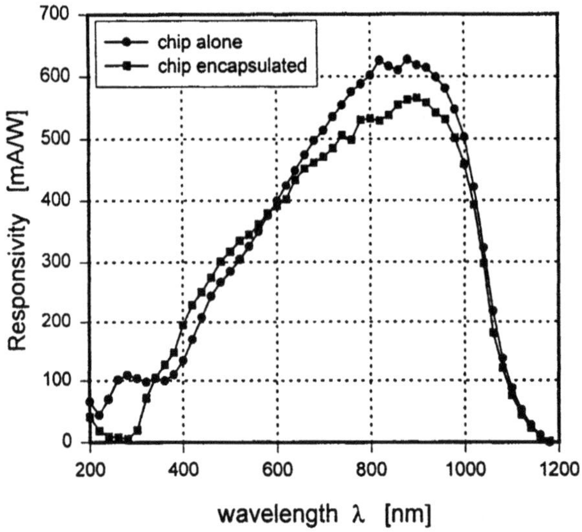

Figure 10. Spectral response of the chip and the chip in the finished module

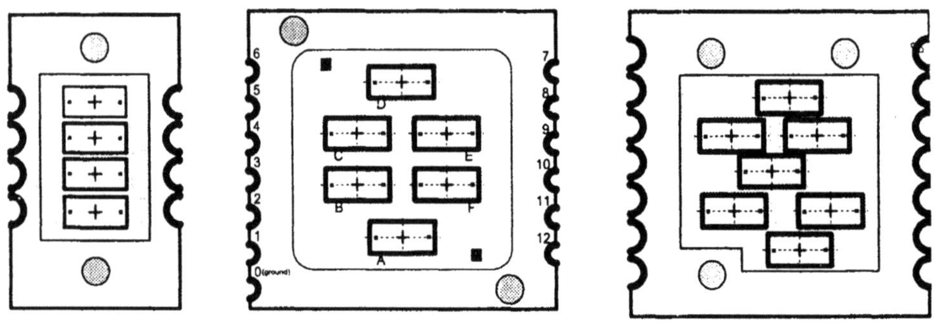

Figure 11. Typical encoder circuits layout with positioning marks

Figure 12. Photograph of fabricated encoder modules

5. Conclusions

Low cost optoelectric PCB based circuits were designed, developed and fabricated. Fabricated miniature circuits with mounted photochips demonstrated good electrical, optical/mechanical and reliability properties. The approach reported is appropriate also for other optoelectric low cost yet reliable circuits with similar constraints as presented encoder circuits.

6. References

1. D. Resnik, D. Vrtacnik, U. Aljancic, S. Amon (1995), "Growth conditions for thin passivation oxides", *Proc. of 23rd International Conference on Microelectronics, MIEL-SD 95*, Terme Catez, Slovenia

2. D. Resnik, P. Popovic, D. Vrtacnik, S. Amon (1994),"Characterization of thermal SiO_2/Si_3N_4 double layer antireflective coating", *Proc. of 22th International Conference on Microelectronics, MIEL-SD 94*, Rogla, Slovenia

MICROWAVE FILTERS MANUFACTURED BY ADVANCED THICK-FILM TECHNIQUES AND NOVEL SILVER PASTES

M.CIEŻ, B.DZIURDZIA

The Research and Development Centre for Hybrid Microelectronics and Resistors, PL. 30-701 Cracow, 39 Zabłocie Str., Poland

St. NOWAK

Department of Electronics Academy of Mining and Metallurgy, PL. 30-054 Cracow, 23 Reymonta Str., Poland

Abstract

With increasing demands on microwave devices of high performance, low cost and high volume of production, thick-film technology has become, for certain frequency range up to 14 GHz, an alternative manufacturing method. In this paper the application of three advanced thick-film techniques such as: ultra-fine screen printing, etching of fired layers and photoimagining of photosensitive layers (FODEL^R technology) for fabricating bandpass microwave filters have been presented. Novel fritless silver paste 9912F by ESL and photosensitive silver paste FODEL^R 4095 by DuPont have been employed to perform filter structures. Compact planar filters of microstrip edge-coupled configurations of centre frequency 12 GHz and 30% bandwidth has been designed and then manufactured in different technological versions. Their insertion and return losses have been measured with a vector network analyzer HP8720C. The results have been listed in forms of tables and plots and then analyzed.

Investigations, carried out at the Research and Development Centre for Hybrid Microelectronics and Resistors in Cracow, have proved, that at comparable good technical results for all used thick-film techniques, taking into account the cost criteria, FODEL^R technology exhibits the most promising prospects for manufacturing microwave hybrids.

1. Introduction

Modern microwave communication systems require, especially in satellite and mobile communications, high performance narrow band bandpass filters having low insertion loss. In order to reduce size, weight and cost, there has been a growing interest in planar structures. Advanced thick-film techniques in combination with novel thick-film conductors offer new solutions in manufacturing cheap planar microwave filters operating up to 14 GHz.

R. Tummala et al. (eds.), Electronic Packaging for High Reliability, Low Cost Electronics, 223–233.

2. Thick-Films for Microwaves - Advantages and Disadvantages

Microwave requirements for conductors are as follows [1,2]:

- high conductivity,
- fine line and space resolution,
- fine edge definition,
- dense, homogenous structure,
- rectangular cross-section,
- smooth surface,
- thickness of several skin depths.

Thick-film conductors manufactured in conventional way exhibit some considerable advantages but also some disadvantages for microwave applications (Fig.1)

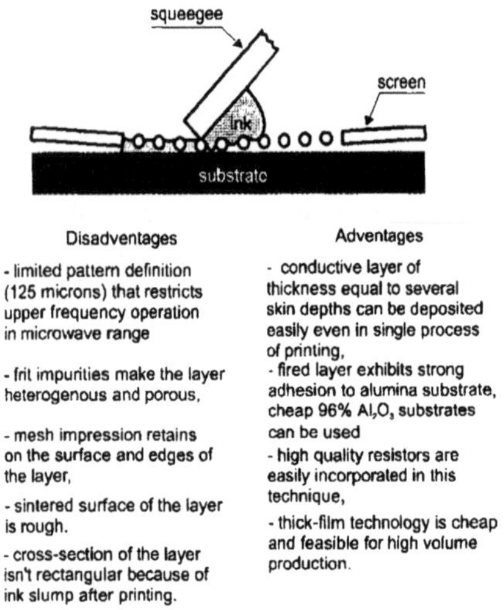

Fig. 1. Advantages and disadvantages of thick-films in microwaves [1].

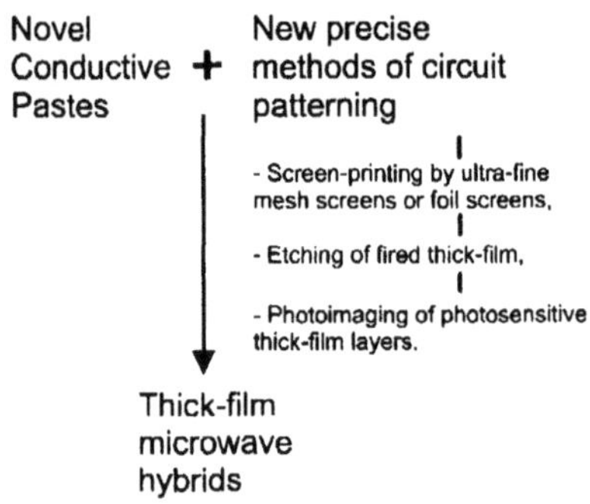

Fig. 2.Prospects of thick-film technology.

Development of novel materials and advanced techniques of circuit patterning diminishes or even overcomes drawbacks of thick-films at high frequencies and opens new prospects for thick-film technology (Fig.2):

- ultra-thin calendered stainless steel screens with novel direct/indirect emulsion system in conjunction with fine line pastes allow to achieve, with using screen printing, line and space resolution comparable to thin-film one,
- photolitographic techniques combined with screen printing provide improvement in edge definition and cross-section of a layer and as well in further increasing of pattern resolution,
- fritless and mixed bonded conductive pastes of high conductivity near that of a bulk metal create after firing dense, uniform layer of smooth surface,

- photosensitive conductive pastes simplify precise photolitographic methods of circuit patterning.

Thin-film technology uses expensive 99.5% alumina substrates for manufacturing microwave hybrids. In the case of thick-films the substrate surface finish seems to play minor role because they tend to be self smoothing at the substrate interface so it is possible to reduce costs by using 96% alumina substrates with a rougher surface. Additionally the increased surface roughness improves conductor adhesion.

Thick-film technology uses also another 'hard' substrates such as sapphire, aluminum nitride AlN, beryllia BeO, silicon carbide, titanates. Fig.3 shows the area of their applications in relation to frequency range and power dissipation [3].

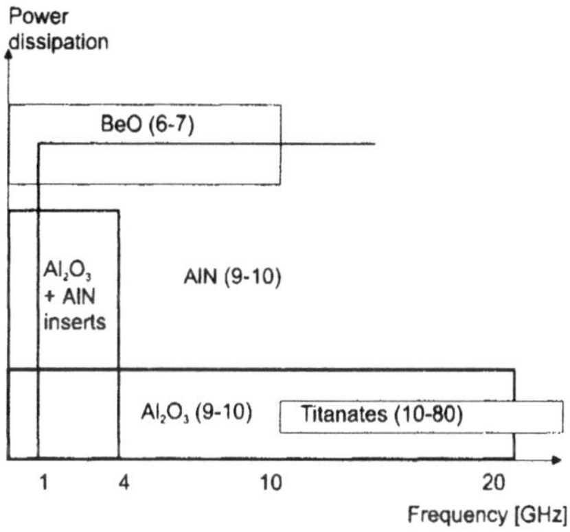

Fig. 3. Substrates used for thick-film microwave hybrids [3].

This paper deals with application of advanced thick-film techniques such as: ULTRA-FINE SCREEN PRINTING, ETCHING and PHOTOIMAGING OF PHOTOSENSITIVE THICK-FILM layer into manufacturing microwave hybrids on the example of band-pass microwave filters. Cheap silver pastes: printable 9912F by ESL and photosensitive FODEL^R 4095 by DuPont were employed in investigations as well as cheap substrates 96% Al_2O_3.

3. Model of a Band-Pass Microwave Filter

A model of a band-pass edge-coupled microstrip filter of a centre frequency 12 GHz and 30% bandwidth has been designed to test (see design characteristics - Table 1).

Fig.4 shows a layout of the model. Fig.5 presented its computed scattering parameters s_{21} (insertion loss) and s_{11} (return loss). TOUCHSTONE computer programme was used for simulation.

Table 1. DESIGN CHARACTERISTICS - filter 12 GHz, 30%

Chebyshev response	
Microstrip	
Coupled quarter-wave sections:	6
Number of poles:	5
Substrate ε_r:	9.6
Substrate thickness:	0. 635 mm
Centre frequency:	12 GHz
Ripple bandwidth:	30%
Passband ripple:	0.2dB
System impedance 50Ω	

Selected geometrical dimensions:

s_1	0.070 mm
s_2	0.184 mm
s_3	0.277 mm

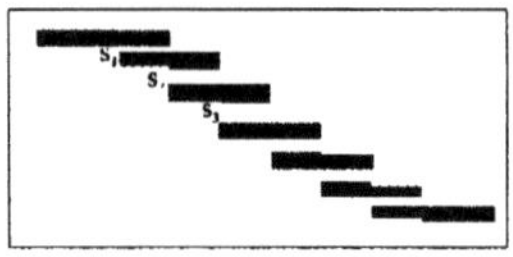

Fig. 4. Layout of a model.

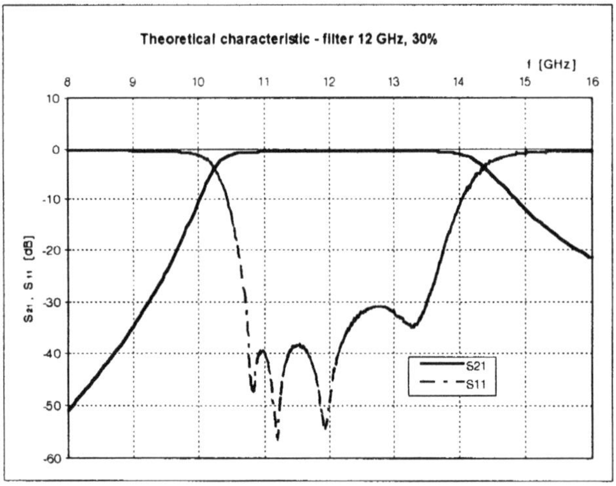

Fig. 5. Theoretical characteristic of filter model.

Models of filters were manufactured in three techonological versions. For each type two electrical parameters: insertion loss (s_{21}) and return loss (s_{11}) were measured using Vector Network Analyzer of type 8720C. The results were presented in form of plots. Also accuracy of geometrical projections for various technologies was checked.

4. Advanced Thick-Film Techniques Used for Manufacturing Models of Filters - Review

4.1. ULTRA-FINE PRINTED THICK-FILMS

Ultra-fine screen printing process was employed to manufacture the designed model of filter. Stencils made of 400 mesh calendered stainless steel wire cloth of wire diameter of 18 μm, mesh orientation of 22.5^0 in combination with capilar foil emulsion system of thickness 15 μm were used to screen-printed the patterns on 96% alumina substrates. Calendering removes knuckles of the weave and makes the printing side of the screen smooth for the emulsion which can create an excellent gasket against the ceramic substrate producing printing lines of excellent edge definition. Mixed bonded silver paste 9912F by ESL of sheet resistivity 6 mΩ/Sq and fired thickness of 8 μm was used for screen printing. Fig. 6 shows definition of edge of a silver screen printed layer. For comparison Fig.7 shows the edge of etched silver layer and Fig.8 the edge of photoimaged silver layer.

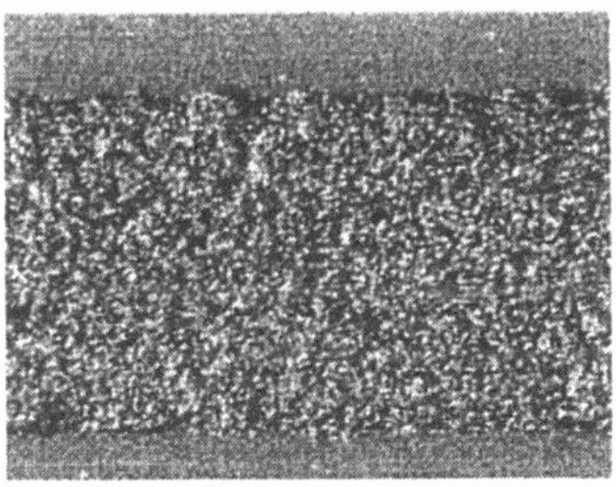

Fig. 6. Edge definition of a screen printed layer (silver paste 9912F) - scan camera magnification 80x

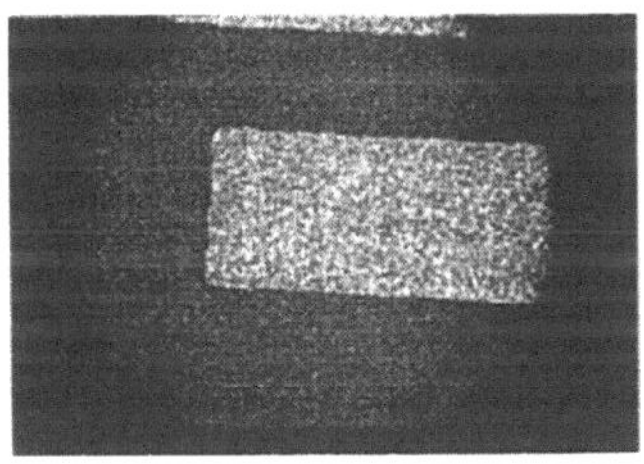

Fig. 7. Edge definition of a thick-film layer etched (silver paste 9912F) - camera, magnification 40x.

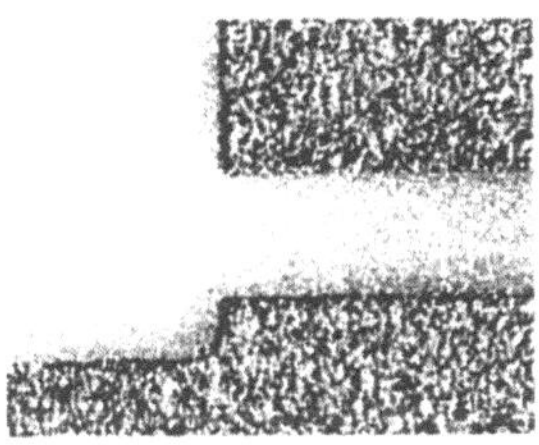

Fig.8. Definition of photoimaged thick-film layer (silver paste Fodel 4095) - scan - camera magnification 80x.

4.2. ETCHED THICK FILMS

The sequence of etching process of a fired thick-film is shown in Fig.9 [2]. It combines a conventional screen printing used for deposition conductive pastes on alumina substrates with photolitography for patterning.

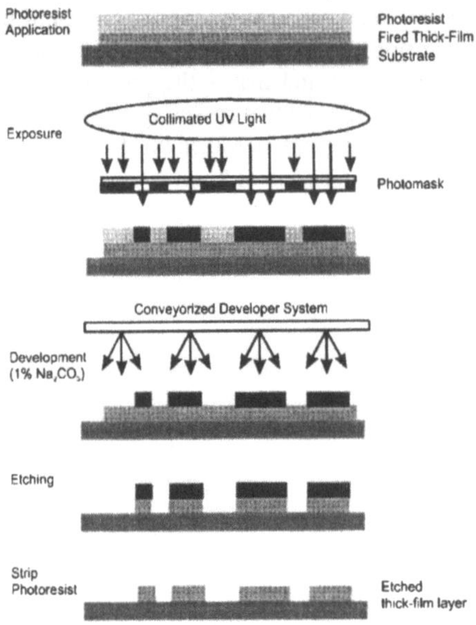

Fig.9. Process of etching of a fired layer.

Conductive paste is blank screen printed over the alumina substrate, dried and fired. The fired layer is spin-coated with photoresist and exposed to ultraviolet light through a glass photomask. The image formed in this way is developed - exposed areas of photoresist remain while the unexposed ones are washed away. Etching in appropriate etching solution removes conductive layer from all places unprotected by photoresist. Patterning of thick-film layer by etching provides excellent edge definition (see Fig.7). In our investigations silver paste 9912F was blank screen printed through 330 mesh stainless steel screen on 96% alumina substrates dried and fired. Negative type of photoresist by Shipley was employed to cover the fired substrate. The etch solution consisting of ferric nitrate and DI water was used.

4.3. PHOTOIMAGED THICK FILMS

The sequence of a process is shown in Fig.10.[4].

Special photosensitive conductor pastes must be used in this process. They combine standard inorganic components used to make thick-film conductors (metal powders, glass powders and metal oxides, refractory powders) with standard organic components used to make photoresist films for printed wiring boards (polymers, photoinitiators, monomers, stabilizers). They are sensitive to UV radiation and should be processed under safe yellow light. The photosensitive pastes are blank screen-printed throuh 330 mesh screens over the total area of the substrate and then dried at temperature 80^0C. Patterns are defined directly in the layer by exposing to ultraviolet light and then developed in Na_2CO_3 solution. Unexposed areas remain soluble so they are washed away. At last samples are fired in a typical for thick-films belt furnace at temperature 850^0C.

In our investigations photosensitive silver paste FODEL^R 4095 of sheet resistivity 4 mΩ/Sq was blank screen printed through 330 mesh screen. It was exposed using typical for thin-films phototools with Hg ultraviolet light source of power 350 Watts.

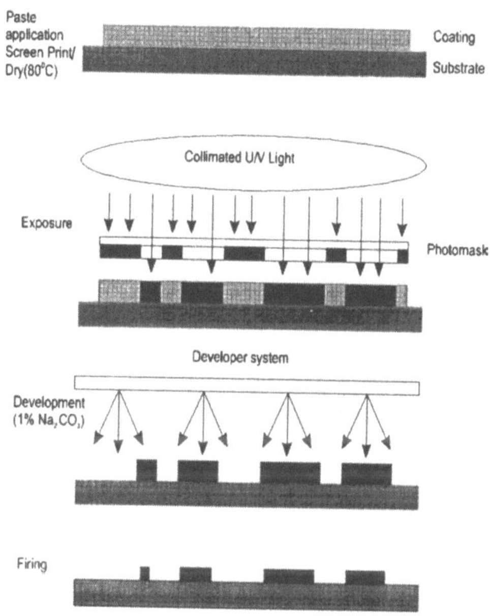

Fig.10. Process of photoimaging of photosensitive thick-films [4]

5. RESULTS OF MEASUREMENTS

Table 2 collects the results of measurements both electrical parameters of filters and accuracy of their geometrical projection.

Table 2

Parameter	Designed	Printed	Etched	FODEL
Centre frequency in GHz	12	12	12	12.1
Ripple bandwidth %	30%	31%	30%	30%
Insertion loss s_{21} in the bandpass	-0.5	-1.1	-1.5	-1.2
Return loss s_{11} in the bandpass (min)	-22	-18	-22	-18
s_1 [mm]	0.070	0.140	0.100	0.120
s_2 [mm]	0.184	0.255	0.216	0.240
s_3 [mm]	0.277	0.325	0.322	0.313

Fig.11, Fig.12, Fig.13 show the measured characteristics of filters manufactured respectively by ultra fine screen printing, etching and photoimaging of photosensitive layer .

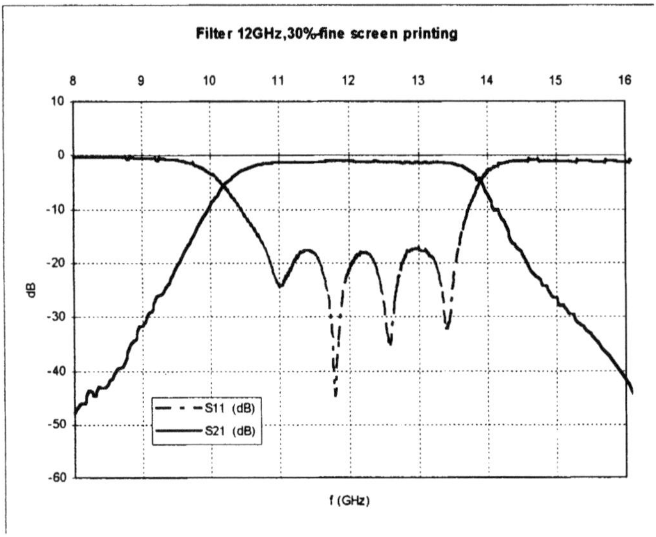

Fig.11. Measured characteristics -filter 12GHz,30% manufactured by screen printing.

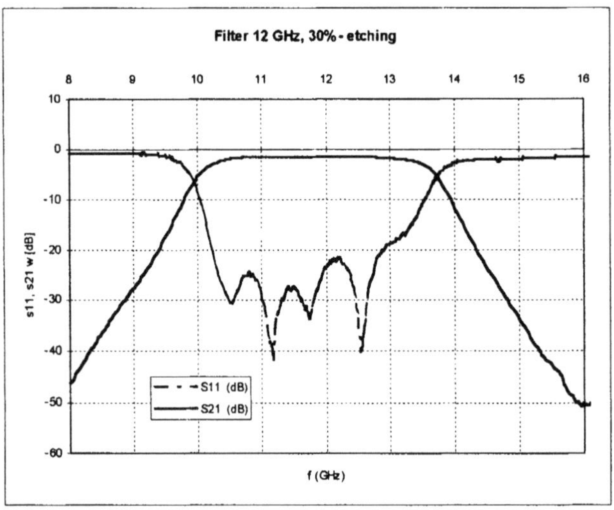

Fig.12. Measured characteristics -filter 12GHz,30% manufactured by etching.

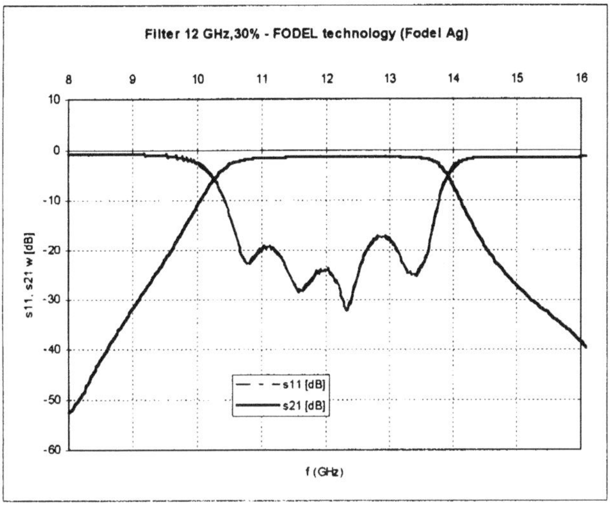

Fig.13. Measured characteristics -filter 12GHz,30% manufactured by photoimaging.

6. CONCLUSIONS

Comparison of capabilities of techniques used for manufacturing microwave hybrids is shown in Table 3.

Table 3. Technology Comparison [3].

	Printed	Photoprintable	Etched	Thin film	Claded
Substrate dielectric constant surface roughness (μm.) loss factor x 10E4 cost	Alumina 96% 8-10 0.4 -0.5 6 1	Alumina 96% 8-10 0.4-0.5 6 1	Alumina 96% 8-10 0.4-0.5 6 1	Alumina 99.5% 10 0.02-0.05 1-2 5-8	PTFE 2 1-2 3-5
Conductors Material Thickness (μm) Resistivity (mohms/sq)	 Au, AgPt,Ag, Cu 5-10 <5 for Au	 Au, AgPt,Ag, Cu 4-8 6-8 for Au	 Au, AgPt,Ag, Cu up to 10 8-10	 Au,Cr,Ni,Cu,Ti ,Mo <4 10	 Au, Cr, Ni, Cu 17 or 35 <2
Line width (μm)	100	≤50	≥25	25	>100
Shape definition	medium	very good	very good	excellent	very good
Multilayer	easy, low cost	more comlex	no	very expensive	expensive
Process complexity cost	 simple very low	 more complex low	 complex medium	 very complex high	 more complex medium

Fig.14 shows the relative manufacturing costs of the various technologies, taking thin film Au technology as a reference [5]. Reduction of labour costs makes FODEL^R technology about 40% cheaper per substrate than thin film technology.

Fig.15 shows the relative raw material costs for each technology, taking again thin film technology as a reference. The cost of FODEL^R materials are 20% lower than those of thin film materials.

It is worth to underline that cost evaluation of FODEL^R technology is based on manual process. The automatization of the FODEL^R process, easier to implement than in thin film, will let a greater cost reduction.

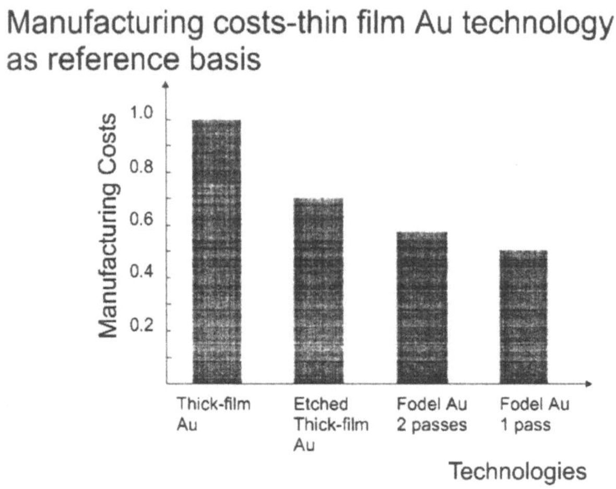

Fig. 14. Cost comparison of technologies involved [5].

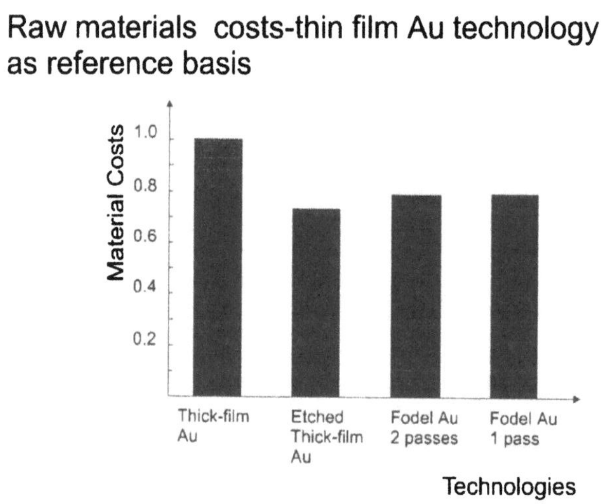

Fig. 15. Raw materials cost [5].

The results of our investigations show that up to 14 GHz electrical parameters of filters manufactured by ultra fine screen printing, etching and FODEL^R technology don't exhibit essential differences. This affect the choice of the most suitable technology for manufacturing microwave hybrids. Taking into account cost of manufacturing PHOTOIMAGED THICK-FILM TECHNOLOGY offers the most promising perspectives.

REFERENCES

1. P.J.Holmes, R.G.Loasby „Handbook of Thick-Film Technology". Electrochemical Publications Limited 1976, p.199,
2. R.Brown „Materials and Processes for Microwave Hybrids", ISHM Publication 1991, p..83,
3. A.Ptak „Thick-Film for RF/HF Communications", Proceedings of XXth ISHM-Poland Chapter Conference, Jurata 1996,
4. E.Polzer „DuPont Application Note: Fodel^R - Photoprintable Thick-Film", April 1996.
5. V.Manier, O.Sockeel et. al. „FODEL^R, Photoprintable Thick-Film Technology in Microwave Aplications"

SYSTEM RELIABILITY

PROFESSOR NIHAL SINNADURAI
TWI, CAMBRIDGE, UK,

Abstract

Modern systems can achieve high reliability cost effectively by selecting components from high quality commercial production lines and by designing circuits to be fault tolerant. Experience gained at British Telecommunications and also by AMSAT shows that this practical approach is more realistic than the outdated prescriptive approach to reliability specification which has not kept up with the developments of high quality and advanced functional products in mainstream applications. The analysis of system life cycle costs shows that the right choices in design and technology must be made at the design and feasibility stage. The outcome is whole-life cost-effectiveness.

Key Words:
system reliability/fault tolerant/fault tree methods/life cycle costs/cost-effectiveness/six sigma components/quality/reliability

1. Modern Requirements for System Reliability

Modern high reliability service providing companies have recognised that expertise in components and equipment lie with those who make the components and equipment and are now moving towards performance based reliability specifications, rather than prescriptive specifications at the component level. The leadership in taking such initiatives came from British Telecommunications in the early nineteen eighties. Such sensible changes provide for cost-effective electronics systems by allowing the systems designer and builder to design for reliability including trade-offs at the system level to compensate for critical components and vulnerable parts of the circuits.

2. System Reliability Versus Component Reliability

The system reliability can be configured to be dependent not only on the failure rates of some components and to overcome the failure of other components and continue to perform in the event of the failure of components in selected parts of the circuits. If a component fails but the system does not, then the system reliability requirement is met.

R. Tummala et al. (eds.), Electronic Packaging for High Reliability, Low Cost Electronics, 235–240.

For high reliability (Hi-Rel) the system reliability requirement is for no failures in 10 or 20 years, which is sometimes translated into a statistical requirement of, for example, 0.2% component failures in 20 years (statistically based on an assumption that 1 failure occurs just after the lifetime of the system).

3. Definitions of Quality and Reliability

A *quality* product or system may be defined to be one that "meets the required functional performance under the stated conditions", while a *reliable* product or system is one that "meets the required functional performance under the stated conditions for the required duration". A *known good* product, may have the same definition as a *reliable* product.

4. Achievement of Quality

Modern electronics components are manufactured in high volumes under strict process control conditions to ±6 sigma control limits, and thereby achieve yield levels at or below parts per million defects. Such controlled manufacturing conditions result in quality levels far better than those attainable by the small quantities manufactured in special manufacturing lines for space and military applications, and contradict the quality factors presumed in MIL-HDBK-217E. The fact that commercial components are delivered with better quality than parts qualified to MIL-STD-883, is borne out by practical experience in telecommunications and space applications. The exception is in radiation tolerance of commercial parts. Commercial parts may also be selected for radiation

Fig.1 - Effect of Increasing "Process Capability"

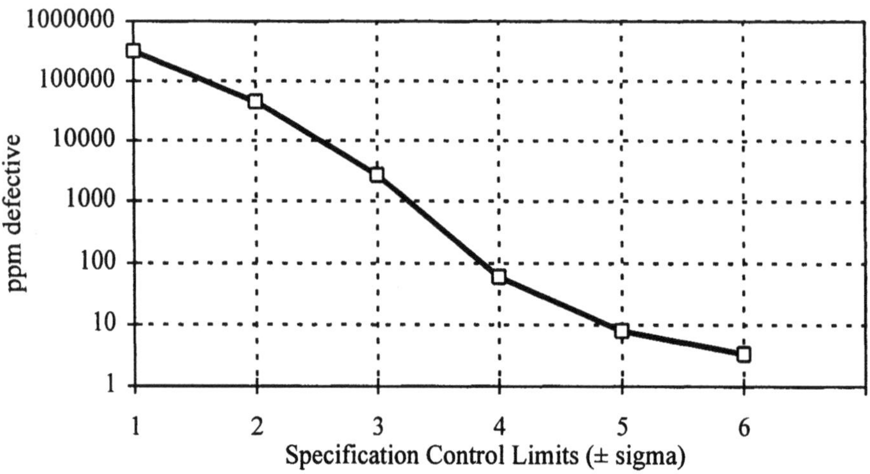

5. Experience Based Approach to Satellite Design

Amateur satellite builders have been forced, by simple lack of money, to take the route of using commercial parts in the design and build of their satellites. This has resulted in the progressive acquisition of information and knowledge of how to buy, which parts to selectively qualify, and how to design for survival. The formal space and military organisations would be horrified at the resultant approach, which they may regard as cavalier, but has a proven and improving track record of success with many years of satellite orbital operation, and failure occurring when the batteries run out.

Amsat used an intelligent approach to satellite design because they found that Hardware Fails in Space because:

- The "Problem Space" of big Satellite Systems exceed the current risk management tools available.
- Poor Designs, because of poor engineering education
- - lack of circuit design courses
- - no hands-on experience
- - students have too narrow an outlook
- - RF? what is RF?
- Space equipment designers are not motivated to provide quality because
- - paperwork kills craftsmanship
- - craftsmen are dying - there are no apprentice programmes
- - "technician" is a dirty word in the USA
- - space is no longer exciting
- Paperwork is applied blindly. Instead use the simple rule " if paperwork will not solve any problem prior to launch, keep it, if not discard it".
- People are the major cause of failures.
- People need improving.
- People can improve people.
- MIL-STDs Cannot improve people.
- The first step to fixing a problem is admitting there is one.

The Amsat Approach to Design and Build of "Little" Spacecraft is:

- Concentrate on single purpose
- The risk versus cost trade-off is important
- The schedule can be short; but do not over qualify the parts
- Reliability is achieved by: KISS, and Use of multiple spacecraft (which are themselves low cost
- and are launched by piggy-backing on the final stage LV.
- Accept that some problems may not be solved

- Satellite is not so important, and the programme manager is not so important
- Build on experience. The Amsat experience disagrees significantly with the MIL-HDBK-217E Quality factor ΠQ. Amsat has observed very low dependence on parts classification compared with MIL-HDBK-217E.

Figure 2. Comparison between Amsat Quality Factor and MIL-HDBK-217E Quality Factor

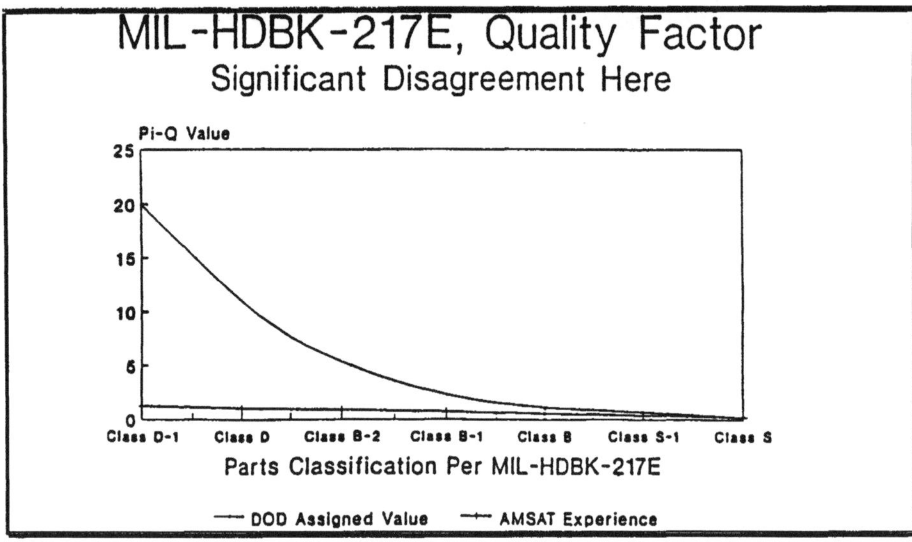

6. System Life Cycle Costs

The costs of systems arises at various stages of the system life cycle, with the predominant costs being borne by the field operations and maintenance. However, that impact of costs are decided much earlier - at the Feasibility (technology choices and operational conditions) and Design stage. For example, designers make the choices of track widths and spaces, the margins of tolerance, package types, junction temperatures, power consumption, system error correction, sub-system replication, and parts and assembly costs, all of which control the costs in all the subsequent life cycle stages.

Figure 3. Causes of Life Cycle Costs

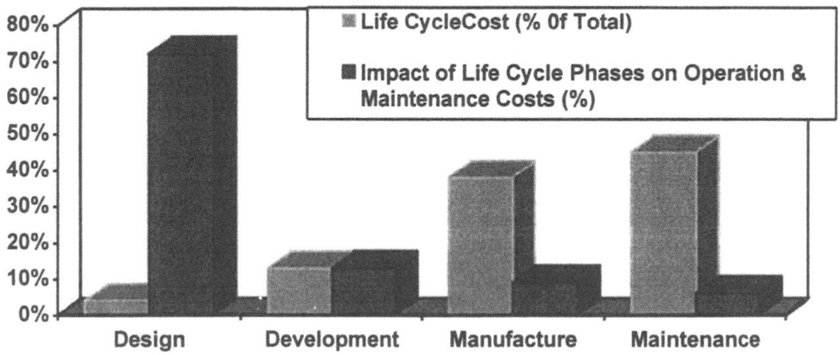

7. System Design Options

The design of systems for reliability require interactive circuit construction with Fault Tree modelling to determine the critical paths for failure. This approach will help the designer identify the most likely components to cause first failures of the system. The design option then is whether to select more reliable components or to design the circuit to avoid the weakness by replication and voting or making the circuit dependent on the more robust components. A useful method in fault tree analysis is to create a Boolean Algebra chart showing the structure of AND and OR gates leading to the top event, which is system failure. It is then possible to reconfigure the circuit to convert the critical path of the circuit into AND gates. It is always necessary to recognise the incidence of common mode faults which are common to the parts of the AND circuit and, if weak, can thereby jeopardise the reliability of the local circuit.

Figure 4. Fault Tree Methods Aid System Design For Reliability

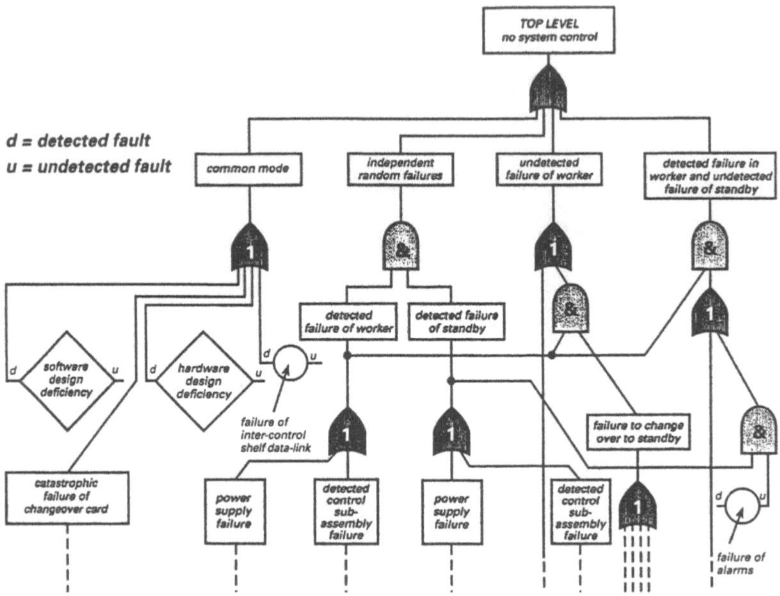

TWI

Example fault-tree with independent, dependent and common-mode faults

Designing circuits and systems for testability and diagnostics is also a necessary part of robust system design.

8. Whole Life Cost Effectiveness

The classic wisdom on life-cycle costing is that reliable components are more expensive, and reliable components give rise to lower product logistics costs (cost of maintenance), resulting in a high whole life system costs. The new message is that choice of commercial components instead of expensive (and possible less reliable) MIL-STD components, together with system Design-for-Reliability, can result in lower procurement costs and deliver whole life costs that are significantly lower than the traditional approach.

Capabilities and Limitations in Microelectronic Packaging and Assembly within the FSU (Former Soviet Union)

Steve Muckett
Moziak Technology Ventures, Ltd. (UK)

The author has had first hand experience during the last 3 years of visiting the various microelectronics facilities in Eastern Europe. The purpose of the presentation is to give a general review and personal impression of the microelectronics packaging and assembly production capabilities in the FSU countries. The countries of Russia, Ukraine, Belarus, Lithuania, Latvia and Estonia are included. The review is limited to those regions of Russia and the FSU to the west of the Ural mountains. The author has no experience of visiting microelectronics facilities farther east although it is known that there are a number of them.

Throughout the 1980's there was a large scale of microelectronics prodcution within Eastern Europe, principally for supply of components to the military complex.The various government Ministries within the Soviet Unionthat were concerned with microelectronics production allocated significant financial resources to the further development and improvements in capabilities across the full range of microelectronics products. During 1989 all financial 'investments' were frozen. The downturn in production capability was exacerbated by the collapse of COMECON and the decline in local 'markets' for electronic production.

The present situation is one of gradual improvements. Certainly within Russia and the Baltic States the electronic assembly and microelectronic industries are growing again (from a low base and with 70% fewer workers) and have even developed a significant export trade (at a commodity level) for their products.Within Russia the government has recognised the importance of the microelectronics sector and through the 'Federal Fund for Electronics Development' is beginning to channel some seed capital, and assistance in the raising of bank finance and credits, into specific microelectronics projects of national importance.

The skills/knowledge levels and training for microelectronic specialists in I.C design. microelectronic components and materials development remain at a high level. Innovative solutions contiue to be explored but lack the funding of equivalent programmes in the West.

Against this backdrop the paper highlights the present capabilities in microelectronic packaging and assembly. The standard packaging options available from the most important semiconductor manufacturers in the region are reviewed.Case studies are presented for some of the innovative activities, notably in the ceramic packaging, thin and thick film metallization and MCM fields.

The paper concludes by identifying specific technology areas where good opportunities exist for western investors, prospective Joint Venture partners and equipment suppliers to be active in the region.

R. Tummala et al. (eds.), Electronic Packaging for High Reliability, Low Cost Electronics, 241–243.

Capabilities Reviewed

Because of the economic crisis caused by the breakdown of Comecon there has been no investment in microelectronics manufacturing since 1990. At the time of writing the situation is changing with some significant investments in silicon wafer processing, wafer fab operations and far sighted research projects. However the situation for packaging is still rather bleak.
There are essentially no packaging technologies more modern than 1989. There is no evidence of ;flip chip, BGA, SO (DIL) packaging with more than 16 pins,no TSOP, no taping and reeling of surface mount or other packages and no high throughput auto wire bonders installed. The author is optimistic that there will be significant market opportunities for the sales of such equipment and processes in the near future.

Innovations in Microelectronics Packaging

Even with the limitations described above there are many examples of highly innovative and practical packaging solutions, which could be utilised by western licensees, collaborators or venture partners. Some examples are given below;

Electrochemical Aluminium oxide/metal substrate technology (ELAT Process)

This technology was originally developed by the RadioEngineering Institute in Minsk. It is highly sophisticated thin film technology whereby the entire substrate and conductor/dielectric multilayer sequence is constructed from aluminium metal and aluminium oxide using an anodic, electrochemical oxidation process.It was used in the Russian space programme and confers good thermal, mechanical and radiation characteristics compared to other substrate technologies.
The minimum track and gap of 10micrometres/10micrometers can be achieved at a conductor thickness of 3 micrometres.Dielectric breakdown voltage is better than 1000V at 250degrees celsius for a thickness of 100 micrometres.No separate steps are needed for via metallisation because it is a planar sequential structure and multilayers up to 5 conductor levels have been produced. Nickel and gold can be sputtered on the top conductor layer for bonding or soldering. Thin film resistors or capacitors can be built into the structure. A special 'sealing' layer is applied to the top anodised aluminium dielectric surface to enhance the dielectric properties.

Photoimageable Thick film technology (Hibridas Process)

This is a photosensitive thick film paste process that was originally developed by the Radio Measurement Engineering Research Institute, Kaunas, Lithuania during the early 1980's. The driving force for developing such a process was the production of low cost microwave devices up to 20GHz frequencies.
A range of paste materials were developed including gold, silver and palladium silver conductors, dielectric and some resistors. Conductor resolution on alumina ceramic is 15micrometres track width and 20micrometres gap for a conductor thickness of 10 micrometres.Dense films with vertical edges are achieved.
This process (a specially designed kit of equipment and the photosensitive materials) is now available for sale to thick film circuit producers through Mozaik Technology Ventures Limited.

Planar MCM Technology

Such a technology has been originally developed by OKB 'MCM Technology', at Zelenograd, Russia. This is a spin off from the research group of 'Angstrem' - the most well known semiconductor producer in Russia.

According to the inventors of the technology it is a new and novel 'Group assembly' MCM-L featuring package free chips. It consists of a planar design of polyimide thin film wiring board with precsion cut recesses for placement of bare die.No wire bonding,welding or solder/adhesive joining of I/O's is needed. A thin film process and photolithography is used to sputter a conductor metallisation and polymer dielectric layers over the top surface of the substrate/die. It is claimed to be a reliable process and allows easy testing of the die. Patents have been applied for and the technology is being implemented for PCMCIA card applications at the sample and pre-production stage.

The three examples outlined above hopefully give the reader some insight into the type of opportunities that are available for genuine collaboration in research and development in microelectronics packaging within the FSU.

APPLICATIONS OF THE INTERCONNECTED MESH POWER SYSTEM (IMPS) SUBSTRATE LAYER REDUCTION TOPOLOGY

LEONARD W. SCHAPER
High Density Electronics Center (HiDEC)
University of Arkansas
W. 20th St.
Fayetteville, AR 72701

Abstract

A significant decrease in interconnection substrate production cost can be achieved by reducing the number of substrate layers from the conventional four or five (power, ground, X signal, Y signal, pad) to two or three. Besides reducing direct processing steps, yield also increases as defect producing operations are eliminated.

The Interconnected Mesh Power System (IMPS) is a new patented interconnection topology which leverages the production technologies of fine line lithography and batch via generation to allow planar power and ground distribution, and dense signal interconnection, on only two metal layers. The IMPS concept is reviewed.

Results of signal transmission measurements (impedance, delay, and crosstalk) and power distribution characteristics on several test vehicles are presented.

The results achieved indicate that the IMPS topology can be applied to a wide spectrum of MCM and other interconnect structures, with considerable savings in manufacturing cost. The IMPS topology is also suitable for multiple layer applications, where its inherent low crosstalk and power distribution flexibility would offer advantages.

IMPS technology is being licensed to Sheldahl, Inc. Other major interconnect suppliers are evaluating the technology. Several designs for various applications are discussed.

What is IMPS?

A significant decrease in MCM substrate production cost can be achieved by reducing the number of substrate layers from the conventional four or five (power, ground, X signal, Y signal, pad) to two or three. Besides reducing direct processing steps, yield also increases as defect producing operations are eliminated.

The Interconnected Mesh Power System (IMPS) is a topology which allows low inductance planar power and ground distribution; and dense, controlled-impedance,

R. Tummala et al. (eds.), Electronic Packaging for High Reliability, Low Cost Electronics, 245–251.

low crosstalk signal transmission in only two physical wiring layers. It utilizes the production methods of fine line lithography and batch via fabrication characteristic of MCM-D and some MCM-L to create a structure not economically possible using standard printed wiring board methods [1].

The derivation of the power distribution structure is shown sequentially in Figure 1. Consider a familiar meshed plane, used in many MCMs for power or ground. Think of this construction not as a plane with holes, however, but as a set of X and Y conductors. In Figure 1a, the X conductors and Y conductors have been placed on two separate metal layers, and at each crossover, a via has been provided to retain the planar characteristics. (Vias typically have low resistance and inductance compared with lines.) This "interconnected mesh" plane is reasonably electrically equivalent to, but topologically different from, the conventional mesh plane.

The mesh of Figure 1a includes enough space between conductors to insert interdigitated conductors of opposite polarity (power or ground) with their own set of interconnecting vias, on the same two physical metal layers (Figure 1b). The resulting "interconnected mesh" structure forms a complete power distribution system, with the necessary characteristics of low resistance and inductance.

In this "dense" mesh of power and ground conductors, there is no room for signals. If the power and ground conductors were narrowed, wiring tracks for signals would be included between every power-ground pair. Where these tracks are needed for signals, they are inserted, allong with the appropriate connecting vias. Where they are not needed for signals, the power and ground traces are widened, to maximize the use of metal. This construction is shown in Figure 2.

The resulting complete MCM substrate, produced on only two metal layers, provides signal interconnection at twice the minimum conductor pitch, and low-impedance, planar power distribution, at a cost far less than a conventional four-layer wiring structure.

Test Vehicles and Results

Several test vehicles in both MCM-D and MCM-L have been built and tested.

The MCM-D test vehicles were fabricated using an aluminum/polyimide process on 5" silicon wafers. The two metal layers are 2 μm thick, with 6 μm of polyimide between them. A 2 μm thick SiO_2 layer separated the bottom aluminum layer from the high resistivity silicon substrate. Each test vehicle has a dimension of 32 mm x 25 mm. The power and ground conductors are 100 μm wide, and signal lines are 20 μm wide. The power and ground lines are on a 160 μm pitch. Signal lines have dual pitch: 80 μm and 160 μm. Figure 3 shows a SEM photograph of such structures.

The upper layer transmission line test structures have an average length of 26 mm and the bottom layer test structures are 18 mm long. The test pads are designed with ground/signal/ground configuration (150 μm pitch) so that they will be compatible with Cascade ACP 40 probes. Signal line measurements were performed using an HP

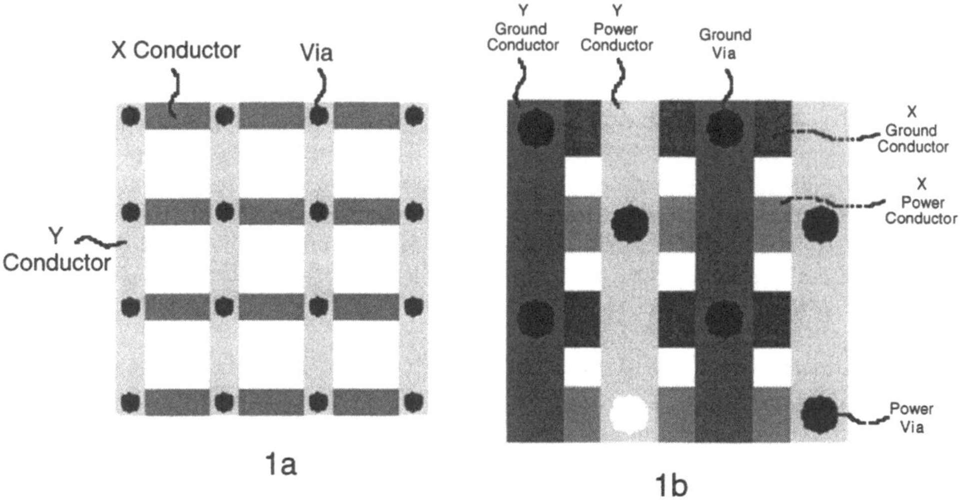

Figure 1. Derivation of the IMPS topology

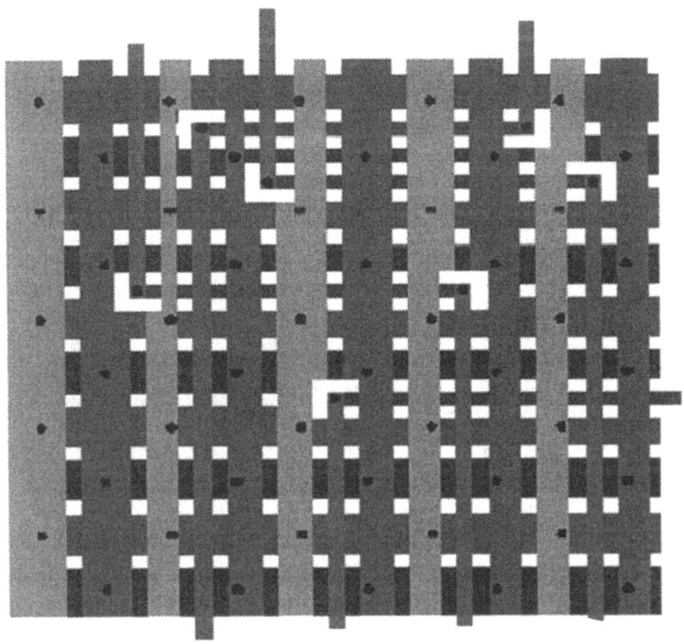

Figure 2. Variable width power distribution

8510B Network Analyzer in the frequency range from 65 MHz to 1.8 GHz (below the skin depth frequency). Line impedance was 48 - 50 Ω on both layers.

Power distribution impedance measurements were also made on this test substrate. Because the power distribution impedance of a bare IMPS substrate is basically capacitive, a simple RLC series equivalent circuit was used as a model. Over the frequency range of 1 MHz - 500 MHz, using the equivalent circuit capability of the HP 4291A, the substrate was modeled as a 2.1 nF capacitor, 200 mΩ resistor, and 140 pH inductor in series. Measurements were an excellent fit to this model. It should be noted that the resistance and inductance parasitics are mainly due to the limited connection area of the probe pads into the mesh structure. Slightly larger pads which would have intersected several mesh lines would have avoided most of these parasitics.

Additional measurements were made on the network analyzer to determine S_{12} of the power distribution system. The capacitive mesh is viewed as a two port network in this case, with probes connected at the center and one corner. These measurements showed no principal inductive resonance below 18 GHz, indicating that a significant portion of the measured inductance with one probe was due to current crowding at the probe. In reality, the performance of the power distribution system will be limited by the parasitics of the decoupling capacitors, not the parasitics of the mesh planes [2].

AC power distribution noise has been measured on the same test substrate, using chip resistors as loads and power MOSFETs as switches. Comparisons were made with a substrate drawing the same .45 A/ns di/dt but with solid power and ground planes. Measured noise on the substrate was 200 mV P-P for IMPS and 150 mV P-P for the solid planes.

Noise has also been measured on a 68040-based MCM, as described elsewhere [3]. Since the noise glitches had frequency components far above the useful range of the standard chip decoupling capacitors, the lesser integral capacitance of the IMPS structure allowed more noise on the power distribution system than the same module built with solid planes, as expected. The use of low-inductance capacitors provides decoupling in the appropriate frequency range. A photograph of a section of this substrate is shown in Figure 3.

An MCM-L IMPS test vehicle has also been built, as previously reported [4]. This substrate was built on Sheldahl ViaGrid® material, with copper conductors on both sides of a polyimide flexible film. Good transmission lines of 95 - 100 Ω were measured, and design rules for 50 Ω lines were developed. A crosstalk measurement on two 32 mm lines, 500 μm apart, with an intervening ground line, indicated below-40 dB under 4 GHz.

Extensive modeling of the IMPS structure has also been performed. Models and measurements are in excellent agreement [5]. The results achieved indicate that the IMPS topology can be applied to a wide spectrum of MCM and other interconnect structures, with considerable savings in manufacturing cost.

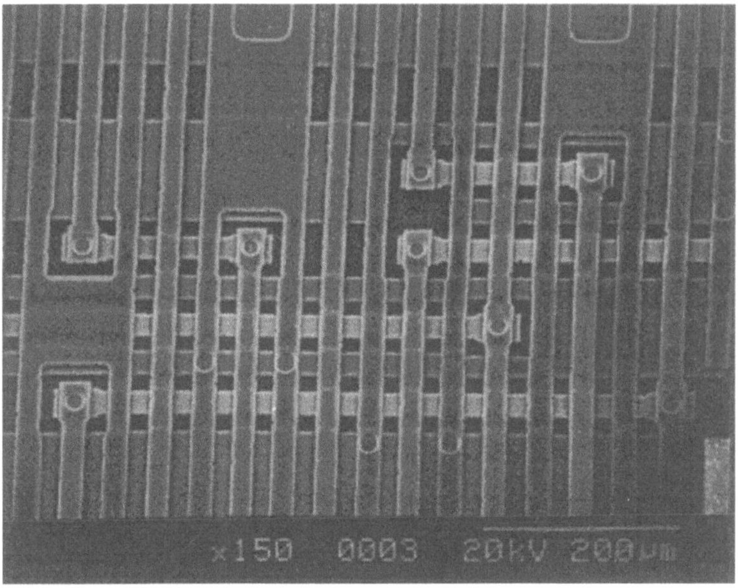

Figure 3. SEM of IMPS MCM-D wiring

Figure 4. Section of an RF module built on IMPS film

Figure 5. MicroModule Systems Gemini™ Pentium® module

Applications

IMPS technology is being licensed to Sheldahl, Inc. Several designs for various applications have already been completed and fabricated, using post-processing software for the Mentor Graphics MCM Station developed at the University of Arkansas [6]. IMPS is particularly attractive for a fine-line flex technology since an entire MCM substrate can be created on only one layer of flex. A photograph of a section of an RF circuit assembled on such a substrate is shown in Figure 4.

AlliedSignal has designed a control system MCM incorporating analog, digital, and power control sections [7]. IMPS provided power distribution and isolation for the digital and analog sections.

The Consortium for Intelligent Large Area Processing (CILAP) has commissioned an IMPS design of the MicroModule Systems Gemini™ Pentium® processor module shown in Figure 5. The IMPS design will serve as the demonstration vehicle for CILAP's 400 mm MCM-D fabrication process.

For high performance applications, where routing density demands more wire than can be provided on two IMPS layers, four layers of IMPS provide advantages in signal integrity and power distribution flexibility compared with four layers of metal in a buried stripline configuration. Test vehicles for MCM-D and MCM-C are being designed at HiDEC for companies which will produce the substrates. Assembly and electrical testing at HiDEC is scheduled for Summer, 1997.

References

1. L. Schaper, S. Ang, Y. Low and D. Oldham, "The Electrical Performance of the Interconnected Mesh Power System (IMPS) MCM Topology," *IEEE Trans. Comp., Packaging., and Manufacturing Tech.*, vol. 18, no. 1, pp. 99-105, February 1995.

2. L. W. Schaper and G. Morcan, "High Frequency Characteristics of MCM Decoupling Capacitors," *Proc. 46th ECTC*, pp. 358 - 364, May, 1996.

3. M. D. Glover, J. P. Parkerson, and L. W. Schaper, "A Signal-noise Comparison of The Interconnected Mesh Power System (IMPS) with a Standard Four-layer MCM Topology," *Proc. 5th International Conference & Exhibition on Multichip Modules*, pp. 216 - 219, April 1996.

4. D. A. Arnn, J. P. Parkerson, L. W. Schaper, and S. S. Ang, "A Low Cost Multichip Module using Flexible Substrate and Ball Grid Array", *Proc. 5th International Conference & Exhibition on Multichip Modules*, pp. 28 - 32, April 1996.

5. Yee L. Low, Leonard W. Schaper, and Simon S. Ang, "Modeling and Experimental Verification of the Interconnected Mesh Power System (IMPS) MCM Topology", *IEEE Trans. Comp., Packaging., and Manufacturing Tech.* - Part B, Vol. 20, No. 1, February, 1997, pp. 42-49.

6. J. P. Parkerson and L. W. Schaper, "A Design Methodology for the Interconnected Mesh Power System (IMPS) MCM Technology, Proceedings of the 1995 IEPS Conference, pp. 192 - 196, September, 1995.

7. Jeff Klein and Armagan Akar, "Multichip Module Offers Ceramic Power", *Electronic Engineering Times*, June 24, 1996, p. 52.

PACKAGING IN THE 21ST CENTURY

Rao R. Tummala
Pettit Chair Professor and Director
NSF ERC Packaging Research Center
Georgia Research Eminent Scholar
Georgia Institute of Technology
Atlanta, GA 30332-0560

ABSTRACT

Next generation of packaging must be consistent with both semiconductors and systems. The packages in the past have been mostly passive containers for semiconductors. Given the needs for extremely compact and ultra low-cost electronic systems of the 21st century and given the wafer fabrication costs that are $1-5B a plant, what should packaging be in the 21st century? Any packaging technology should also be built upon the learning of current state-of-the-art packaging. Highly integrated and low-cost packaging as a single level and single component that is capable of improving performance, reducing cost and size of about 10x in each is proposed.

INTRODUCTION

Dramatic changes are underway in the computer, telecommunication, automotive, and consumer electronics industries. In computers two major changes are underway: (1) computing power is doubling every 15 months reaching billions of instructions per second within 5 years as illustrated in Figure 1 and (2) the cost per unit of computing has come down from $100,000/MIP (millions instructions per second) with mainframes to less than $10 with PCs and some workstations, as illustrated in Figure 2. In addition, to increase performance, computing is increasingly being performed in hand-held next-generation computers. There is a similar trend in the telecommunications industry, multimedia products with voice, image, video, text, and other functions, are expected to be commonplace within the next decade, as illustrated in Figure 3. Major changes in the automotive industry, both in the amount of electronics in automobiles and the functions needed to support the intelligent vehicle highway systems (IVHS), are occurring in this decade. The common and pervasive requirements in all of these electronics are: (1) ultra-low cost, (2) thin, light, and portable, (3) very high performance, and (4) diverse functions involving a variety of semiconductor chips.

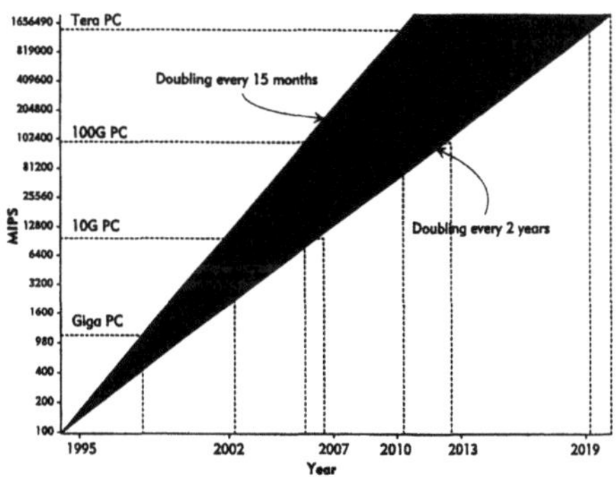

Figure 1. Exponential Growth Trends in Computer Performance

R. Tummala et al. (eds.), Electronic Packaging for High Reliability, Low Cost Electronics, 253–258.

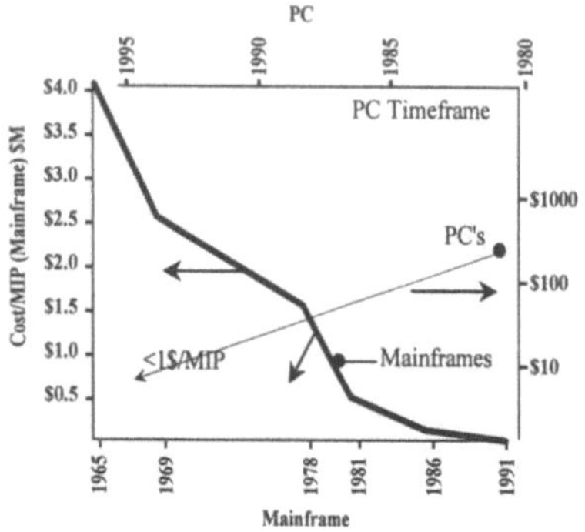

Source: Fortune, June 1993

Figure 2. Cost Per Unit of Computing

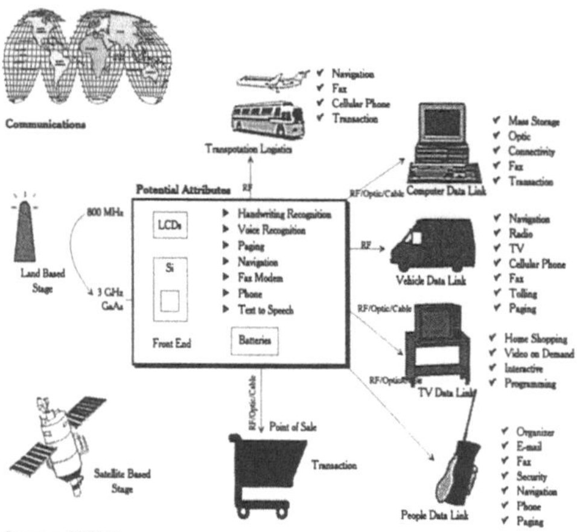

Source: NEMI

Figure 3. Future Portable Electronics

Semiconductor Trend

Next generation of packaging must be consistent with these needs and with the semiconductor roadmap. The semi-conductor roadmap is much clearer than in the past, almost totally directed to CMOS technology, in contrast to CMOS for low performance and bipolar for high performance applications. The reasons for this paradigm shift are understood from Figures 4 and 5 where in cost and integration are the base for CMOS and the performance is beginning to approach bipolar. The roadmap for the semiconductor technology is indicated in Figure 6 for the next decade. It appears that this roadmap is technically doable but the economics as discussed later is of major concern. The key question for next generation of packaging, therefore, -- can it provide its traditional role of interconnecting, powering, cooling, and protecting semiconductor ICs and at the same time address the economic concern of larger wafer fab costs, and high costs of large and fine-line ICs which will invariably be low-yield and high cost. In addition, can the next generation of packaging address the ultra-compact portable needs of tomorrow? This paper begins to address these challenges.

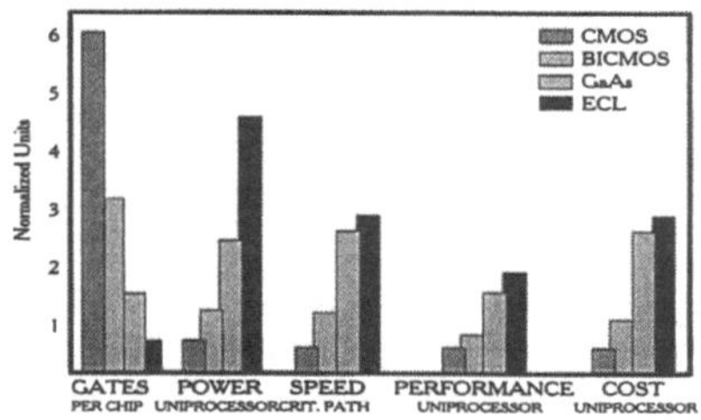

Figure 4. Comparison of Semiconductor Technologies (Hannemann)

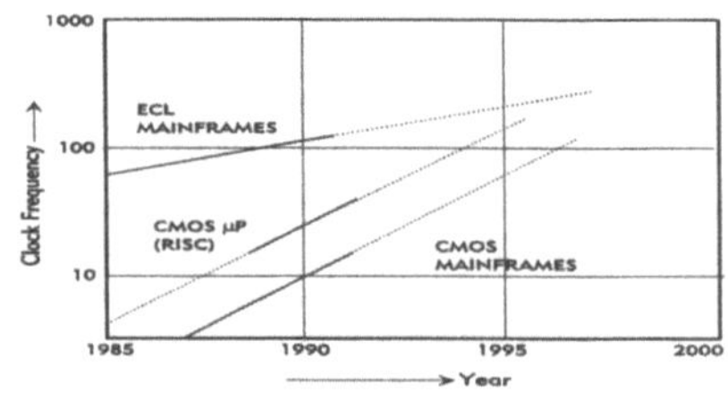

Figure 5. Performance Comparison of CMOS and BiPolar (Hannemann)

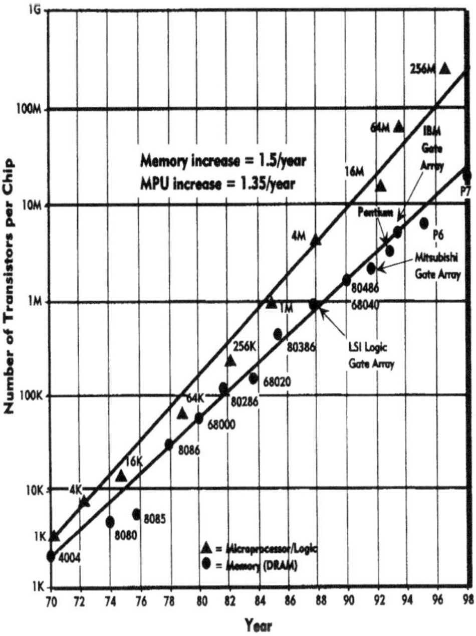

Figure 6. IC Density Trends

PACKAGING TREND

The overall packaging trend is indicated in Table 1, which illustrates how packaging has evolved during the last three decades, starting with DIP and wirebond in the 1970's, QFP and SMT in the 1980's, and BGAs in the 1990's. There is similar progress in miniaturization of passive components and in organic board using drilling and lamination technologies as described below. All these technologies, however, provide only about 10% silicon packaging efficiency, a factor of 10x from the theoretical limit possible. This table indicates the next generation of packaging as direct chip attach to organic board with fine line photolith wiring and with integration of passives and optoelectronics. These developments which are described in the next generation packaging section are aimed at reducing the number of packaging levels to 1 from the current 3, eliminating 5-10 types of active and passive components to 1 and simultaneously reducing size or improving silicon efficiency by a factor of 5-10x from current levels.

Table 1. Packaging Evolution & PRC Strategy

	← Past	→	← Current	→ ← PRC Strategy	→
	1970's	1980's	1990's	2000	2005
Chip Connection	Wirebond	Wirebond	Wirebond	Flipchip	Low-cost Flipchip
Package	DIP	P-QFP	P/C-BGA		
Pack. Assembly	PTH	SMT	BGA-SMT		
Passives	C-Discretes	C-Discretes	C-Discretes	C-Disctretes	
Board	Organic	Organic	Organic	DCA to Board	DCA to Int. Board
No. of Levels	3	3	3	1	1
No. Types Components	5-10	5-10	5-10	5-10	1
Si Efficiency %	2	7	10	25	>75

CURRENT PACKAGING

Single Chip Packaging

The fact that portable electronics are the prime drivers of packaging technologies is illustrated in Figure 7. This need is forcing the packaging technologies in terms of reducing packaging size by improving the I/O pitch and by reducing the passive component size. These developments have resulted in improving component density to about 25/cm^2 currently and with fine pitch QFP or BGA, this may be increased to 50. The overall packaging strategy leading to this density is illustrated in Figure 8 again indicating that QFP and BGA packages will lead the parade. The only newcomers are those that will enhance these such as VSPA, a new revolutionary simple and elegant

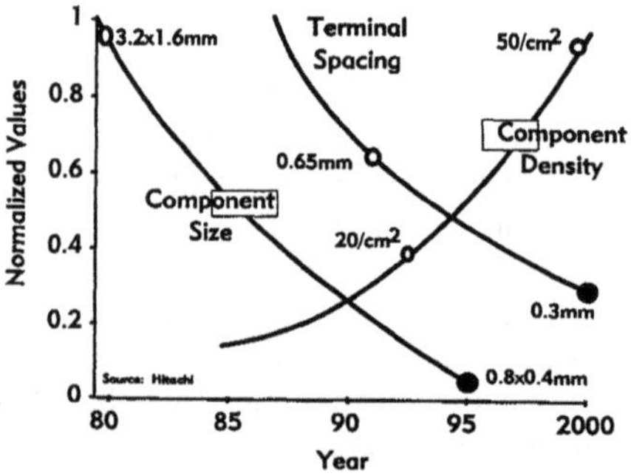

Figure 7. Portable Electronic Density

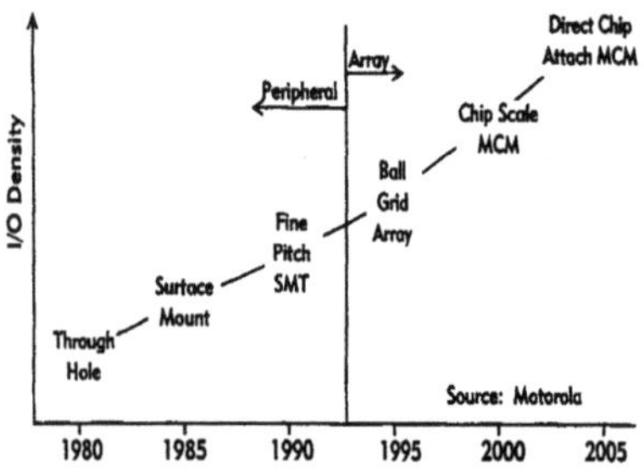

Figure 8. Overall Packaging Evolution

package from the Panda Project and a number of chip scale packages. Another reason for the selection of these limited

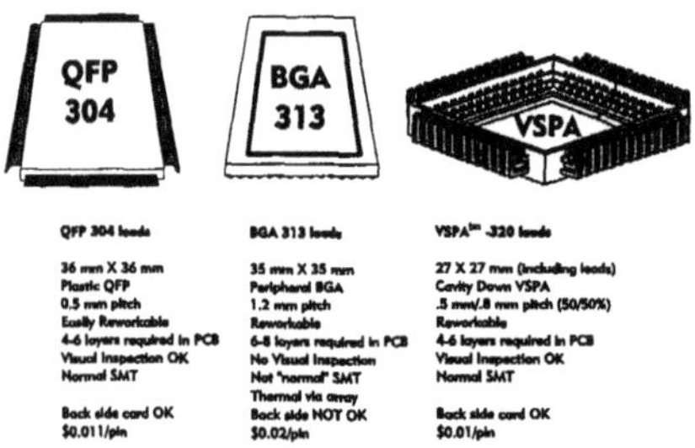

Figure 9. QFP, BGA and VSPA^{tm} Package Comparison

packages is their low cost as illustrated in Figure 9.

It is apparent that while the dominant packages are QFP, BGA and chip scale packages, TAB, PGA and VSPA will continue to be used in selective markets. Figure 10 illustrates the I/O ranges in which they are expected.

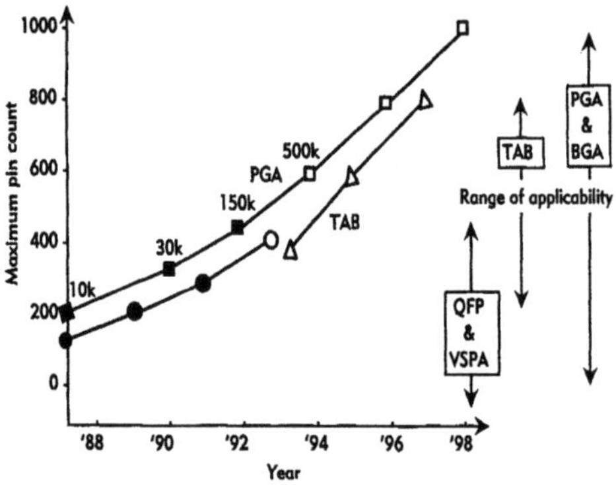

Figure 10. High Pin Count Packages (Hitachi)

Multichip Packaging

Multichip technologies are generally divided into three groups based on ceramic, thin film, and printed wiring board technologies. As illustrated in Figure 11, the wiring density per layer or J/O is the highest for thin film layers, followed by ceramic and least for printed wiring. The cost per I/O connection in Figure 11 is not consistent, however, with the wiring density, favoring printed wiring as the most appropriate MCM technology if this can be extended. It appears that this technology can be easily extended with thin films based on large area processing, combining the best aspects of each into one MCM technology.

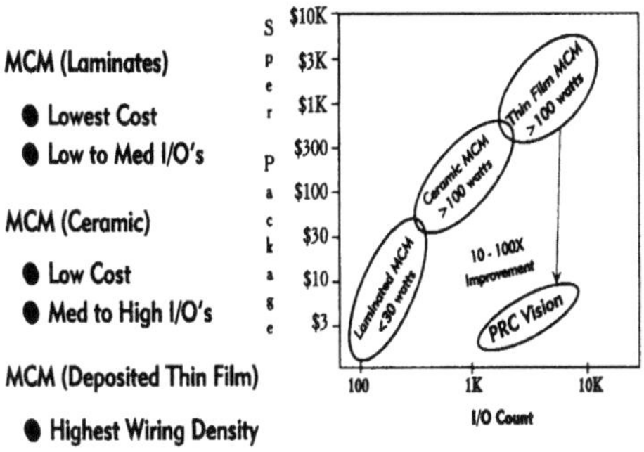

Source: IBM

Figure 11. MCM Options and Costs

NEXT GENERATION OF PACKAGING

There are several fundamental limits that are being imposed by both semiconductors and packages described above toward the system needs of the 21st century. The semiconductor limits are in cost and performance as the sub, sub micron dimensions are approached. The cost of wafer fab currently is at $1-1.5B and can be expected to increase to $5B for next generation as illustrated in Figure 12. The performance limits on chip are due to

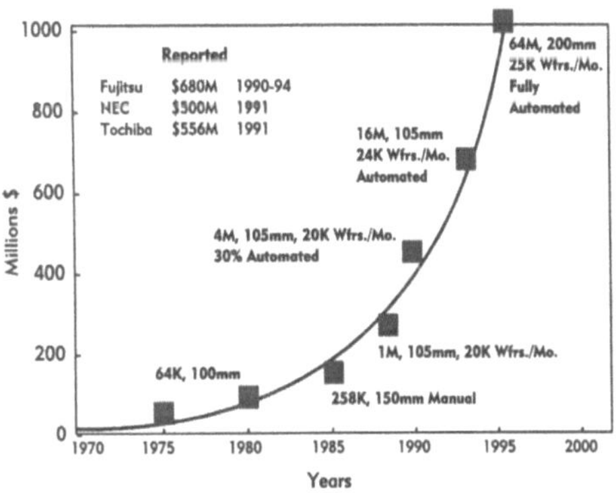

Figure 12. IC Wafer FAB Capacity/Cost

interconnection delays as the complexity of the interconnections grow. Three semiconductor plants costing $1B or more are already up and running and 10 more sites (Figure 13) have been announced in the US, making the US back on top (Figure 14).

The package limits are due to the poor silicon packaging efficiency, cost, and the low reliability associated with the number of levels required to complete system-level packaging.

The packaging strategy being pursued by the Packaging Research Center at Georgia Tech addresses these limits by concentrating on ultra low-cost, very high-efficiency and highly integrated thin film on PWB. The integration includes MCM-compatible resistors, inductors and capacitors as well as optoelectronics. Such a package is compared with current packages as illustrated in Figure 15 for a number of important attributes. The overall vision of the research is to reduce cost by a factor of 10 while simultaneously improving performance and reliability and reducing size by a similar order of magnitude.

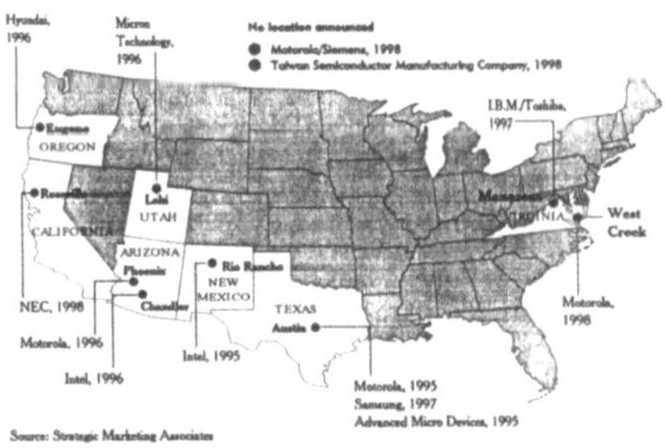

Figure 13. State of the Art - Semiconductor Plants and Opening Dates in US

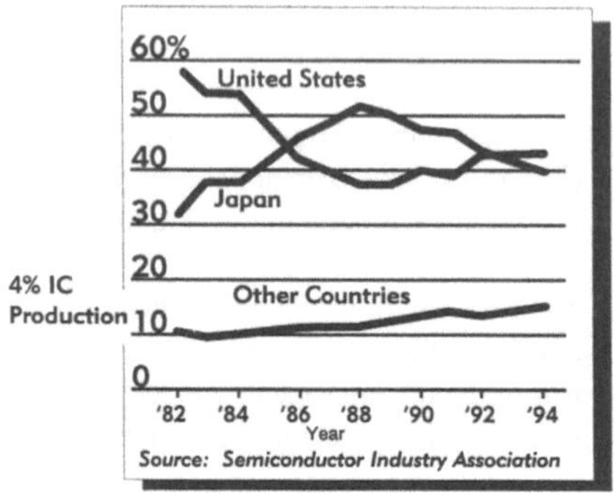

Figure 14. Market Share of Worldwide Semiconductor Sales

	Current Packaging	Next Generation Packaging	Improvement Factor
Size (nominal)	10	1	10X
Package Efficiency	8%	80%	>10X
Performance	50-100 MHz	200-1 GHz	5-10X
Cost ($/sq.in.)	30	3	10X
Reliability (PPM)(	10	1	10X

Figure 15. Next Generation of Packaging at Georgia Tech

Acknowledgments: The Packaging Research Center is support by the National Science Foundation as one of its Engineering Research Centers. The author wishes to acknowledge this support and the support of the State of Georgia.

Plastic Ball Grid Array (PBGA) : A Rising Package Solution

K.KURZWEIL, X.SAINT-MARTIN, Y.STRICOT
Bull SA
Rue Jean Jaurès B.P. 68
78340 Les Clayes Sous Bois
France

Abstract

Plastic Ball Grid Array (PBGA) packages are in industrial world now, for memories and some types of integrated circuits. Their application will continue to grow rapidly and will be also increasingly used to package complex VLSIs in the range of 500 I/Os, showing increased operating frequency and high thermal dissipation. PBGA implementation into manufacturing requires a substantial investment in technology development. To comply to the specifications of such complex VLSIs, PBGA design shall address the die electrical and thermal characteristics while taking cost considerations into account. Packages with controlled impedance in cavity up and down configuration are then implemented, incorporating high density Printed Wiring Board technology refinements. Process assessment dedicated to PBGA's technology shall be carried out to achieve a high reliability of the package as a stand-alone and the package as mounted on PCB. This paper will review PBGA packages for housing VLSIs in the range of 15 x 15 millimetres, 10 to 20 Watts power dissipation developed through the two European projects CHIPPAC and JESSI.

Keywords : CHIPPAC, JESSI, PBGA, multilayer, organic substrate, cavities

1. Introduction

First level packaging trends are driven by the evolution of the dice complexity and cost reduction pressure. For some electronic systems, the device itself represents a whole function. The advances in the semiconductor technology (e.g. size of the devices, number of I/Os, switching speed, thermal dissipation, frequency), require adequate effort supporting the developments in the field of

R. Tummala et al. (eds.), Electronic Packaging for High Reliability, Low Cost Electronics, 259–266.

packages, in such a way as to insert advanced semiconductor devices in an appropriate packaging environment.

Since many years, these trends mainly focus on package compactness and standardisation, electrical performances, thermal dissipation capability and board assembly with a high yield.

To comply to these trends, the pitch of the common packages (PGA, QFP) has been decreased, thus generating constraints at board level which came beyond the limits of the technology. The BGA technology has appeared as a possible alternative to package complex devices with respect to cost [1]. BGA technology has been primary introduced by Motorola and IBM, which implemented the Solder Ball Connection (SBC) based on a ceramic body and dedicated to high performance devices like the power PC. Nowadays the plastic family is becoming the main stream of high pin count BGA packaging, due to cost and reliability potential. the table 1 below shows the BGA technology trend from 1995.

Table 1. World IC packages types 1995-2005

Pieces (Bn)	1995	2000	2005
DIP	16.3	12	7.6
SIP/ZIP	4.5	3.8	3.3
PGA	0.5	0.8	0.7
total	21.3	16.6	11.6
SO	17	33.6	44.6
PLCC	3	1.9	0.8
PQFP	5.9	17.3	21.6
CSP	0	0.3	2.8
BGA	**0**	**0.7**	**2.0**
total	25.9	53.8	71.8
TAB	0.8	1.6	3
COB	2.8	5.8	12.9
Flip-chip	0.4	1.9	6
total	4	9.3	21.9
TOTAL	51.2	79.7	105.3

Source : BPA

Major silicon founders (Intel, IBM, NEC, SGS-Thomson and others) and sub-contractors (ASAT, Amkor, OSE, Hana, OIT and others) have now PBGA in their road-maps. To take part in this emerging technology, the European Community decided 3 years ago to promote Consortia to develop it in Europe [2]. BULL participated in the two following projects as leader :

- CHIPPAC (Cost effective High Performance Packaging for Electronic Systems) a 2,5 year project (Nov. 1993- April 1996) , part of the Esprit III programme, gathering 12 European partners ;
- JESSI (Joint European Submicron Silicon Initiative) AE 82, a 3 year programme (1994-1996), in association with SGS-Thomson

2. JESSI and CHIPPAC PBGA objectives

Part of these two programs was to design, build and test functional demonstrators using PBGA packages with cavities, dedicated to high performance dice. In the CHIPPAC project, developments of the basic technologies to manufacture multilayer cavities PBGA were carried out by BULL for high performance telecom and computer ASIC's application. The JESSI program was focused on :

- the design of the PBGA to handle high speed signals (high speed data processing and transmission)
- the process developments to assemble dice in organic package.

The cost of these manufactured demonstrators is as important as the electrical and/or thermal performances. The printed wiring board industry proposes an industrial experience for the manufacturing of boards in large quantities, thus leading to possible cost reduction of the PBGA. Nevertheless, the manufacture of cavities in a board using the standard industrial equipment was a technical challenge which has been set in these two programs.

3. Substrate designs and associated Printed wiring Boards technologies

3.1 : PACKAGE OUTLINES

Packages are based on multilayer substrate. The chip is connected to the PBGA on a double tier. Bonding fingers are connected to the balls through plated thru holes (PTH). A cavity is routed to allow the chip to be protected by a glob-top. The multilayer is built from a high tetrafunctional epoxy or BT resin. For cost reduction purposes, the packages were designed for the housing of dice of different dimensions and pinouts. Ground and power rings on the tiers of the packages have been conceived.

3.1.1 : *JESSI AE 82 PBGA demonstrator*

This package (figure 1 below) is a cavity-up structure with a heat spreader for the thermal dissipation. The thermal dissipation can be up to 20 Watts. The body size is 40 mm SQ according to the Jedec MO 151 standard and shows 480 balls on a 29 x 29 matrix at 1.27 mm pitch on 5 rows.

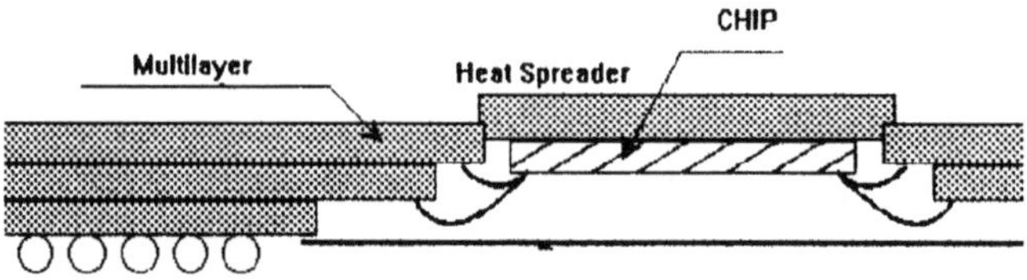

Figure 1 : Cross-section of the PBGA

3.1.2 : *CHIPPAC cavity PBGA demonstrator*
The structure of this demonstrator (figure 2) is a cavity-up without spreader dedicated to dice with a thermal dissipation less than 3 W. The size is 42.5 mm with 604 balls on a 31 x 31 matrix at 1.27 mm pith on 6 rows.

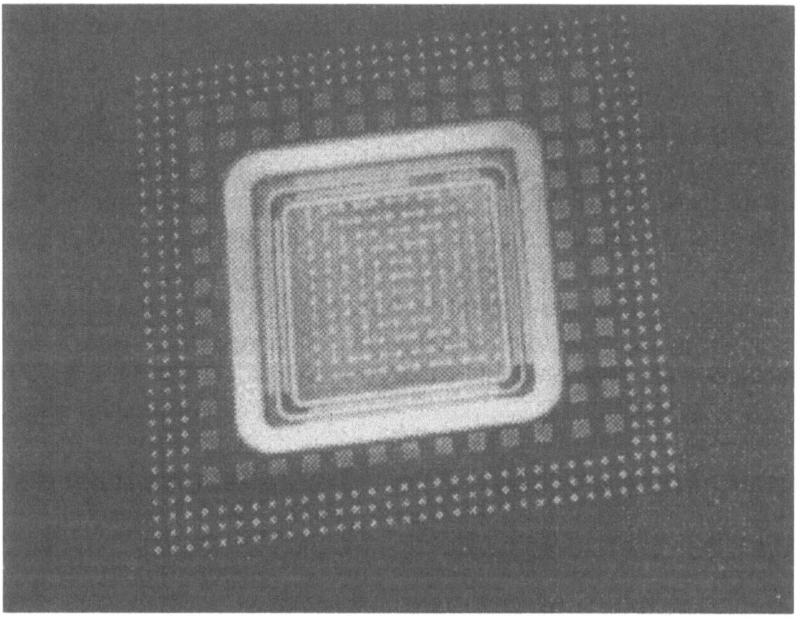

Figure 2 : CHIPPAC demonstrator, cavity side

3.2. PRINTED WIRING BOARD TECHNOLOGY ISSUES :

In order to design a PBGA structure with cavities, specific design rules differing from the current PCB technology were implemented :

3.2.1. *Pre-preg flowing control*
When laminating the multilayer, flowing occurs onto the bonding fingers. To minimise this flow, a solder resist layer is laid on the metal layers prior to lamination. Measurements by a non contact machine show that the pre-preg flows in the range of 200 to 400 μm.

3.2..2. *Misalignment between bonding fingers tiers*
The misalignment due to the lamination process yields difficulties for wire-bonding and can generate crosses between wires (not allowed by the Mil. 883 Std.). The design shall take into account the bonding finger length tolerance when the wedge connection technique is selected. Typical tolerance is 300 µm (difference between the nominal and the minimum value of the distance form bonding finger tip to cavity wall).

Fig. 3 : View of the tiers and the bonding areas - JESSI PBGA

3.2..3. *Pads openings*
The ball pads underneath the package are defined by circular openings in a solder resist layer exhibiting a gold plated surface. Diameter opening tolerance must be better than 50 µm, in order to limit ball height variations.

3.2.4. *Coplanarity*
The Jedec standard MO 151 requires a ball tip coplanarity less than 150 µm. Measurements have been carried out using 2 different methods : the best plane method and the 3 points method. The first one calculates the best plane to minimise the distance between the lowest and the highest ball tip [3]. Measurements are in the range of 50 to 100 µm. The second method refers to a plane defined by 3 pads chosen at the package periphery. Results are between 100 and 150 µm.

3.2.5. *Heat spreader attachment*
In the case of the PBGA developed in the JESSI program, the heat spreader is epoxy glued to the multilayer. The width of the fillet is 3 mm. Mechanical tests using 3 bending points method have shown that attachment strength is around 70 Kgs after humidity and thermal cycling.

3.2.6. *Metallisation*
Packages for complex dice running at high frequencies require to minimise the inductance and the capacitance on the signal lines. Antennas on the tracks are not allowed. For that purpose, the electroless gold has been selected for the CHIPPAC PBGA. With this technique, 0.1 μm of gold can be deposited. The JESSI PBGA has used a 0.5 μm thick electrolytic gold. One external metal has been used as a tie-bar for that purpose.

4. Process Developments

4.1.PACKAGE ASSEMBLY

In both programmes, the process assessments were roughly the same. The die assembly flow-chart is : die sawing, die attach, wire-bonding, glob-top and ball attach.

4.1.1. *Die attach*
A silver filled epoxy was selected for improving the thermal path from the die and to connect its backside to the ground. This epoxy was selected to comply with the low expansion coefficient of the chip. The curing profile of the adhesive has been optimised to achieve a low stress assembly.

4.1.2. *Die connection*
Techniques and constraints to connect a chip in a PBGA are identical to the COB (Chip On Board) technology. Aluminium wedge and ball bonding were evaluated. The aluminium wedge technique is more flexible than the gold one with respect to the gold thickness and the gold pollution. Electroless gold is more dedicated to the aluminium wedge and the electrolytic gold to the Ball bonding. A precise control of the wire loop heights is required for coating. Specific development has been conducted to take into account the very small gap between the rings and the bonding fingers, to avoid short-circuits with respect to the loop height of the wire.

4.1.3. Glob-top
The chip protection is carried out by a coating dispensing which completely covers the chip and the wires. This coating is selected to minimise the stresses on

the wires, especially at the bond heel. Material and process deposition have been implemented in order to reach a good resistance of the assembled package against moisture.

4.1.4. Ball attach

Eutectic solder balls can be used in the case of PBGA (same range of TCE as the board). High reliability of the soldered joints is then achieved. The two advantages provided by eutectic balls are the tolerance at balls coplanarity level and the simplification of the ball mounting.

Flux is deposited on the PBGA prior to the balls placement. A specific tool places balls onto the grid array of solder mask resist openings of the PBGA substrate. Balls connection to the PBGA is done by infrared reflow. The strength of balls, measured after thermal aging, goes over 1 Kg. The gold thickness has no impact on the ball strength nor on the soldered joints reliability. Fig. 4 shows a PBGA populated with solder balls.

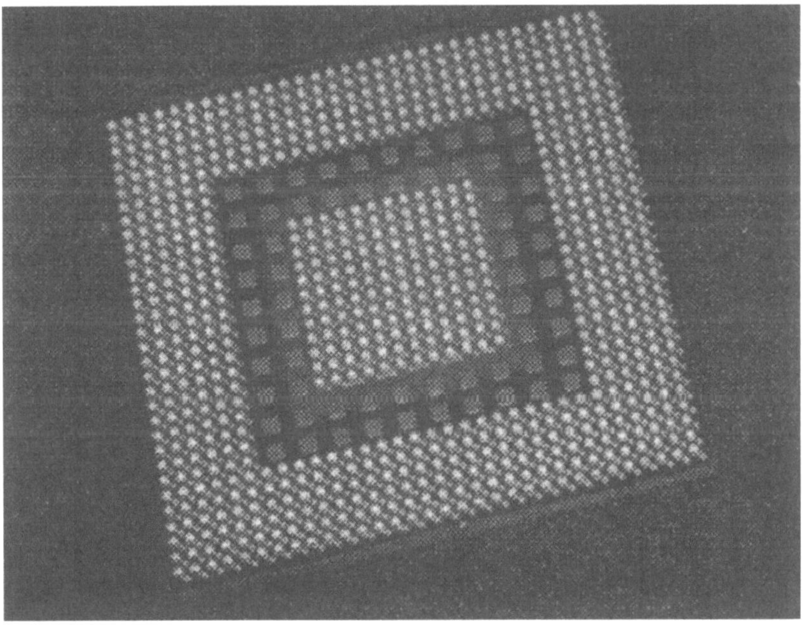

Fig. 4 : CHIPPAC PBGA with balls

4.2. MOUNTING ON BOARD

The process of board assembly has been developed internally, repair included. No-clean solder paste is first deposited on the PCB. The control of its thickness is one of the key parameters to achieve a good yield. Typical thickness is 150 µm. Another parameter is the weight of the PBGA. 50 milligrams per ball is the maximum value to avoid short-circuits. The reflow is performed in an oven.

Solder paste thickness deposit, as well as reflow thermal profile shall be compatible with the other components on the same PWB [4]. Control of the solder joints can be done by X rays to detect the short-circuits. Visual inspection of the external row of balls gives a good idea of the joints aspect.

5. Conclusion

The PBGAs developed in the CHIPPAC and the JESSI programs show this technology as a good solution to package complex and high pin count dice in a cost/performance approach given by the organic substrate and the manufacturability aspects of the PWB industry. Adaptation of the current PWB technology is necessary. Good reliability has been achieved regarding to the expected applications. Electrical performances of these demonstrators have shown the capability of complex PBGA to handle high frequencies up to 500 MHz. The PBGA package will be increasingly used by the industry, even for packaging high performance devices. It can be expected that alternative PBGA constructions will appear with design simplifications and resulting in cost reduction.

6. Acknowledgements

The authors would like to acknowledge all partner companies involved in the CHIPPAC and the JESSI AE 82 projects. The work has been carried out thanks to the financial support of the European Union and the French Ministry of Industry.

7. References

1. Romenesko B.M. (1996) BGA and flip-chip Technologies. Their histories and Prospects , *Microcircuits and electronic packaging*, Vol 19, Number 1, p 68
2. Kurzweil K. (1995) Advanced packaging in european cooperative programs, *Semicon-West, San-francisco, USA*
3. Saint-Martin X. Stricot Y. (1997) BGA : what does flat mean , *European Microlectronics Conference , Venice, Italy*
4. Saint -Martin X., Stricot Y., Auray M. (1995) Are BGAs a concern in SMT, *European Microelectronics conference, Copenhagen, Denmark*

LOW COST FLIP CHIP ON GLASS-EPOXY FR4 SUBSTRATES FOR PERSONAL ELECTRONICS

N. VAN VEEN

Nederlandse Philips Bedrijven CFT
P.O. Box 218
5600 MD Eindhoven
The Netherlands
+31-40-2733364, fax 2736815, veen_nja@cft.philips.nl

ABSTRACT

The drive towards lightweight electronics has led to the development of a low cost flip chip process on fine line FR4 substrates. As the flip chip mounting is to be embedded in a fine pitch reflow soldering line the assembly process was designed in such a way that maximum compatibility was assured. The process development was performed as part of the ESPRIT Project LOCOP.

1. Introduction

In the Esprit project LOCOP, which started at the end of 1993, was dedicated to the development of low cost multichip modules. Participants were Siemens (D), Dassault (F), Thomson (F), Philips (NL), FhG-IZM-TU Berlin (D), ICT (D), INESC (P) and IMS (D). Philips was awarded the task of developing a process for flip chip on rigid organic glass epoxy FR4 board in a no-clean environment. The driving force for starting development in flip chip technology can be found from Table 1. Starting point of the project was that flip chip provides the highest degree in miniaturisation in printed circuit board assembly and the aim of the project to bring flip chip mounting to the level of pilot production in a real product within three years. It was evident from the beginning that flip chip would have advantages especially in the field of high frequency behaviour. Compared to SMT packages the capacitance of a solder bump is one tenth at 0.1 pF and the inductance is only one hundredth at 0.01 nH.

However as one starts it is appropriate to realise the reasons for packaging:
- environmental protection
- adaptation to the PCB process, the package functions as pitch converter.
- improvement of electrical/ functional testability compared to wafer level test

R. Tummala et al. (eds.), Electronic Packaging for High Reliability, Low Cost Electronics, 267–274.

PACKAGE	PITCH [MM]	SUBSTRATE [MG]	DIE [MG]	PACKAGE [MG]	ASSEMBLY [MG]	WEIGHT % DIE
PLCC 68	1.27	1450	160	4500	6100	3
VSO 40	0.76	450	35	600	1100	3
QFP 80	0.80	1100	80	1300	2500	3
SO-L 24	1.17	400	32	500	900	4
BGA 169	1.27	700	100	1500	2300	4
SSOP 24	0.50	100	17	200	300	5
QFP 80	0.50	400	80	500	1000	8
COB	0.30	370	80	400	850	9
Flip Chip	0.30	100	80	10	190	40
μBGA	0.50	100	80	10	190	40

Table 1: An indicative indication of the efficiency of packaging die into different packages and subsequent assembly on printed circuit boards. All die sizes are maximum allowable size and have thickness 0.6 mm. Printed circuit board thickness varies for different components.

- marking; it is easy to write on the package what is inside the package.

At the beginning of the project the choice had to be made between the different possibilities for flip chip:

2. Bump structures

The interconnection is made of either:

i. High melting solder bumps (95/5 lead/tin) on the bond pads and eutectic solder on the footprint of the printed circuit board. Bump pitch can be as low as 175μm.
The bumps are made by evaporation or, as is presently more pursued, by electroplating or even stencilling, provided the bond pad pitch is around 250μm and up.
The counter part on the board side is a solderable pad supplied with flattened eutectic solder. Specifically this local solder is expensive in PCB technology. This system is exercised by IBM and subsequently by Motorola. However due to some of the arguments mentioned here, the choice was made to go for the following system:

ii. Low cost flip chip systems contain eutectic solder bumps, made by electroplating or stencilling. In this case there is no need to apply solder onto the board. Care has to be taken in the design of solder pad on the substrate in relation to the solder volume in the bump to assure a proper stand-off for the application of the underfill. There are four factor that determine the final stand-off:

1. The volume of the solder.
2. The surface energy of the solder.
3. The dimensions of the pads on both the IC and the printed circuit board.
4. Thc wcight of thc IC.

In a previous paper the relation between these parameters has evaluated analytically and it has been demonstrated that using a simple analytical trial function [Van Veen, 1995] for the bump shape results can be obtained that are equivalent to more laborious numerical solutions [Goldman, 1994]. The process flow for mounting of flip chips is chosen such that maximum overlap with the existing reflow soldering process is achieved:

- printing of solder paste for the SMDs, followed by inspection of print result.
- placement of the SMDs followed by inspection of the placement result.
- picking the flip chip and inspection of presence of the bumps.
- fluxing the flip chip followed by placement.
- reflow of flip chip together with the SMDs.
- electrical testing, followed by repair or
- underfilling process with subsequent curing.

3. Layout

The layout of bond pads on the printed circuit board has been a matter leading to discussion. The major obstacle is provided through the misregistration of the solder resist, which in present printed circuit board technology is about 75 micron. The pads have to be in the order of 100 micron in length and width. Definition of bond pads by having rectangular openings in the resist leads to openings of 250 micron width and 100 micron length. This provides the option for achieving finest pitch on the board. However for ICs having bond pads on all four sides this may lead to skewed interconnections.

In order to be more independent of the misregistration of the resist it was chosen to use round openings in the resist projected on a copper pad of 250 micron minimum. This systems allows for bond pad pitches of 350 micron and up but has the advantage that all bump structures are nicely vertically aligned. Confinement of the solder is achieved by pad design through overlay of the solder resist over the edges of the pads. Through this confinement a proper control of pad size on the substrate side is achieved and therefore this part of the height control is assured.

4. Flux

Placement of the flip chip is done in a flux, which has to fulfil the following conditions:
1. It has to allow for a no-clean process
2. It has to provide sufficient adherence between the bumped chip and the substrate such that transport to the reflow oven will not cause serious displacements.
3. It is supposed not to leave residues. The residues that remain are to be compatible with the chemistry of the underfill as otherwise cleaning would be necessary for the underfill step.

The best approach is to apply the flux only locally where it is needed. This can be achieved through tampon printing, dispensing (using multiple dispensers) or printing using ink-jet technology. A simple technique is to dip the IC with its bumps into a thin layer of flux, just before placement. Disadvantage of this technique is that it happens inside the pick and placement machine and adds to the cycle time.

5. Thermal mismatch

The strain induced by thermal mismatch between IC die and substrate is proportional to the die dimensions. With increasing size, the shear forces on the die increase and lead to premature crack formation at the places of highest strain concentration in the construction. It is evident that due to the spherical nature of the bumps the highest concentrations appear at the IC bond pad-bump and the bump-PCB pad interfaces. The

whole construction can be regarded as two plates of different expansion coefficients with wheels in between.. The cracking starts at the place where the rolling occurs.
After soldering, the structure is therefore underfilled with a filled epoxy based encapsulant. At present viscosity is optimised for underflow behaviour and the expansion coefficient of the cured material is made to match the bump material through filling the underfill with solid particles. Care has to be taken that little or no voids occur in the cured underfill. Through the underfilling the assembly transforms into a bi-layer construction and the shearing forces are predominantly transformed into bending forces.

6. KGD

As with all bare die delivery, flip chip generally is hampered with a deficient knowledge of the 'goodness' of the die.
There is at present still some reluctance by semiconductor manufacturers to release wafers for bare chip mounting applications. Requests to deliver bumped wafers to external customers are not always honoured. Also IC manufacturers who want to deliver bumped dies will have to set up their own bumping facility as they have to do final testing after bumping. At present some contract bumping facilities are being set up. However the discussion on who has the final responsibility for the achieved quality level has not come to an end yet.
This implies that flip chip at present is limited to a few larger vertically integrated companies, who have the complete processing in their own hands. If this situation will continue for a longer period this will also be lethal to the concept of flip chip as it will effectively prevent the spreading of this technology to small and medium size companies.

7. Chip Size Packages

This type of IC package compares favourably with respect to fine pitch QFPs, with respect to occupied area and assembly weight (IC + package + substrate) in a ratio of about one to three or four. Furthermore the electrical performance of flip chip constructions is excellent due to short interconnects. Together this provides excellent means for the successful realisation of small sized, lightweight and high performance electronics as portable computers, pagers and handheld phones or camcorders.

In recent developments, the total size of the BGA package is reduced to the dimensions of the die itself. In this way a 'chip-size package' is obtained. This package has characteristics in common with a flip chip, but the die-testing problem (KGD) has been solved, as the die can be adequately tested in the package. In a period of one to two years time, a lot of companies have published work on chip size packages. These packages make the miniaturisation achieved by flip chip assessable to everyone.

A lot of companies have already gone a part of the way towards realisation of such a package. An overview and categorisation as done by TechSearch is given in Table 2. All these companies have either their own developments or have acquired licences. The packages solve the 'known good die' problem, and are therefore very well acceptable for semiconductor manufacturers. On the other hand they very well fit into the regular reflow line as a normal SMD component. However, only if they are equipped with sufficient strain relieve for the bump-interconnect, rendering underfill superfluous.

8. The separation between FC and CSP

In a recent study (October 1995) BPA has projected the number of chip size packages for the year 2005 at 3 billion compared to 6 billion flip chips, 2 billion BGAs and 20 billion PQFPs on a total of 100 billion packaged ICs. In their study they find a tendency in CSP towards higher pincounts, whereas the lower pincounts have better coverage from flip chip. This can be understood as follows: At regular time intervals the IC manufacturers perform so-called die shrinks, reducing the die surface, available for placing bumps. Therefore the number of IOs per unit surface increases to values, which cannot be handled with the printed board technology. In this case (chip size) packages are needed.

This approach also determines the transition between the use of micro BGAs (fan-in systems) and MiniBGAs (fan-in/fan-out systems). The number of IOs accommodated by micro BGA can be estimated as follows: Along the edge of the die an area of approximately 0.5 mm is needed for making the connections to the IC pads. Therefore the number of bumps N over the rest of the active area equals:

$$N = \mathrm{INT}((a-1)/p) * \mathrm{INT}((b-1)/p)$$

where a and b stand for the length and width of the IC in mm and p represents the ball pitch. INT stands for the integer function. If the required number of bumps cannot be accommodated a fan-out system has to be considered.

A system in between the μBGA and miniBGA is Shellcase's Slimcase which through its special processing using glass plates and semiconductor type processing can accommodate one additional row of bumps.

9. Substrate issues

A flip chip has bumps on the bond pads and bond pads can be as close 150 μm apart and still have sufficiently high bumps. In fine line printed circuit boards the lines and spacings are 100 μm. In case the pads are defined by overlays of solder resist with a registration error of 75 μm, the minimum pad pitch is 400 μm. In case all connections

TECHNOLOGY	COMPANY

Flex circuit interposer:	Fujitsu, General Electric, NEC, Sharp, Tessera (Amkor/Anam, Hitachi, Mitsui, Read-Rite, Shinko)
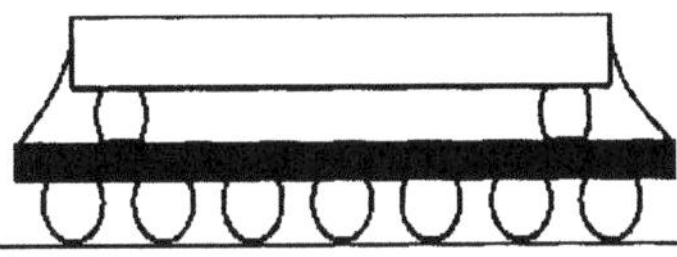	
Rigid substrate interposer:	Fujitsu, Matsushita, Motorola, Sony, Toshiba

Custom leadframe (for memories):	Fujitsu, Hitachi Cable, LG Semicon, Mitsubishi Electric, Rohm

Wafer level assembly:	ChipScale, Motorola ShellCase, Sandia

Table 2. Some CSP package types and manufacturers. Source: TechSearch International, reprinted with permission

cannot be realised on the periphery of the chip, staggered configurations can be used at pitch of 600 μm, or bumps have to be realised in the inner circle. These can be connected to the board to pads which can be routed out through vias. Vias, at present, mechanically drilled plated through holes, can be made at a pitch of 0.75 mm . If with this board technology the necessary number of interconnects cannot be realised, then FC is not an option.
A solution can be found in CSP, where the pitch and spaces between the bumps is stretched too such an extent that the maximum number of interconnects can be realised in accordance with the available printed circuit board technology.

10. The future

MCM-L is very well acceptable as a low cost solution, however the line widths, via size and via land size are, at present, such that the performance (low power dissipation, and low electromagnetic emissions), and packing density suffers. For rigid substrates the solution can be found in the use of Sequential Build Up boards (SBU), which combine both narrow line width, small vias and via land sizes and low cost.
In this technology, a two- or four layer standard fine line PCB is used as a base for deposition of dielectric layers. Via holes can be made in the dielectric layer with three different technologies. After this the holes and the dielectric surface is metallised and structured. Now the process of applying a dielectric layer can be repeated until the desired number of additive layers is reached.

11. Conclusions

Although technological problems in the mounting of flip chip on FR4 can be solved, it is questionable whether a sufficient number of companies will pursue flip chip on board.

12. Acknowledgement

Financial support from the EC for activities undertaken in the LOCOP project is gratefully acknowledged.

13. References

Veen N.J.A.; Proceedings TAP '95 San Jose, **5**, 146 (1995)
Goldmann L.S., Proceedings ECTC **43**, 1120 (1993)

Chip Stacking using Flip Chip Technologies

Maja Amskov
DELTA Danish Electronics, Light & Acoustics
Venlighedsvej 4
DK-2970 Hoershom
Tel:. +45 45 86 77 22
Fax. +45 45 86 58 98
e-mail: mam@delta.dk

Introduction

For miniaturisation purposes the flip chip technology has for a long time been known as the smallest package yet seen and lot has been written concerning different flip chip technologies and their reliability [1,2,3,4].
The motivations for putting this effort in the area of flip chip mounting are many fold. Depending on the technique that are chosen the flip chip mounting are both smaller, more reliable, cheaper etc.

As also the micromechanic technique mature we will see all different kinds of mechanic transducers and actuators made out of silicon. Now combining the conventional IC's with the micromechanical components, extremely small systems can be made with only a few additional components.
For many microsystem applications the size and weight are very important and that is why the idea of stacking the two (or more) silicon parts has come up.

The most severe objection to the stacking idea is that it is necessary to have electrical access to both parts from the substrate. There exists technologies that can take care of this problem [5], and the subject will not be pursued further here.
Instead we will look into the challenges of stacking two silicon parts, where at least one of them could be rather fragile.

Driving forces

Some of the driving forces for using the stacking technique for the microsystems are:
Concerning size:

> The entire module will take up less substrate space, which depending on the type of substrate will mean a decrease in cost, weight or both.

R. Tummala et al. (eds.), Electronic Packaging for High Reliability, Low Cost Electronics, 275–279.

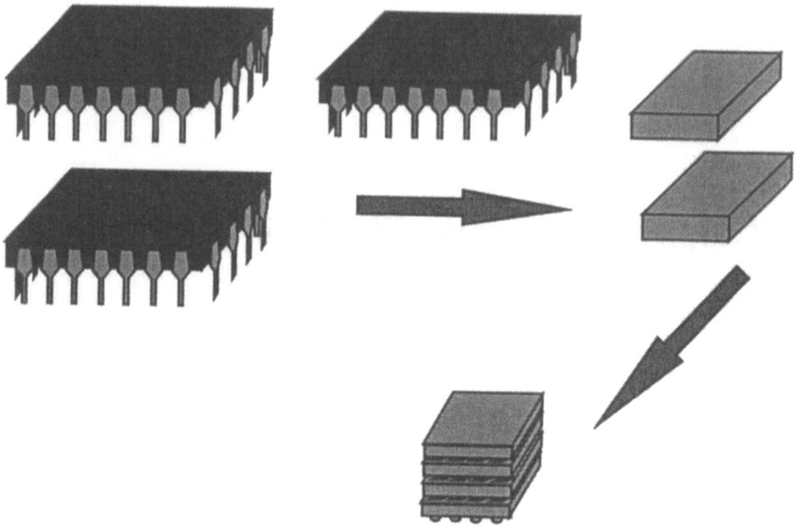

Figure 1 By stacking the different parts of a microsystem both weight and size can be tremendously reduced

Concerning electrical connections:

The electrical routing form one part to the other is the shortest possible, and that will be important for the speed of the system. As most micromechanic devices deal with a transformation of some kind of measurand to an electrical signal or vice versa, also the sensitivity towards EMI will be minimised.

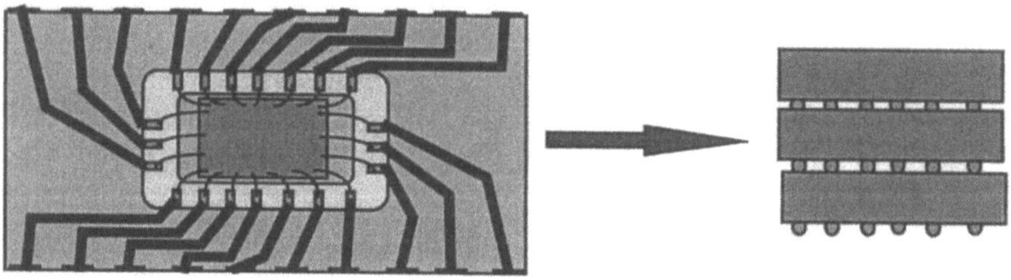

Figure 2 In the stacked version of a microsystem the electrical path is the shortest possible and furthermore the parasitic loads and thus the noise sensitivity will be minimised.

Concerning reliability:

It is well known that one of the worst problems that the flip chip mounting face are the fact that most substrates have a Coefficient of Thermal Expansion (CTE) that are larger than that of silicon. So during thermal cycling the flip chip connection see a great stress, that might induce cracks or opens in the connection. When stacking the chips this crucial mounting to a substrate other than silicon are made only once, thereby decreasing the risk of failures.

Conductive adhesive systems for stacking

When it comes to performing the different flip chip mountings in the stack it is important to keep in mind that the processes must be compatible with respect to the process conditions, that is: the pressure and temperature etc.
In other words, the system that requires the highest temperature should be utilised first.

Now concentrating on the conductive adhesive flip chip techniques, there are a number of methods that could be used for chip stacking as well as for the conventional flip chip mounting.

Stud Bump Bonding:

Each connection consists of the ball from a gold ball bond (a stud) and conductive adhesive. This method is a low cost, low investment method that requires only a ball bonder and the conductive adhesive [4]. The adhesive can be applied by dipping the stud into the conductive adhesive, alternatively the adhesive can be applied to the other part (the one without the studs) by dispensing or by stencil printing. After curing the conductive adhesive, the underfill is applied.

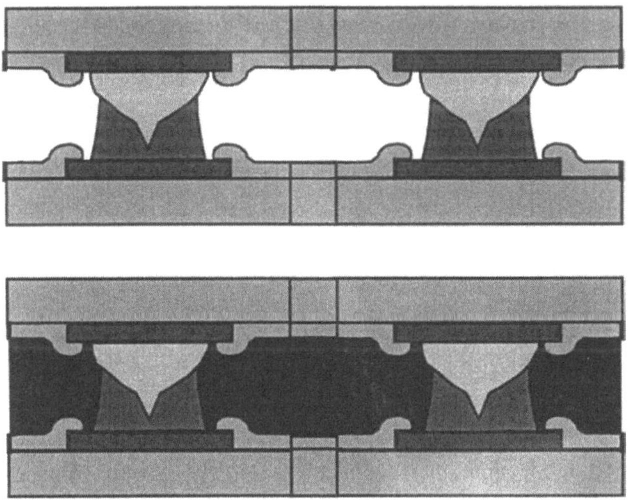

Figure 3 Stud Bump Bonding

Polymeric bonding:

The electrical connection is made entirely out of conductive adhesive. This method requires a way of positioning a small amount of conductive adhesive with a high precision (at least 1/3 of the pad size), either by stencil printing or by dispensing. Underfilling is still required.

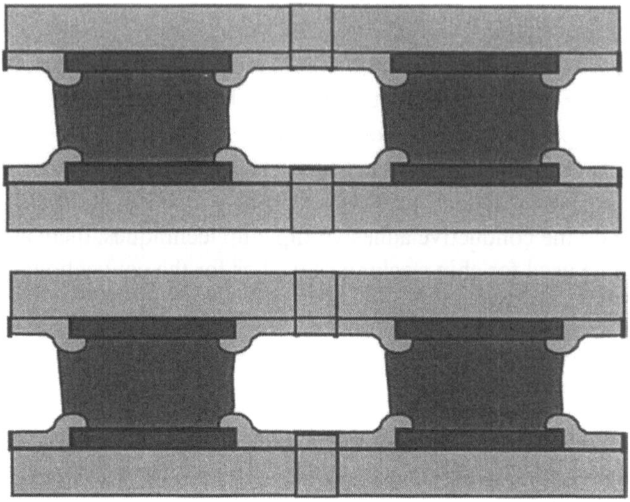

Figure 4 Polymeric flip chip bonding

Non conductive adhesive:

Here the electrical connections consists only of a gold stud. The studs are then pressed through a non conductive adhesive into the metal pads on the other part. No underfilling is needed.

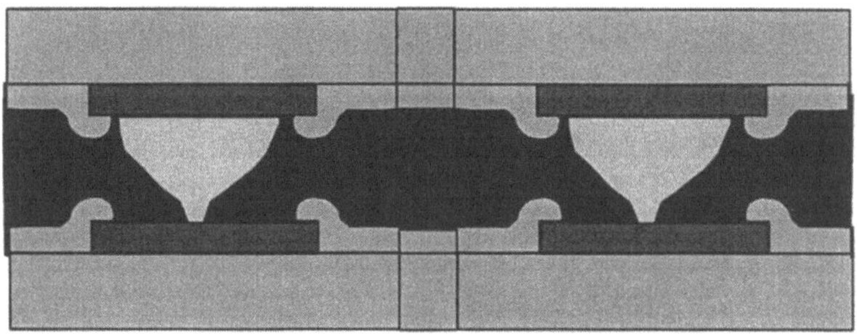

Figure 5 Non conductive adhesive flip chip bonding

From our initial work with these methods it seems most straight forward to handle a method where a gold stud is involved, that is the stud bump bonding and presumably also the non conductive adhesive technique. The great advantage is that the real fine pitch or precision work is done by the use of equipment that is already well known and in many cases already standing in the companies. Also the stud methods are Al pad compatible. Different conductive adhesives and underfillers work well with this method and thus it is possible to chose systems that are compatible with each other in terms of thermal budget.

References:

[1] J. Liu, K. Boustedt & Z. Lai. "Development of flip chip technology on flexible circuitry using anisotropically conductive adhesives and eutectic solder", Circuit world Vol. 22, No 2, 1996.

[2] F. Kriebel, K. Melzer, E. Meusel & Th. Seidowski, "Electrical and mechanical properties of flip chip connections using Isotropic conductive adhesives", Porc. Adhesives in Electronics 1996, Stockholm.

[3] Outi Rusanen, V.Pennanen & A. Björklöf, "The use of thermosetting and thermoplastic isotropically conductive adhesives in flip chip bonding", Proc. ISHM Nordic konf. sep. 1996.

[4] K. Tsunoi, T Kusagaya & H. Kira, "Flip chip mopunting using stud bumps and adhesives for encapsulation", Flip Chip Technologies, McGraw-Hill.

[5] C. Christensen, S. Bouwstra, J. W. Petersen & P. Kersten, "Wafer througholes interconnection for high vertical wiring densities", EuPac '96

FLIP CHIP JOINING TECHNOLOGY APPLICATIONS

J. VÄHÄKANGAS, O. RUSANEN and T. JAAKOLA
VTT Electronics
P.O.Box 1100, FIN-90571 Oulu, FINLAND
Tel. +358 8 551 2111, Fax +358 8 551 2320

1. Introduction

Flip chip technologies offer today many alternative methods to accomplish dense packaging of bare chips. Naturally most of the research and development work has been devoted to mounting of silicon integrated circuits. Much less flip chip joining has been applied to quite new micro machined sensor components or surface acoustic wave (SAW) components. Many micro machined sensors are very fragile due to their structures and their performance is sensitive to thermomechanical stresses. Here adhesive flip chip bonding seems to offer a proper mounting solution.

Yatsuda *et al.* [1] and Onishi *et al.* [2] have developed SAW filter stud bumping techniques and used both thermosonic and adhesive bonding to connect the chips on the substrate metallisation. Small size and good electrical performance have been obtained by the methods.

In this paper we show results of flip chip joining experiments made for three quite different applications, namely a smart card, a miniaturised optical spectrometer and a SAW filter. The aim is to see how well the special needs of these applications are achieved using flip chip joining techniques.

2. Materials and Methods

Table 1 summarizes the component materials and types, the substrates and flip chip joining methods used in the experiments. More detailed experimental parameters for each application are described in Section 3.

TABLE 1. Components, substrates and joining methods

Material	Component type	Substrate	Joining method
Si	ASIC	Polyimide, alumina, FR-4	Adhesive bonding
Si	Micro machined	Silicon detector	Adhesive bonding
Lithium niobate, lithium tantalate	SAW filter	Alumina	Thermocompression, thermosonic,adhesive bonding

R. Tummala et al. (eds.), Electronic Packaging for High Reliability, Low Cost Electronics, 281–286.

3. Results from Different Applications

3.1. SMART CARD

A smart card was manufactured utilising flip chip interconnection technology. The use of this technology results in increased packaging density especially in vertical direction. Thus the flip chip components can be processed into a smart card that has a thickness of less than 0.8 mm. Isotropic adhesive bonding was chosen instead of soldering.

The use of isotropically conductive adhesives in connecting flip chip components makes the bonding process simple and inexpensive. In addition, the reliability of a product does not decrease even when the rather expensive process of flip chip soldering is replaced with adhesive bonding. Therefore, cost sensitive products can also benefit from the advantages of flip chip technology.

Four isotropically conductive adhesives were used in flip chip bonding. Two adhesives were epoxy based and two adhesives had a thermosetting polymer matrix. An epoxy underfill material was used with all tested structures. Three different substrate materials were used; alumina, polyimide and FR-4.

The manufactured flip chip modules were subjected to 2000 thermal cycles between -40 °C and +125 °C. Thermal cycling broke many of the tested structures and the most influential factor affecting reliability was the modules substrate material. A switch from a small TCE substrate material (alumina) to a large TCE material (FR-4) decreased the structures thermal cycling reliability with all tested adhesives. The effect of damp heat was studied in 85%RH/85°C-ageing for 1000 hours. This testing had only a minor effect on the modules reliability, since the component and substrate metallisations (Au, Ag/Pt or Sn) were not especially sensitive to moisture [3].

The flip chip modules were laminated to smart cards. The smart cards were tested with bending and torsion tests. The cards showed to be very durable. The first cards did not break until 17 500 bendings when PVC plastic fractured. Even at that point the cards were electrically functional. The results show that the use of isotropically conductive adhesive offers a viable option for soldering of flip chip components [4].

3.2. MINIATURISED SPECTROMETER

The completion of a miniaturised near infra-red range spectrometer at the end of 1996 was an important technological breakthrough by VTT Electronics. The spectrometer consists of an IR source, an interferometer, a detector and an IC. The interferometer is fixed on top of a silicon detector by a flip chip technique utilising an isotropically conductive adhesive. The whole spectrometer assembly is housed in a ceramic DIL package. The dimensions are only 12 mm x 23 mm x 5 mm, so that the device can be integrated into a great number of systems.

Infrared analysers often use a rather bulky IR-blackbody source and a motor driven rotating filter wheel. Often a dispersive bulk optics grating is included. It is quite clear that such a spectrometer cannot be truly miniaturised. In this miniaturised spectrometer the IR source is an electrically modulatable thin film structure replacing the conventional black-body radiator and the rotating tooth or filter wheel. The analyser is an electrically tuneable pass band filter of Fabry-Perot interferometer type mounted on top of a bolometer or thermopile IR detector. All components are made of silicon using micromachining and other semiconductor technologies at VTT Electronics in Espoo [5].

The assembled spectrometer is shown in Figure 1. On the left there is the monolithically integrated thin film infrared source, in the middle the IC and on the right the pass band filter on top of the monolithically integrated IR-detector. The necessary electrical wiring is made by thin film metallisation on the silicon substrate.

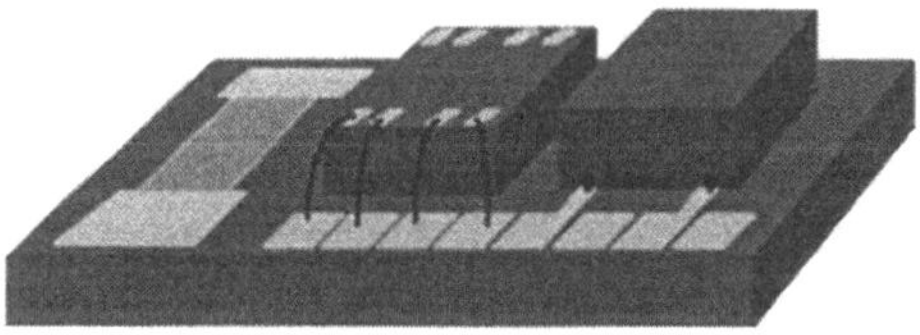

Figure 1. Silicon micromachined miniaturised IR-spectrometer.

The Fabry-Perot interferometer consists of two parallel mirrors that occupy most of the surface area. For reliable operation the mirrors must remain free of strain. Flip chip bonding with adhesives is a plausible interconnection method since soldering could easily have damaged the rather fragile component.

Manufacture of the spectrometers proceeded in a reasonable manner. A proto-series of fifteen spectrometers was produced and fourteen of them passed the operational testing. Adhesive flip chip bonding proved a successful interconnection method for a micro-machined component.

3.3. SAW FILTER

Surface acoustic wave (SAW) devices are the key components for miniaturization of wireless applications such as the mobile phone, whre these components are used in several kinds of filters in, for example, front-end duplexer, receiver and transmitter stages.

The inteconnection method normally used for SAW chip bonding is the conventional epoxy die bonding and wire bonding. Electrically better (lower inductance) and smaller structure would result if flip chip bonding could be employed to make the electrical connections to substrate or package.

Conventional bumping processes based on sputtering and chemical plating of metals are not well suited for SAW chips due to the very sensitive surface with thin (0.2 μm) and narrow Al electrodes and the pyroelectricity of the chip materials leading to electrode burning during temperature changes. Of concern is also the high TCE of the materials (14 ppm/°C for lithium niobate), which results in high TCE difference for substrates like alumina and silicon (TCEs 6 and 3 ppm/°C, respectively).

In this work SAW chips (size 1x1.6mm^2, 8 bumps) were bumped by using the stud bumping method. A slightly modified conventional gold wire ball bonder and AuPd1 wire (diameter 25 μm) were used in making the bumps. The bump shear strength was measured to be between 30 ..40 gm. As substrates alumina with thick film metallization (Ag/Pt) and silicon with aluminium metallization were employed. In order to obtain a reliable contact between the aluminium pad and the gold bump the pad was plated with 5 μm of electroless nickel and flash gold.

The flip chip bonding methods experimented were thermocompression, thermosonic and adhesive bonding. Table 2 shows comparison of the typical bonding parameters used for these methods. The benefits of thermosonic bonding as compared to thermocompression bonding are shorter bonding time, lower force and lower temperature. However, this method is more sensitive to cracking of material under the pad areas of SAW chip or substrate, so special methods are needed to avoid this problem. For adhesive bonding two methods were employed. In the transfer method before placing the chip on the substrate the bumps were coated with adhesive by dipping onto a thin layer of adhesive. In the second method the bumps of chip were bonded on thermoplastic bumps stencil printed on substrate pads.

TABLE 2. Comparison of Candidate Flip Chip Bonding Processes

Process	Typical Bonding Temperature (°C)	Typical Bonding Load (g/bump)	Remarks
1. Au-Al TC	280-350	100-200	Gross deformation of bumps, slow bonding
2. Au-Al TS	150-200	50-100	Concern about damage
3. Au-Au/Al adhesive transfer	150-200	5-30	Low strength
4. Au-Au/Al thermoplastic adhesive	150-200	5-30	Low strength

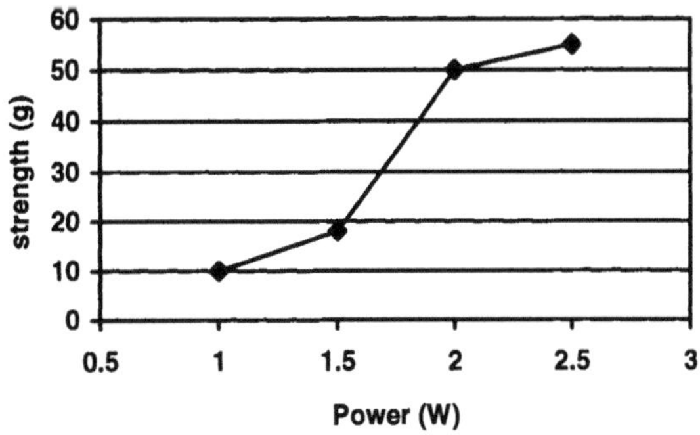

Figure 2. Die shear strength per bump as a function of ultrasonic power. Substrate temperature 150°C and bonding time 500ms.

In thermosonic method the bonding between gold bumps and substrate metallization is accomplished by applying a force and ultrasonic power to the back of the chip. In addition to force and power important parameters effecting bond quality are time and substrate temperature. Figure 2 shows an example of die shear strength as a function of ultrasonic power applied. For good strengths/bump a power of 2 W was required. Applying ultrasonic power greater than 2.5 W lead to increasing cracking of components under the pads, an example of this is shown in the SEM-image of Figure 3.

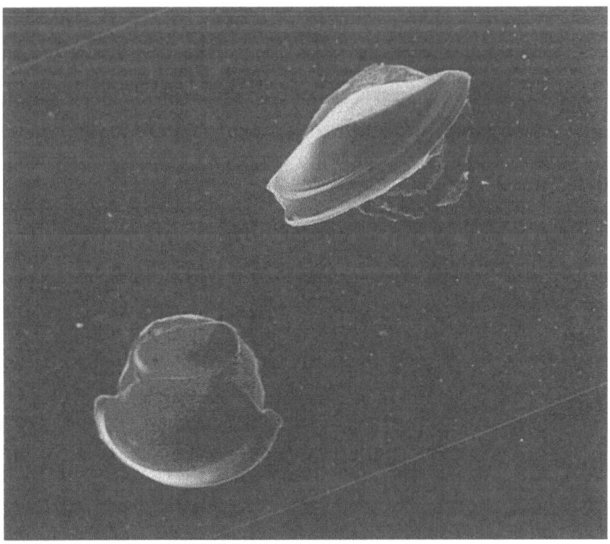

Figure 3. SEM image of fractured surfaces of thermosonicly bonded samples after die shearing (high ultrasonic power).

In thermocompression bonding the parameters affecting the bond quality are force, temperature and bonding time. Die shear strengths of 10 to 50 g/bump were measured. Quite a high force of 150 g/bump was needed for attaining the higher die shear strengths.

Adhesive bonded chips showed a much lower die shear strengths of 3 to 5 g/bump. Consequently, additional insulating resin sealing around the edges to support the chip is needed as used by Onishi *et al* [2].

The electrical response between 800-2000 MHz of test structures manufactured on alumina and silicon substrates was measured to be nearly identical with wire bonded structures. Preliminary thermal shock testing (32 cycles between -40...85 °C) have shown no changes in electrical properties. This is accordance with the results of Yatsuda *et al* [1], who in fact have reported 1000 thermal shocks between -30...+85 °C without failures.

4. Conclusions

We have applied various flip chip joining methods to a smart card, a miniaturised spectrometer and a SAW filter applications. Flip chip technology shows great potential as a joining method, although further work is needed especially as to reliability, electrical performance and productivity.

5. References

1. Yatsuda, Y., Horishima, T., Eimura, T., and Ooiwa, T. (1996) Miniaturized SAW Filters Using a Flip-Chip Technique. *IEEE Transactions on Ultrasonics, Ferroelectrics, and Frequency Control*, Vol. 43 (1) , pp. 125-129.
2. Onishi, K., Seki, S., Taguchi, Y., Bessho, Y., Eda, K., and Ishida, T. (1993) An Application of a Flip-Chip-Bonding Technique to GHz-Band SAW Filters for Mobile Communication. *IEICE Trans. Electron.*, Vol. E76-C, No. 6, pp. 993-999.
3. Pennanen, V. and Rusanen O. (1977) Reliability testing on flip chip joining with isotropically conductive adhesives.. To be published in the *Proceedings of the 11th European Microelectronics Conference 1997*, Venice.
4. Pennanen, V. , Rusanen, O. and Björklöf, A. (1996) The mechanical strength of smart cards made by isotropic adhesive bonded flip chip. *Proceedings of the Adhesives in Electronics '96* , Stockholm , pp. 410 - 413.
5. Blomberg, M., Lehto A., Rusanen, O. , and Keränen, K. (1997) A Silicon Microsystem - Miniaturised Infrared Spectrometer, *Transducers 97*, Chigago 1997.

NOVEL UNDERFILLS: EFFECT OF THE FILLER FOR FAST-FLOW UNDERFILLS AND LATENT CATALYST FOR NO-FLOW UNDERFILLS ON THE PROCESSING OF FLIP-CHIP DEVICES

C.P. WONG, S.H. SHI, & M.B. VINCENT
School of Materials Science & Engineering
And
Packaging Research Center
Georgia Institute of Technology
Atlanta, Georgia 30332-0245 USA
Phone: (404) 894-8391
Fax: (404) 894 – 9140
Email: cp.wong@mse.gatech.edu

Underfill encapsulant is critical to the reliability of the flip-chip solder joint interconnects. Two underfilling techniques presently used are the "Fast-Flow" and "No-Flow" processes. A key factor in the processing of flip-chip devices using the "Fast-Flow" underfill technique is the flow rate. The flow rate is mainly affected by the filler loading, particle size, and particles size distribution. A key factor in the processing of flip-chip devices using the "No-Flow" underfill technique is the selection of the latent catalyst. In the underfill formulation, this catalyst should give a minimal curing reaction before and a rapid curing reaction shortly after the solder joint reflow. In this paper, we present the preliminary experimental results of the above factors on underfill materials.

1. Introduction

Underfill encapsulant is placed between the die and substrate to extend the flip-chip solder joints life by 10-100 times. Currently, two types of underfill materials called "Fast-Flow" and "No-Flow" are under development for corresponding with the "Fast-Flow" and "No-Flow" processes, respectively. The "Fast-Flow" process reflows the solder bump of the flip-chip at first, underfills the gap between the chip and substrate, and then cures the underfill material. Whereas the "No-Flow" process reflows and cures the underfill material simultaneously. Since the underfill resin in itself has a high coefficient of thermal expansion (CTE), in order to match the CTE of the underfill materials with that of the solder joints and substrate, silica filler is added to the resin.

A key factor in the "Fast-Flow" process using the "Fast-Flow" underfill materials is the flow rate. Capillary flow drives the underfill under the chip to completely fill the chip/substrate gap. Many factors can play a role in the rate of flow of the underfill encapsulant between the chip and substrate. These factors can be placed into four categories: filler particles, underfill material, chip/substrate surfaces, and obstructions. The filler particle factors that can affect rate of flow include loading, size, size distribution, surface morphology, shape, material, density, and dispersion. The underfill material factors include rheology, density, and surface tension. The chip/substrate surface factors include surface energy and roughness. Obstruction factors can include solder bump pattern, particle size to gap height ratio, and contamination. The work presented in this paper shows the filler loading and particle size effect on CTE and flow properties.

A key factor in the "No-Flow" process using the "No-Flow" underfill materials is the availability of the successful underfill materials. Proper selection of latent catalysts plays a very important role in developing "No-Flow" underfill materials. A successful "No-Flow" underfill material should meet primary requirements such as: (1) Minimal curing reaction occuring at the temperature below solder bump reflow

R. Tummala et al. (eds.), Electronic Packaging for High Reliability, Low Cost Electronics, 287–294.

temperature, (2) rapid curing reaction taking place at the maximum solder bump reflow temperature, (3) good adhesion to passivation layer, chip, substrate, solder mask, and solder joints, (4) reasonable shrinkage of the material during curing, (5) low CTE, and (6) reasonably high moduli to minimize the residual thermal stress resulted from the curing process and consequent cooling.

2. Experimental

2.1 CHEMICAL INGREDIENTS OF UNDERFILL FORMULATIONS

The epoxy resin used as the underfill encapsulant is 3,4-epoxy cyclohexyl methyl-3,4-epoxy cyclohexyl carboxylate, manufactured by Union Carbide under the trade name ERL-4221D (used as received). The molecular weight is 252.3 g/mol and the epoxy equivalent weight is 133 g. The hardener used is hexahydro-4-methylphthalic anhydride (HMPA), purchased through Aldrich Chemical Company, Inc. (used as received). The HMPA has a purity of more than 97% and its molecular weight is 168.2 g/mol. The curing catalyst used is a metal acetylacetonate (used as received). The filler used is spherical silica with average particle diameters of 8, 15, and 30 microns.

2.2 PREPARATION OF UNDERFILL FORMULATIONS

The specified quantity of epoxy resin and hardener were stirred together for approximately 2 hours. A base formulation was prepared by adding a specified quantity of catalyst to the resin/hardener mixture and stirring for an additional 2 hours until the catalyst was homogeneously dispersed. Then the base formulation was stored in a freezer for later use. Before a filled underfill formulation was prepared, the base foumulation was warmed up to room temperature. The filler was placed in a vacuum oven at 120 °C for approximately 24 hours. This is to ensure the filler is free of absorbed moisture. Then a specified amount of filler was added to the base formulation in a mixing cup. This mixture was then mixed using a Waring commercial blender at low speed while pulsing. After mixing, the sample was placed in a vacuum chamber and the vacuum pressure was ramped slowly to 27 inHg over a period of 1 hour. This was done to remove the entrapped air introduced into the sample during mixing. The samples were stored in a freezer at -40 °C while not in use.

2.3 MEASUREMENT OF FLOW RATE BETWEEN PARALLEL GLASS PLATES

The flow rates of the filled underfill formulations were measured using a parallel plate flow test. This test involved flowing the underfill between two glass plates separated by a 3 mil (app. 75 microns) gap at a 12 mm width. The glass plates were pre-cleaned using isopropanol. They were then preheated to 80 °C to aid in reducing the underfill viscosity, to increase the repeatability of the test, and to mimic the actual production underfill dispensing process. A graduated ruler with millimeter markings was placed on top of the top plate as a tool for measuring the flow distance. The distance of flow was timed cumulatively at regular distance intervals.

2.4 MEASUREMENT OF FLOW RATE BETWEEN DIE/SUBSTRATE TEST VEHICLE

A test vehicle was used to test flow rate in-situ. This test vehicle was provided by National Semiconductor Corp.(Santa Clara, CA). It consists of a copper substrate with a solder (eutectic Pb/Sn) pad grid and a 6x6 mm transparent quartz simulated chip with a solder bump grid. The substrate and chip were cleaned with isopropanol. The vehicle was preheated to 80 °C and a drop of underfill encapsulant was placed along one edge of the test chip. The time taken to underfill the entire chip was recorded.

2.5 COEFFICIENT OF THERMAL EXPANSION

Coefficient of thermal expansion (CTE) readings of the filled underfill formulations were measured using a thermo-mechanical analyzer (TMA) from TA Instruments (Model 2940). Specimens were made for TMA by dispensing the underfill samples into aluminum pans that were 7mm dia. X 2.5mm ht. These pans were then placed into a convection oven at room temperature, heated to 250 °C, held at 250 °C for 15 minutes, and then removed to cool at room temperature. The TMA was programmed to heat the samples to 280 °C at a rate of 3 °C/min. The coefficient of thermal expansion was obtained from the slope of the thermal expansion vs. temperature plot. The inflection point of thermal expansion was defined as TMA Tg.

2.6 RHEOLOGY MEASUREMENTS

Rheology of the underfill encapsulant was measured using a Rheolyst AR-1000 from TA Instruments. The tests were conducted at 80°C at varying shear stresses and strains. Results were provided by TA Instruments.

2.7 DSC STUDY ON CURING PROFILE AND Tg

A modulated DSC instrument made by TA Instruments (Model 2920) was used to study the curing profile and Tg of the underfill formulations. The underfill formulations used for this study are all unfilled. A sample of ~10 mg of a formulation which had been warmed up to room temperature from -40 °C was placed into a hermetic DSC sample pan. The sample was then heated in the DSC cell to around 300 °C at 5 °C/min to obtain the curing profile. The cured sample was left in the DSC cell and was cooled to room temperature at 5 °C/min. Then the sample was reheated to 280 °C at 5 °C/min to obtain the Tg of the cured sample.

2.8 DMA STUDY ON DYNAMIC MECHANICAL PROPERTIES

A DMA instrument made by TA Instruments (Model 2980) was used for the investigation of mechanical properties of cured underfill materials. The underfill formulations used for this study are all unfilled. A sample of ~4 grams of a formulation was placed in an aluminum pan ($1^1/_2$ inch diameter), then was transferred to a preheated convective oven at 80 °C and heated to 250 °C at a heating rate of about 3 °C/min, and then was left in oven at 250 °C for 15 minutes. It was then removed from the oven to cool at room temperature. The sample was then polished to obtain two parallel surfaces. A diamond saw was used to cut the polished sample into strips with dimensions of about 32×11×3mm. The measurement was performed in single cantilever mode under 1 Hz sinusoidal strain loading. Storage moduli E', loss moduli E", and tanδ were calculated by the preinstalled software. The peak temperature of tanδ was defined as the DMA Tg.

3. Results

3.1 FLOW RATE AS A FUNCTION OF FILLER LOADING

As expected, Figure 1 shows that the filler loading increase results in the decrease of the flow rate. This is because higher filler loading results in higher viscosity of the underfill formulation as shown in Figure 2.

3.2 FLOW RATE AS A FUNCTION OF FILLER PARTICLE SIZE

Work conducted by Farris[1] with monodispersed particles showed that particle size has no effect on relative viscosity. Also, Van Der Werff[2] reported that particle size has no effect on relative viscosity through the whole range of volume fractions of particles in fluid. The same conclusion was made by Goto and Kuno.[3] Figure 3 shows that faster flow is obtained with a sample of 8 micron average particle size compared to that

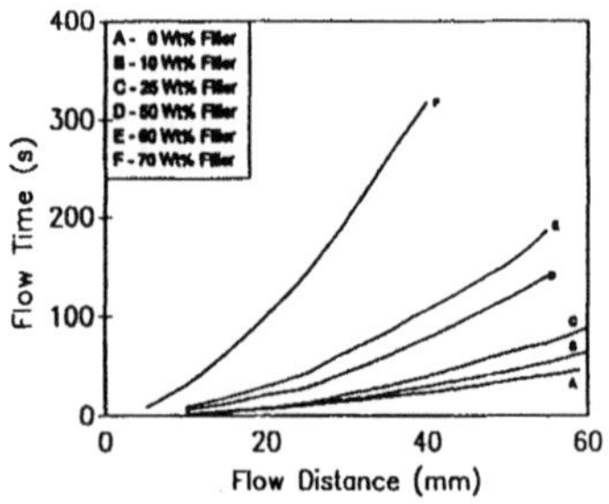

Figure 1: Flow Rate as a Function of Filler Loading (15 micron average particle size used for each sample.)

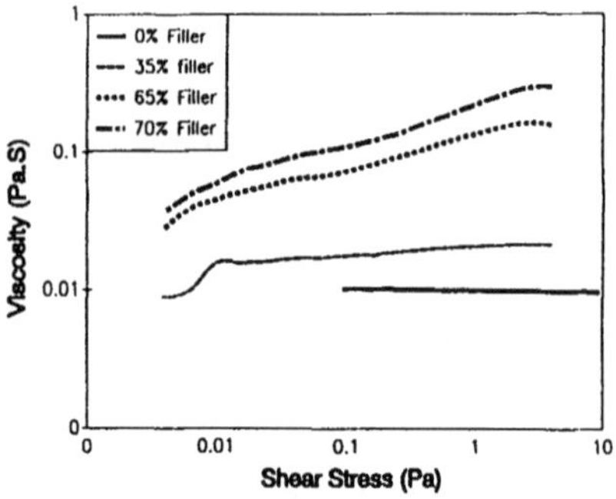

Figure 2: Viscosity as a Function of Filler Loading.

of a 30 micron average particle size. The discrepancy in these results compared with the reported literature could be from obstruction effects due to the small gap to particle diameter ratio. The average particle size for the 8 micron sample is 8 microns but the distribution taken from particle size analysis shows that there are particles present as large as the gap size (75 microns) of the test. This can cause obstruction effects during the flow. The filler materials used in this study were analyzed by a Mastersizer Particle Analyzer (Malvern Instruments, Inc.) and showed to have a broad distribution (Figure 3). A broader distribution was used because the flow is improved with mixtures of particle sizes rather than with monosized particle sizes. It has been reported for particles during capillary flow that multimodal particle size distributions can be used to enhance the flow properties of filled systems.[3,4] Adding a small amount of the smaller particles brings about a decrease in viscosity of the suspension of larger particles.[5,6,7,8]

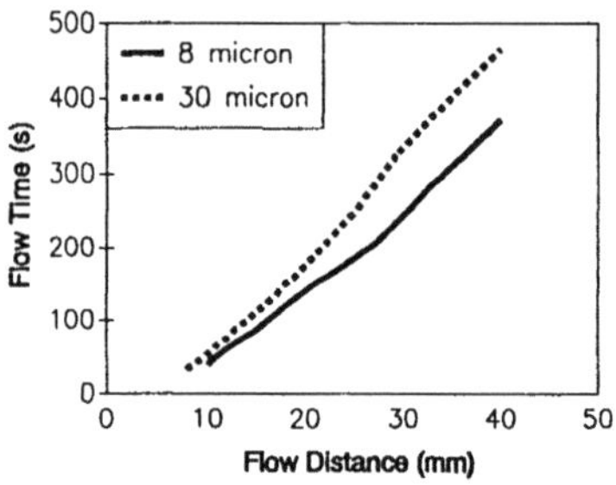

Figure 3: Flow Rate as Function of Filler Particle Size (70 wt% filler loading).

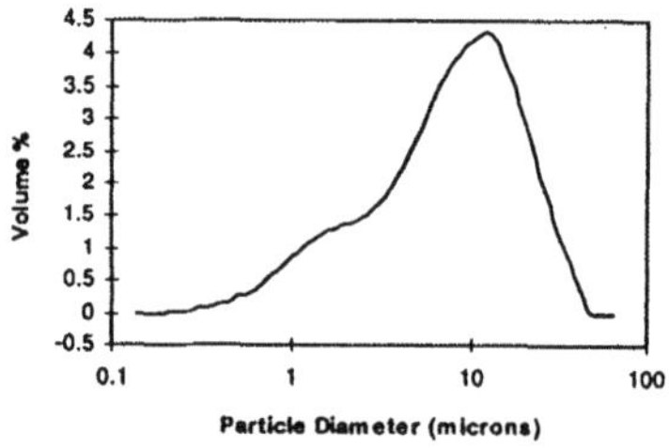

Figure 4: Particle Size Distribution (avg. size app. 8 micron)

3.3 VISCOSITY AS A FUNCTION OF FILLER LOADING

It is shown in Figure 2 that the viscosity increases with increasing filler loading. It has been reported in literature that an increase in viscosity occurs as the fraction loading of particles increase.[1,2] Also, the effect of increasing loading of filler in underfill encapsulants has been shown to decrease the flow rate.[7] This graph also shows that as the filler loading increases, the underfill becomes more and more dilatant (shear thickening).

3.4 CTE AS A FUNCTION OF FILLER LOADING AND FILLER PARTICLE SIZE

The epoxy resin/HMPA hardener mixture has a CTE of 87 ppm/°C and the silica has a CTE of 0.5 ppm/°C. As the ratio of filler to resin increases, the CTE will decrease due to the lower CTE of the filler compared to that of the resin. This decrease can be seen in Figure 5 which shows CTE as a function of filler loading. This graph also shows that the effect of a unit change in loading at high loadings decreases CTE more than a unit change in loading at smaller loadings. Figure 6 shows the relationship between the CTE and the filler particle size. As the filler particle size decreases (at a constant volume % loading), the CTE also decreases. This can be explained by looking at the total surface area of the filler particles. As shown before, the CTE of the filler is much lower than that of the resin. The interface of the filler particles and the resin matrix constrict the expansion of the matrix. As the surface area of the filler particles increase due to the decrease in particle size, there is an increase in the interface between filler and resin. Therefore, an increase in the constriction of the matrix due to increased surface area allows a decrease in the expansion of the matrix. If a large particle size filler and a small particle size filler (each with a narrow size distribution) were mixed in proportions progressing from 100% large particle size to 100% small particle size, the CTE values will decrease due to the increasing surface area.

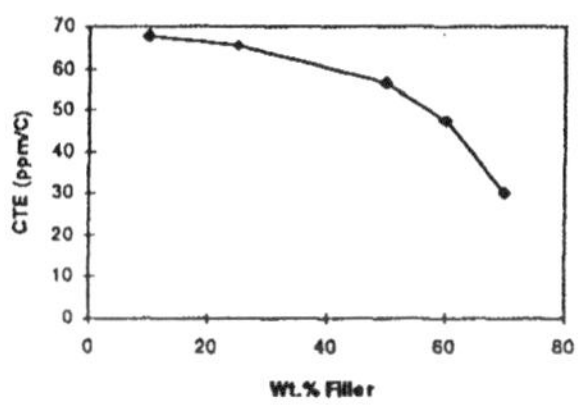

Figure 5: CTE as Function of Filler Loading (15 micron average particle sizeused for each sample)

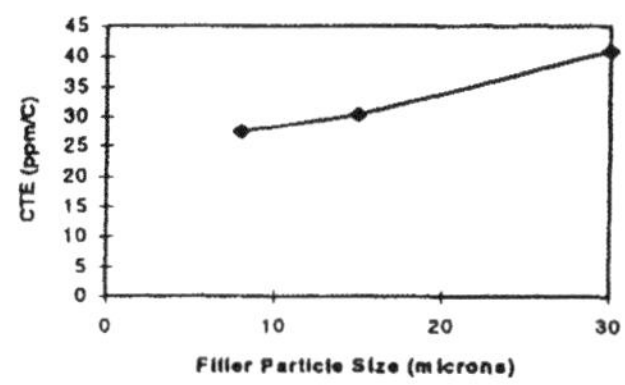

Figure 6: CTE as Function of Filler Particle Size

3.5 CURING PROFILE OF "NO-FLOW" UNDERFILL MATERIALS

Figure 7 shows the typical curing profile of one series of our "No-Flow" underfill materials which is catalyzed by metal acetylacetonate. On one hand, the curing peak temperature of this system is around 213 °C which matches the reflow profile of eutectic tin/lead alloy solder bump. On the other hand, some curing reaction occurs before the solder bump reflow which is desired to prevent the underfill material from thinning out at reflow temperature. The Tg of the underfill is more than 200 °C, which is not only good for underfill applications, but also are good for other applications such as multi-layer dielectric substrate and conductive adhesives.

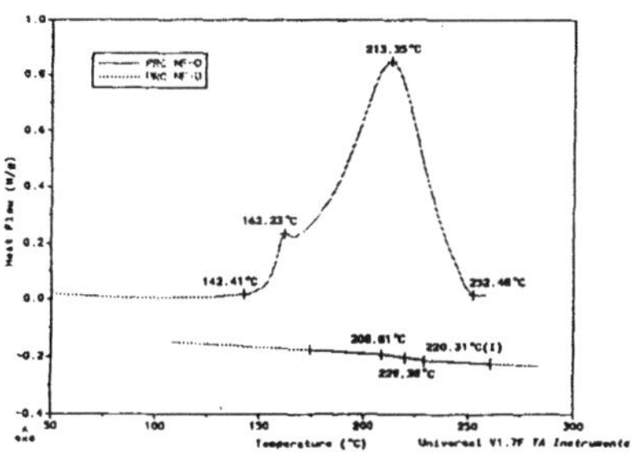

Figure 7: Curing Profile and Tg of A Metal Acetylacetonate Catalyzed "No-Flow" Underfill

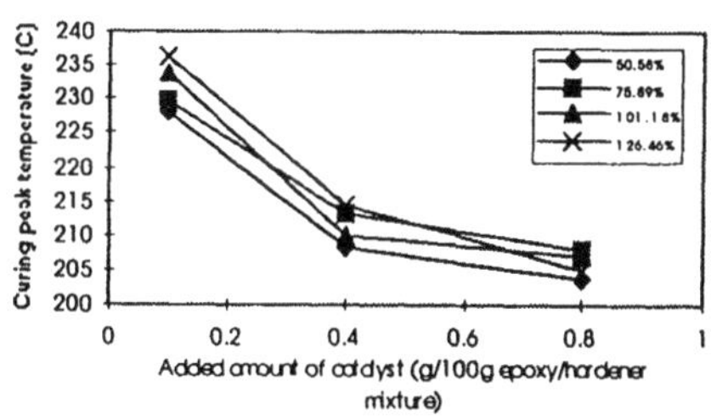

Figure 8: Effect of Catalyst Concentration on Curing Peak Temperature

3.6 EFFECT OF CATALYST CONCENTRATION ON CURING PROFILE

Figure 8 shows that increasing the concentration of catalyst significantly decreases the curing peak temperature. This effect appears more obvious in the low catalyst concentration region. This experiment is reasonable. The catalyst (i.e. metal acetylacetonate) must react with anhydride at first to form active species which open the epoxy ring since metal acetylacetonate or anhydride itself can not effectively open cycloaliphatic epoxy rings even if the temperature reaches 230 °C[9,12]. Considering the fact that the added amount of hardener is far beyond the added amount of catalyst, increasing the catalyst concentration will increase the concentration of active species, and consequently decrease the curing peak temperature.

3.7 EFFECT OF CATALYST CONCENTRATION ON Tg AND CTE

Figure 9 and Figure 10 show the effect of catalyst concentration on Tg (TMA data) and CTE of the cured formulations. It is found that catalyst concentration has little effect on the two properies. This is because the catalyst concentration does not noticeably affect the crosslinking density of the cured formulations (see Figure 11).

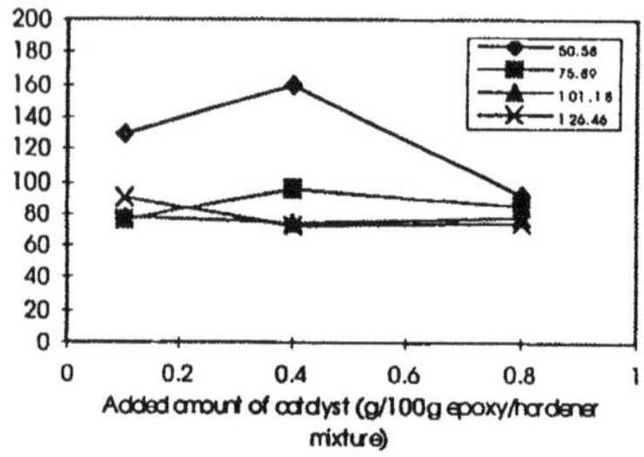

Figure 9: Effect of Catalyst Concentration on CTE of The Cured Formulations.

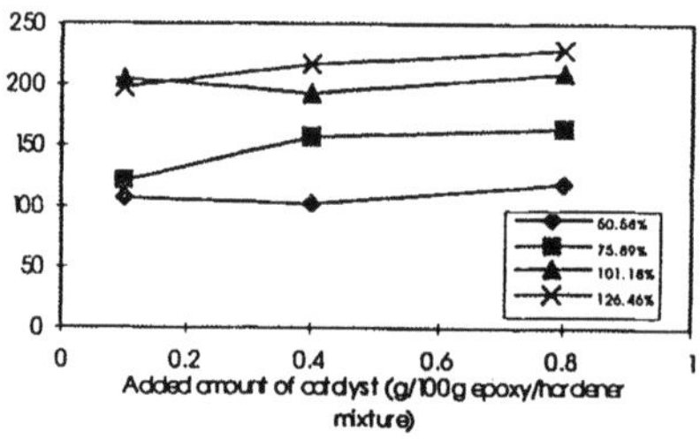

Figure 10: Effect of Catalyst Concentration on Tg (TMA data) of The Cured Formulations.

3.8 EFFECT OF CATALYST CONCENTRATION ON CROSSLINKING DENSITY

Crosslinking density can be calculated using kinetic theory of rubber elasticity[10]. The crosslinking density $\rho_{(E')}$ can be determined by equation (1) as follows.

$$\rho_{(E')} = E'/3\phi RT \qquad (1)$$

Where E' is the storage elastic modulus of the cured samples at peak temperature of tanδ + 40 °C, ϕ is a front factor (assumed as $\phi = 1$), R is the gas constant, and T is the absolute temperature. The assumption behind this calculation is that the cured epoxy resin is in the real rubbery state at the temperature Tg (DMA data) + 40 °C. Figure 11 shows the calculated $\rho_{(E')}$ vs. catalyst concentration. It is seen that the crosslinking density of the cured formulations has little affect by the catalyst concentration within the investigated concentration range.

3.9 CURING MECHANISM OF EPOXY/ANHYDRIDE/METAL ACETYLACETONATE SYSTEM

The curing mechanism of epoxy/anhydride/metal acetylacetonate system is by no means clear at this time. However, the research work in this field can go back as early as 1981 or even earlier[11]. Based on the studies on the bisphenol A/anhydride/metal acetylacetonate system, Smith, et al.[11] thought that the decomposition fragments of metal acetylacetonate is the most likely active species responsible for the initiation of polymerization in epoxy/anhydride resins. And they further pointed out that the more likely initiation mechanism is the one involving electron transfer between carboxylic anhydride and the liberated metallic cations to give a reactive initiating species. But this initiation mechanism can not explain the experimental result that Co^{2+} and Co^{3+} acetylacetonate show same catalytic reactivity, and the characteristic color of Co^{2+} and Co^{3+} did not change before and after curing. Therefore we proposed another initiation mechanism here to explain all the experimental results we obtained, which can be schematically shown in

(I) and (II) in Figure 12. The acetylacetone anion is first released from the complex when it is heated to high enough temperature. The enolate type acetylacetone anion is a very strong base and can effectively attack the carboxylic carbons of the anhydride and form a carboxylic anion which can further open epoxy ring or another anhydride molecule to initiate anionic polymerization and form crosslinked networks.

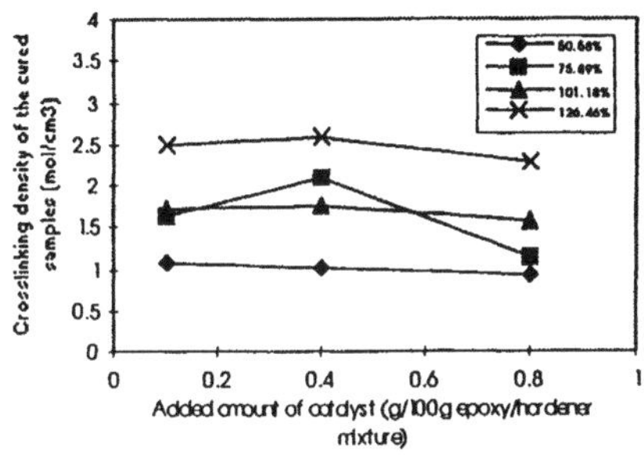

Figure 11: Effect of Catalyst Concentration on Crosslinking Density of The Cured Formulations.

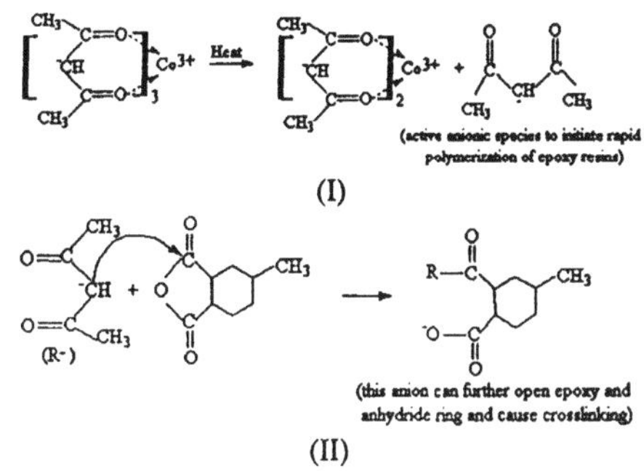

Figure 12: Possible Curing Mechanism for Epoxy/Anhydride/Metal Acetylacetonate System.

4. Conclusions

It has been presented that a low underfill encapsulant CTE value is needed for the flip-chip assembly. It was proven that as filler loading is increased, a lower CTE value is reached. This decrease in CTE is more significant at higher loadings than at lower loadings. This work showed that an increase in filler loading decreases the flow rate of the underfill encapsulant between parallel plates as well as increasing the viscosity. The particle size was also proven to have an effect on CTE and flow rate. As the surface area of the filler increased (i.e. smaller particles at constant loading), the CTE decreased. The proposed reason for this is due to the increased constrictive forces on the matrix from the increased surface area of the filler. This work showed that as the filler particle size decreases, the flow rate increases. More work is needed to determine the real factors that affect the flow between parallel plates with a small gap to filler particle size ratio. Metal acetylacetonate is a potential curing catalyst for epoxy/anhydride based "No-Flow" underfill formulations. The curing reaction peak temperature can be manipulated within the range from 195 to 230 °C by changing the catalyst concentration. The concentration of catalyst mainly affects the curing reaction peak temperature, but has little effect on Tg, CTE and crosslinking density of the cured formulations. The curing reaction is most likely initiated by enolate type acetylacetone which is released from the metal acetylacetonate during heating.

5. Acknowledgment

This work is partially supported by NSF and DARPA through the USAF Wright Petterson Laboratory

6. References

1 R.J. Farris, "Prediction of the Viscosity of Multimodal Suspensions from Unimodal Viscosity Data," *Transactions of the Society of Rheology*, Vol. 12(2), p. 281-301, (1968).

2 J.C. Van Der Werff and C.G. De Kruif, "Hard-sphere Colloidal Dispersions: The Scaling of Rheological Properties with Particle Size, Volume Fraction, and Shear Rate," *Journal of Rheology*, Vol. 33(3), p. 421-454, (1989).

3 Goto and Kuno, "Flow of Suspensions Containing Particles of Two Different Sizes through a Capillary Tube," *Journal of Rheology*, Vol. 26(4), p. 387-398, (1982)

4 Goto and H. Kuno, "Flow of Suspensions Containing Particles of Two Different Sizes through a Capillary Tube. II. Effect of the Particle Size Ratio," *Journal of Rheology*, Vol. 28(3), p. 197-205, (1984).

5 C.E. Park, "Evaluation of Epoxy Underfill Materials for Use in 'Chip-on-Board' Method of Packaging Silicon Integrated Circuits," *ANTEC*, p. 2871-2876, (1995).

6 K.B. Wun and G. Margaritis, "The Evaluation of Fast-Flow, Fast-Cure Underfills for Flip Chip on Organic Substrates," *Electronic Components and Technology Conference*, p. 540-545, (1996).

7 M.K. Schwiebert and W.H. Leong, "Underfill Flow as Viscous Flow Between Parallel Plates Driven by Capillary Action," *International Manufacturing Technology Symposium*, p. 8-13, (1995).

8 S. Han, K.K. Wang and S.Y. Cho, "Experimental and Analytical Study on the Flow of Encapsulant During Underfill Encapsulation of Flip-Chips," *Electronic Components and Technology Conference*, p. 327-334, (1996).

9 C.P.Wong, S.H.Shi and G.Jefferson, "High Performance No Flow Underfills for Low-Cost Flip-Chip Applications", *IEEE Proceedings of The 47th Electronic Components and Technology Conference*, P. 850-858, (1997)

10 A.V.Tobolsky, in "*Properties and Structure of Polymers*", Wiley, New York, (1960)

11 J.D.B. Smith, "Metal Acetylacetonates as Latent Accelerators for Anhydride-Cured Epoxy Resins", *Journal of Applied Polymer Science*, Vol. 26, 979~986 (1981)

12 S.H.Shi, C.P.Wong and G.Jefferson, "High Performance Underfills for Low Cost Flip Chip Application", *3rd International Symposium and Exhibition on Advanced Packaging Materials*, Page 42-45, Chateau Elan, Braselton, Georgia, March 9-12, (1997)

Authors Index

Achmatowicz, Selim, 142
Amon, Slavko, 211
Amskov, Maja, 275
Belavic, Darko, 161, 197
Caravasile, D., 61
Ciez, Michael, 223
Ciszkowski, W., 9
Codreanu, N. -D., 9
De Mey, Gilbert, 1
Dobay, Robert, 133
Dziurdzia, B., 223
Ganciulescu, S., 61
Garrou, Philip, 107
Golumbeanu, V., 9
Gramc, Slavko, 197
Harsanyi, Gabor, 133
Hrovat, Marko, 161
Illyefalvi-Vitez, Zsolt, 69, 133
Ionescu, Radu Mihai, 61
Jaakoola, T., 281
Jakubowska, M., 142
Jones, W. Kinzy, 99
Kolesar, Frantisek, 169
Kurzweil, Karel, 259
Leonescu, D., 61
Leppavuori, Seppo, 157
Lernout, Jo, 179
Liu, Yanqing, 99
Mach, Pavel, 17
Makogon, Yuri, 91
McNeilly, K., 79
Muckett, Steve, 241
Nemesz, W., 142
Nowak, Stanislaw, 223
Osieczkin, S., 142
Pavlin, Marko, 197
Philippov, Philipp, 151
Pinkola, J., 69
Plut, Janeta Fajfar, 161
Reczey, Maria, 133
Resnik, Drago, 211
Rocak, Dubravka, 161, 197
Rusan, O., 281
Ruszinko, M., 69

Saint-Martin, X., 259
Schaper, Leonard, 245
Schlag, Stefan, 79
Sexton, Peter, 79
Shah, M., 99
Shi, S.H., 287
Sidorenko, S.I., 91
Sikula, Josef, 53
Sinnadurai, Nihal, 235
Sladic, Tone, 211
Soba, Stojan, 197
Somora, M., 169
Stricot, Y., 259
Suhir, Ephraim, 25
Svasta, Paul,, 9, 61
Szabo, P.J., 69
Tol, O., 61
Trontelj, Janez, 205
Trontelj, Lojze, 205
Tummala, Rao R., 253
Urbancik, J., 169
Uusimaki, A.K., 152
Vahakangas, Jouko, 281
Van Calster, A., 179
Van Loan, Paul R., 187
Van Veen, N., 257
Vanfleteren, Jan, 179
Vasina, Petr, 53
Vincent, M.B., 287
Visy, Csaba, 133
Vladescu, M., 9
Vrtacnik, Danilo, 211
Wong, C.P., 287
Zwierkowski, E., 142

MIX
Papier aus verantwortungsvollen Quellen
Paper from responsible sources
FSC® C105338

If you have any concerns about our products,
you can contact us on
ProductSafety@springernature.com

In case Publisher is established outside the EU,
the EU authorized representative is:
Springer Nature Customer Service Center GmbH
Europaplatz 3, 69115 Heidelberg, Germany

Printed by Libri Plureos GmbH
in Hamburg, Germany